The Maya Calendar

The Maya Calendar

A Book of Months, 400–2000 CE

Weldon Lamb

UNIVERSITY OF OKLAHOMA PRESS : NORMAN

Published through the Recovering Languages and Literacies of the Americas initiative, supported by the Andrew W. Mellon Foundation.

Library of Congress Cataloging-in-Publication Data

Name: Lamb, Weldon, 1953– author.
Title: The Maya calendar : a book of months, 400–2000 CE / Weldon Lamb.
Other titles: Book of months, 400–2000 CE
Description: Norman, OK : University of Oklahoma Press, [2017] | Includes
    bibliographical references and index.
Identifiers: LCCN 2016031030 | ISBN 978-0-8061-5569-2 (hardcover : alk. paper)
Subjects: LCSH: Maya calendar. | Months—Terminology. | Mayan languages—
    Terminology.
Classification: LCC F1435.3.C14 L34 2017 | DDC 529/.32978427—dc23
LC record available at https://lccn.loc.gov/2016031030

RECOVERING
LANGUAGES & LITERACIES
OF THE AMERICAS

The Maya Calendar: A Book of Months, 400–2000 CE is published as part of the
Recovering Languages and Literacies of the Americas initiative. Recovering Languages
and Literacies is generously supported by the Andrew W. Mellon Foundation.

The paper in this book meets the guidelines for permanence and durability of the
Committee on Production Guidelines for Book Longevity of the Council on Library
Resources, Inc. ∞

1 2 3 4 5 6 7 8 9 10

Contents

Tables

Introduction

This study collects, defines, and extensively relates all the Maya month names. By some 1,700 years ago speakers of proto-Ch'olan, the ancestor for three Maya languages still in use, had developed a calendar of eighteen twenty-day months plus a set of five days and begun to record in stone numerous dates of dynastic and cosmological history. Throughout the Classic period (200–900 CE) the original hieroglyphic set remained unchanged, with only minor and brief variants. But across the long subsequent history (900–2000 CE) only two of the original names eluded modification, while the rest altered their form, their meaning, or both, developing into fourteen distinct traditions. Altogether these year counts encompass 250 standard forms, variants, and alternates, with some 570 meanings among the cognates, synonyms, and homonyms. Nearly 75 percent of these new names derived ultimately from the hieroglyphs.

This work expands and intensifies the research on the Maya calendars accomplished by my mentor Munro Edmonson with his 1988 opus *The Book of the Year*, a compilation and discussion of the representative sacred (260-day) as well as civil (365-day) calendars for all Mesoamerica. In the present volume the attention to zenith passage dates as candidates for the start of either the year or the five-day period follows from the research by Anthony Aveni, John Carlson, Horst Hartung, and Edwin Krupp. By 1984 the data from astronomy, ethnography, linguistics, and epigraphy had become more than abundant enough to make the project represented here feasible: first as a dissertation and now as a book.

The corpus of Maya month names consists of fifteen distinct sets, with the hieroglyphs being the earliest system and to a large extent the source of the rest. Each set or "tradition" is specific to one ethnic group and goes by the name of its language. The definition of the standard forms and their variants delineates the major themes but also facilitates the comparison of meanings and forms. The lists of cognates, synonyms, and homonyms within and across traditions permit a tentative outline of the names' descent and diffusion.

The start dates for the year and the "month" of five days in counts fixed or frozen to a correlation with a European calendar point to agricultural, astronomical, ritual, and even political motivations. They sometimes uncover the history of particular changes. More specifically, the winter solstice, the vernal equinox, and the two solar zenith passages apparently motivated initial dates for the year and the five-day month in several fixed calendars. Tribute schedules imposed by Nahua militarists might explain names and dates in a few others. Ritual, however, proved the major factor in the origin of most of the new names.

Over the past five hundred years clerics, historians, travelers, archaeologists, ethnographers, and linguists have documented the traditions fashioned since the era of the hieroglyphs. Only a few Maya shamans have offered names, definitions, or lore, so deeper insights into Classic iconography and cosmology depend on relating the names to texts and archaeological finds.

Chapter 1, "The Ancient Maya Hieroglyphic Calendars," treats the hieroglyphic set, a nearly monolithic uniformity marred by merely sporadic, short-lived variants. Chapter 2, "The Ethnographic Calendars," presents the normal forms and significant variants in each tradition. This highlights synonyms, alternates, replacements, and unique forms as well as shifts in position for the First Month and the Five Days. Chapter 3, "The Maya Month Names: Forms and Meanings," defines each member of every tradition, then groups the meanings by themes. Chapter 4, "The Maya Month Initial Dates," shifts the focus to annual occurrences. This analysis establishes the frozen traditions, then explores rationales for the initial dates of their First Months and Five Days. It closes with a contrast between seasonality and ritual. Chapter 5, "Continuity in Sound and Sense," presents stemmas that order the cognates, synonyms, and homonyms into a descent and diffusion history of unexpected connectedness and longevity.

Author's Note

Several conventions distinguish proper names from their related dictionary entries. Entries appear in lowercase italic (*pop*), while proper names and their translations are lowercase roman with initial capitals (Pop "Mat"). Hieroglyphic names in appendixes, captions, and tables also appear in lowercase roman with initial capitals (Pop). In the text itself, however, glyphs that represent a whole word or a grammatically meaningful part of a word appear in all capitals and bold (**K'AYAB** "Drum," from **K'AY** "sing" **+** **-AB** "instrumental suffix"). Signs that record just sounds with no meaning appear in lowercase roman and bold (**YAXK'IN**, from **YAX + K'IN + ni** or **POP** from **po + po**). The orthography employed by Horst Nachtigall (1978: 306) to record his Ixil names has been simplified, omitting most diacritics. This does not affect the substance of the present work.

The single asterisk precedes hypothetical or reconstructed forms of words (*poop). The double asterisk marks even earlier ancestral versions, based on single-asterisk forms (**pohp*). For consistency in style, all glottal stops before vowels at the beginning of words are omitted.

Please notice that figure numbers in the text are not always cited in order. Varying discussion topics make this inevitable. The illustrations of both the standard names and the innovative forms are grouped by convention in the sequence of the Yukatek calendar and under the Yukatek names.

Acknowledgments and Dedication

It delights me to credit here those whose faith and encouragement have helped compose and conclude this never-ending story, for this is a list of friends. It must always remain incomplete, however, as I never dreamed the project would last all these years or involve that many people, so when I finally realized their names should be set to paper, some had already faded beyond recall. To these, I apologize and ask forgiveness.

In this hunt for quality data and the bigger pictures my beaters flushed out countless articles and books from the thickets of the nation's libraries. These include Holly Reynolds, Fran, Norice, and Sandra of Interlibrary Loan at New Mexico State University; Gene, Pat, Dolores, and Nancy at Thomas Branigan Memorial Library in Las Cruces, New Mexico; the late Mrs. Martha Robertson, at the Latin American Library of Tulane University; and the indefatigable Mr. Dennis Rowley of Brigham Young University.

Bearers and guides on this expedition have been Professor Brian Stross, my friend Professor William Ringle, and my fellow lover of star lore Professor Susan Milbrath. Particularly valuable maps were the manuscripts generously provided by the late Dr. George Stuart and my mentor, the late Professor Munro S. Edmonson. Along the way reinvigorating words frequently came from my close friends Gilbert Sarabia, Keith and Kathy Austin, and especially Don Kurtz and Anne Gutierrez.

The game, once brought to ground, required considerable gutting, dressing, and cooking prior to serving. The members of my dissertation committee demonstrated prodigious skills here, and should the fare prove palatable, the thanks go to the late Professor Munro S. Edmonson, Professor Judith Maxwell, Professor E. W. Andrews V, and Professor Robert C. Hill II. The same looms true, too, for the many involved with the actual presentation of this feast: Theresa May, who referred me to the University of Oklahoma Press; Alessandra Jacobi-Tamulevich, acquisitions editor for American Indian, Latin American, and Classical Studies; Emily J. Schuster, manuscript editor; Amy Hernandez, marketing assistant; and Kathy Lewis, copyeditor *par excellence*. I must take the blame for any mangled cuts or poorly seasoned portions.

Despite my many mistaken calls and misdirected steps, the beloved elders managed to bring the enterprise and its captain more or less intact to a safe haven and a happy end. Thanks, then, to those who watched over the expedition from above, namely Professor Donald Robertson, Professor Marion Hardman, the Most Reverend S. M. Metzger, my aunt Agathe Jackson, my grandmother Käthe Klein, my virtual grandfather Professor C. H. Stubing, and my mother Eva K. Lamb. May you all look upon this work and find it good.

Finally, above all others, credit and gratitude are due my wife, Mary Kerwin. More than anyone else, she made it happen. To her I dedicate this book.

The Maya Calendar

CHAPTER ONE

The Ancient Maya Hieroglyphic Calendars

For most of their precontact history (200–1500 CE) the Maya recorded dates with month names virtually unchanged in form, sound, and meaning. Occasionally scribes altered a compound or set of two to five signs constituting a name, but the variant or innovation nearly always proved short-lived, employed only rarely in one or two texts at a handful of sites. Since the late 1500s, natives and foreigners alike have recorded the remarkably diverse descendants of the original set of month names. Of the fourteen traditions derived from the Classic texts, only three significantly resemble the inscriptions in form, though most reflect their hieroglyphic ancestors in meaning.

The sparse data prior to 800 BCE do not allow a fine fix in time for the origin of the three basic Mesoamerican counts. First is the cycle of twenty named days, often called a "month," perhaps derived from tallying toes and fingers. Next comes the divinatory or sacred almanac, a period of 260 days, the permutation of the twenty days with the numbers 1 through 13. This count is an apparent estimate for the length of human gestation; scholars designate this cycle the *tzolk'in* "order of days." The final count is the *haab* or "Vague Year," a round of 365 days (composed of eighteen "months" plus a span of five days), an approximation for the tropical year of 365.2422 days. Beginning with the eighth century BCE, artifacts record dates that allow inferences significant to the history of the glyphic calendar.

Sages within the geographic triangle delimited by Cuicuilco in the Valley of Mexico, Monte Albán in the Valley of Oaxaca, and San Andrés Tuxtla in the Valley of Tuxtla, Chiapas, meshed the count of 260 days with the count of 365 days to fashion the Calendar Round of 18,980 days, apparently for the summer solstice of 22 June 739 BCE (Edmonson 1988: 115, 124). This cycle consists of 73 sacred almanacs and 52 Vague Years, stating a date in terms of both tallies. The first record of the Calendar Round might perhaps survive on an ear spool from Cuicuilco; its day count of "2 Lord" at least presents the earliest example of the almanac, from 691 or 679 BCE (Edmonson 1988: 20, 167). An Olmec sun priest etched the next suspected instance of the double cycle onto the face of an ax from Simojovel in 667 BCE. The first unimpeachable Calendar Round date, complete with a distinct month name, distinguishes Stela 17 at the Zapotec capital Monte Albán, from 528 BCE (Edmonson 1988: 20–22).

The earliest Maya month name appears on the diminutive Hauberg Stela of 199 CE (Edmonson 1988: 31; fig. 1 Xul in appendix C, which contains all the figures cited below). The month name is known by the Yukatek label **XUL** "Planter Stick." Within the monumental corpus the latest name endures on Stela 6 at Itzimté, erected in 910 CE (fig. 2 Xul). It too might depict a variation of Xul.

After 1000 CE literacy barely manifested itself on monuments. Writing survived on portable objects in dates and in tags that function as brief statements of ownership. The disappearance of writing from the ceramics, though, remains a quandary. Why did Postclassic rulers no longer exchange calligraphic vessels as prestigious gifts like their Classic predecessors? Early accounts and confiscated or extant codices attest that writing on the painted page operated during the Postclassic on a modest scale in Highland Chiapas and Guatemala but flourished in northern Yucatán. Here codices on numerous topics actually proliferated into libraries (Tozzer 1941: 38, 282–85). The Spanish suppressed hieroglyphic literacy through book burnings and the promotion of the alphabet, eradicating it only in the early 1700s.

The Four Sets of Maya Month Names

Four sets of glyphic compounds encompass the major variations of Maya month names in form, sound, and sense. By source they are designated Classic Period Hieroglyphs; the indigenous Codex Dresden; the records of Friar Diego de Landa; and the eighteenth-century Book of Chilam Balam of Chumayel, one of a dozen or so colonial native collections of religious and historical texts written in Yukatek Maya with the Roman alphabet and named for the town where it surfaced. Table 1.1 displays the first three series, omitting the Chumayel set as meaningless because its glyphs present only scribblings with little or no resemblance to their earlier counterparts.

The Classic Period Hieroglyphic Set: Standard Forms

The Classic period hieroglyphic signs were used from about 200 to 900 CE and remained little changed in appearance through 1566; with the demise of the writing system by about 1750 they disappeared from the historical record. Table 1.1 displays their standard forms, while table 1.2 gives their translations. These hieroglyphs appeared in dates on stelae, altars, and panels in temples, palaces, and tombs, on stone and plaster, wood and bone, and shell and paper.

Scholars have long debated the language of the glyphs. The consensus began with Yukatekan then shifted to Ch'olan (Lacadena and Wichmann 2002: 279–81). Later it favored bilingualism. In the past twenty years researchers have detected the early forms of several languages represented in the inscriptions: Western Ch'olan, Eastern Ch'olan, Yukatekan, Tzeltalan, and even an undetermined member of the Greater

Table 1.1

TABLE 1.1

Three standard month sets by Yukatek names

Month Name	Classic 200–950 CE	Dresden 1450 CE	Landa 1566 CE
Pop			
Uo			
Sip			
Sots			
Sec			

(continues)

TABLE 1.1 Three standard month sets by Yukatek names (*continued*)

Month Name	Classic 200–950 CE	Dresden 1450 CE	Landa 1566 CE
Xul			
Yaxkin			
Mol			
Chen			
Yax			
Sac			

Month Name	Classic 200–950 CE	Dresden 1450 CE	Landa 1566 CE
Ceh			
Mac			
Kankin			
Muan			

(continues)

TABLE 1.1   Three standard month sets by Yukatek names (*continued*)

Month Name	Classic 200–950 CE	Dresden 1450 CE	Landa 1566 CE
Pax			
Kayab			
Cumku			
Uayeb			

Three versions of the Maya months: from the Classic Period (drawings by Huberta Robinson in Thompson 1971, © 1971 University of Oklahoma Press, used with permission, all rights reserved); Codex Dresden (drawings by Carlos A. Villacorta, in Villacorta and Villacorta 1977); and Landa's *Relación de las cosas de Yucatán* (courtesy of George Stuart and the Biblioteca Real, Madrid).

TABLE 1.2

Standard hieroglyphic month names with translations

#	Yukatek	Translation	Hieroglyphic	Translation
1	Pop	Woven Mat	K'än Jäl-ab'	Yellow Loom/Weaving
2	Uo	Toad	Ik'K'ät/Tahn	Black Wicker/Center
3	Sip	Miss; Brocket Deer	Chäk K'ät/Tahn	Red Wicker/Center
4	Sotz	Bat	Sutz'	Bat
5	Sec	Skull/Small Bat	Kasew	Pacaya Palm
6	Xul	Planting Stick	Tz'ikin	Bird
7	Yaxkin	Green Days	Yäx K'in	Green Days
8	Mol	Pile; Harvest	Mol	Pile; Harvest
9	Chen	Cave	Ik' Ha'/Hab'/Sihom	Black Rain/Year/Soapberry
10	Yax	Green	Yäx Ha'/Hab'/Sihom	Green Rain/Year/Soapberry
11	Sac	White	Säk Ha'/Hab'/Sihom	White Rain/Year/Soapberry
12	Ceh	Deer	Chäk Ha'/Hab'/Sihom	Red Rain/Year/Soapberry
13	Mac	Enclosure	Mak	Enclosure
14	Kankin	Yellow Days	Uniw	Avocado
			Tz'utz'uw	Coatimundi
15	Muan	Hawk	Muwan	Hawk
16	Pax	Upright Drum	Paax	Upright Drum
			Paax-il	Drumming
			Pá'ax-ab'	Upright Drum
			Pa'-tun	Horizontal Slit-Drum
17	Kayab	Horizontal Slit-Drum	K'än Aj Si'	Yellow Woodchopper
			K' än Asij	Yellow Vulture/Raptor
18	Cumku	Thunder God	Ol Waj; Ol	Ball of Food; Portal
19	Uayeb	Bed	Way-ab'	Bed
			Way-hab'	Poison/Vision of the Year

Note: The Yukatek forms come from the Chilam Balam literature.

K'iche'an family. Together designated as "Classic Ch'olan," the first two languages by far dominate the texts.

While acknowledging occasional traces of "vernaculars," the local dialects or languages, some scholars have proposed that a form of Eastern Ch'olan functioned as the one prestige dialect for all the Classic texts (even those of northern Yucatán), the ceramics, and the codices. They refer to this lingua franca as "Ch'olti'an" or "Eastern Ch'olti'an" (Lacadena 2011: 344) or "Classic Ch'olti'an" (Houston et al. 2000: 322–25). Specific linguistic features, many related to verbs, distinguish this prestige script. Ch'olti'an marks transitive verbs with -*Vw* (V = vowel), forms some passives with -*w-aj*, derives passives from transitives with -*h-* . . . -*aj* or -*n-aj*, and derives antipassives from transitives via -*Vw* or -*Vn*. It forms mediopassives with -*Vy* and uses -*laj* to create positionals and converts positional roots into causatives with -*b'u*. *Winal*

"20-day month" and *ajawil* "lordship" prove diagnostic for the dialect (Lacadena and Wichmann 2002: 307). The phonemic contrast between long and short vowels (VV:V) and between the glottal and velar spirants (*h:j*) also characterizes Ch'olti'an, although both distinctions fade away between about 600 and 900 CE (Lacadena and Wichman 2002: tables 8, 9). Its last defining feature is its lack of split-ergativity, a system that relates classes of verbs and pronouns to time (Law et al. 2006).

Nonetheless, the prestige dialect theory has its doubters. Some argue that the "Ch'olti'an" grammatical markers are in fact retentions from the earlier proto-Ch'olan and not later innovations, so they cannot count as evidence for a specifically Eastern Ch'olan affiliation of the inscriptions. Major problems with the prestige dialect hypothesis remain unresolved, especially in regard to its implications for the nature of the script. Nor have any new data in its favor come to light (Wichmann 2006:282).

Debate also continues about descent. Does Eastern Ch'olan exhibit enough distinctive features to define it as "Ch'olti'an" or "Ch'orti'an"? Or did it instead develop into these dialects, which then evolved into separate languages: Colonial Ch'olti' (now extinct) and Modern Ch'orti'? Some scholars propose that Eastern Ch'olan morphed into Classic Ch'olti'an (the language of the hieroglyphs), then Colonial Ch'olti', then Modern Ch'orti' (Houston et al. 2000: 322–25; Hull 2003: 13–23; Lacadena and Wichmann 2002: 292 n. 13; Robertson 1992: 169; Robertson 1998). Some dispute parts of this chain, rejecting the idea that Ch'orti' descended from Ch'olti', for example (Wichmann 2006: 281).

Many epigraphists agree on qualifying numerous Classic texts as clearly Western Ch'olan (such as Yaxchilán, Lintels 29–31) and plenty of others as indisputably Eastern Ch'olan (for instance, Tikal, Temple IV, Lintel 2) (Lacadena and Wichmann 2002: 305–11). Though the present evidence might slightly favor Eastern Ch'olan, given the unresolved issues in the prestige dialect theory, some experts propose the term "Classic Ch'olan" as a neutral label for the language behind the southern lowland inscriptions, though the labels "Hieroglyphic Ch'olan" or "Epigraphic Ch'olan" would serve equally well (Wichmann 2006: 281). More data should eventually resolve this debate as well as the squabble over "dialect" versus "language." Meanwhile it appears safe to assert that Eastern versus Western Ch'olan differentiation had set in by 400 CE at the level of dialects, which had likely become distinct languages by around 600 CE, the start of the Late Classic, with Eastern Ch'olan beginning to split into early forms of Ch'orti' and Ch'olti' and Western Ch'olan into Ch'ol and Chontal (Wichmann 2006: 283).

Appendix A presents the phonemic inventories for the languages relevant to the Classic Maya script, while appendix B illustrates the uncertain affiliation between these languages and major sites. Details of verbal morphology and grammar should eventually resolve the issue.

Two kinds of hieroglyphs constitute the month compounds. Logographs or morpheme signs, marked by bold capital letters, represent complete words and any meaningful part of a word. Devoid of sense, phonetic signs, in bold lowercase letters, spell out morphemes. Many glyphs can function in either capacity.

Four names are always spelled out with phonetic syllables: **SEC** as the Classic **CASEU** with **ka-se-wa**; **MOL** as **mo-lo**; **MAC** with **ma-AK-(ka)** or **ma-ka**; and **PAX** as the putative Classic **PA'AX-AB'**, written **PA'AX-HAB'**. **XUL** presents a likely fifth case: as the Classic **TZ'IKIN**, it spells as **tz'ik-(ki)-ni** (only one text, though, supplies the **-ki-**).

Three names regularly consist of a logograph plus one or more phonetic syllables. First comes **KAYAB** in its Classic interpretations **K'ÄN AJ SI'** and **K'ÄN ASIJ**, both from **K'AN ASIJ-(si)-ya**. Second is **UNIU**, recorded with **(u)-UN/UNIW-(ni)-wa/wi**. Last comes **UAYEB** when analyzed as **WAY-AB'** and rendered as **WAY-HAB'**.

The next group consists of ten names written only as logographs, with optional phonetic complements: **POP**, **UO**, **SIP**, **YAXKIN**, **CHEN**, **YAX**, **SAC**, **CEH**, **PAX**, and **UAYEB**. **PAX** seems to read **PA'AX-AB'**, consisting of **PA'AX** + **HAB'** + **(ma)**. This is a hybrid: a logograph plus a logograph used as a homonym for a suffix or **HAB'** "year" for **-AB'**, the instrumental suffix. The optional **-ma** perhaps indicates **-ham**, based on the word-final variation of *b'* with *m'* in Poqom and Tzotzil. The compound reads **PA'AX-ab'** (literally "tap-instrument, drum"). **UAYEB** as **WAY HAB'** is scripted **WAY-HAB'-(ma)**.

The final class is composed of **SOTZ**, **MUAN**, and **CUMKU**, the Classic **SUTZ'**, **MUWAN**, and **OL**. They are nearly always written as logographs with optional complements, but very rarely they are recorded with the fully phonetic spellings **su-tz'i** (St. Louis Columnar Monument C:1, in Liman and Durbin 1975: 316, fig. 2), **mu-wa-ni** (Quiriguá M.23:F′2c; I′4b, in Schele and Looper 1996: 151, 152, 207; Dresden 46b:B1) and **o-la** (Copán Hieroglyphic Staircase fragment, fig. 99 in appendix C). No month name is written only logographically, but all those rendered as logographs with complements do occasionally appear without them. All the names could have come down in spellings, logographs, and logographs with complements. Why they did not remains unclear.

Some prefixes and suffixes perform as "semantic" phonetic complements because they determine one of several known values for the main sign. With both **UO** and **SIP**, for example, the **-ta** cues to **K'A̱T** and the **-na** to **TA̱N**. A glyph qualifies as a prefix if it appears in front or on top of a main sign, as a suffix if under or right after it.

Other syllabic affixes might serve as "dialect" phonetic complements signaling the pronunciation of the main sign's value in the local or prestige idiom. For example, **su-** marks the bat head of **SOTZ** at Chichén Itzá as the Ch'olan *sutz'* in contrast to the Yukatekan *sòotz'* (the asterisk denotes a reconstructed or hypothetical earlier form of a word). Either **b'u** or **wa** is suffixed to **POP** "woven mat" for **K'AN JAL-AB'** "yellow weaving/command instrument." They hint at either a language or a dialect difference, as between the Ch'ol *halaw* and the Yukatek *haleb'* "paca," or between the Yukatek variants *haleb'* and *halew* (Aulie and Aulie 1978: 62; Barrera Vásquez et al. 1980: 172).

A few syllabic suffixes might perform as "redundant" phonetic complements, indicating the last phoneme in the main sign's unique value. The **-ni** on **MUAN** and **YAXKIN** and the -**ka** on **MAC** do this. Linguistically superfluous, these suffixes perhaps satisfy calligraphic aesthetics, balancing with each other across a composition of several glyph blocks (Justeson 1978). Current research suggests that such syllables might instead indicate the nature of the root vowel. It is short (V) if it and the vowel of the complement are the same, but either long (VV) or "complex" (that is, followed by *h* [Vh] or a glottal stop [V']) if they differ (Houston et al. 1998: 276).

THE CLASSIC PERIOD HIEROGLYPHIC SET OF MONTH NAMES: INNOVATIVE FORMS

Innovations are major changes in the meaning or pronunciation of a name signaled by the addition, deletion, alteration, or rearrangement of logographs or phonetic elements. Rare in the Classic corpus, they seem to appear at random. Except for Pop, Yaxkin, and Muan, all the month names each develop at least one other version. Two (Cumku, Uayeb) do this with just logographs; five (Sotz, Sec, Xul, Ceh, Mac) exclusively via phonetic complements; eight (Uo, Sip, Chen, Yax, Sac, Kankin, Pax, Kayab) by either method; and three (Uo, Mac, Pax) by both means at the same time (see appendix C). Tables 1.3 to 1.8 summarize the types of innovations, by name and set. Angle brackets (<>) indicate uncertainty regarding tone for Yukatek and proto-Yukatek. Tone denotes the pitch that distinguishes words that otherwise sound the same but have rather different meanings; *kan*, for instance, means "four" or "snake" or "sky" in Yukatek, depending on the tone: *kan*, *kàan*, and *ká'an*, respectively (Bricker et al. 1998: 122–23).

Uo

In one instance of this month name (fig. 5 Uo) a glyph of unknown meaning replaces the logograph for "black." In another (fig. 6 Uo) the sign for **i-** suppresses the color prefix to give **i-K'AT** for **IK' K'AT**. On a Classic codex-style vessel the spelling **wo-ji** (fig. 7 Uo) indicates *wooj* "glyph" (Houston et al. 1998: 281). At Chichén Itzá (fig. 8 Uo) the glyph for **wo** (Justeson 1984: 321; Houston 1988: 133; Berlin 1987; Stuart 1990: 220) immediately follows the month compound. It prefixes the variant drawn for Landa (fig. 9 Uo). The name's crossed bands can appear inside the sign for "black" (figs. 10, 12 Uo) or bear the suffix -**ma** (fig. 11 Uo), an apparent cue to the proto-Yukatek *táam*, a dialect variant of *táan* "dissolved dough or flour put into stews to thicken them" (Barrera Vásquez et al. 1980: 769; Bastarrachea Manzano et al. 1992: 119) (**TAN** is an accepted reading for the main sign). The Dresden scribe occasionally exchanges the bands for a knot with two darkened loops (fig. 13 Uo); its contexts indicate **TAN**.

Sip

On several similar vases perhaps from the El Mirador basin, scribes replaced the usual "red" prefix with the sign for "green" (K1005: A6, K1371: L3, K1372: J3, K6751: H4)

TABLE 1.3
Classic hieroglyphic month names: innovative forms

Reference name	Color prefix	Translation	New name	Meaning
Uo	Ik'	Black	Wooj	Hieroglyph
			Tahn/Táan	Center
			Táam, Táan	Dissolved Dough
Sip	Yá'ax	Green	K'at	Wicker; Clay
	Yáax	First	Táam, Táan	Dissolved Dough
			Tahn/Táan	Center
			Tan/Tá'an	Ashes
			Sib'ik/Sab'ak	Soot
			Paw	Net Bag
Sotz			Sutz'il	Bat Time; Rains
			Sek/Sik	Small Bat
Sec			Tsèek'	Skull
			Kaas	Scepter
			Seew	?
Xul			B'ah	Gopher, Mole
			Chik	Bird
			Chi'ik	Coati
			Cho'o'	Rat
			Halaw	Agouti
			Tzub'	Paca
			Tzuk	Mouse
			Tz'i'	Dog
			Tz'ijk	Animal, Frog
			Tz'ikin	Bird
Mol			Molol	Gatherer
			Molom	Gatherer
			Molaw	?
			Molow	?
			Mow	?
Chen			Sihoom	Soapberry
			Ha'	Soapberry; Water
			Ha'al	Rain
			Kúum	Thunder
			Ku'um	Stone
			Tuun	Stone
Yax			Ha'al	Rain
			Kúum	Thunder
			Ku'um	Stone
			Tuun	Stone

(continues)

TABLE 1.3   Classic hieroglyphic month names: innovative forms (*continued*)

Reference name	Color prefix	Translation	New name	Meaning
			Pawob'-tuun	Net-Bag Stone
			Winik	20 Days; Person
			Kun	Seat
Sac			Sek	Bat; White
			Ha'al-al	Rainy Season
Ceh	Chäk	Red	Ku'um	Stone
	Chäk	Red	Kúum	Thunder
Mac			Mam	Grandfather
			Ma	Grandson; Old
			Ma Majk	Elder Turtle Shell
Kankin			Un	Avocado
			Uniw	Avocado
			K'an K'ìin	Yellow Days
			Tz'utz'uw	Coatimundi
Pax			Paax	Drum
			Pàax-il	Music
			Pas-hab'	Sprout Season
			Pa'x-ha'ab'	Break-Open Season
			Paax-ab'	Upright Drum
			Tum	Drum
			Tunkul	Slit-Drum
Kayab			Asih	Cicada, Raptor
Cumku			Ha-waj	Tortilla Gourd
			Kumk'u	Pot God
			Ol	Ball; Heart; Hole
Uayeb			U Way Hab'	Poison of the Year
				Vision of the Year
				Co-essence of the Year
				Bed of the Year
			Koch Ajaw	Kick Lord
				Lord of Inevitable Completion
			Kol Ajaw	Loose or Left-Over Lord
			Lok	Bend, Fold

TABLE 1.4

Codex Dresden innovative hieroglyphic names

Reference name	Color prefix	Translation	New name	Translation
Uo	Éek'-éek'	Black Black	K'at	Crisscross; Wicker
	Éek'	Black	Táan	Center
Tzec			Tzek'	Skull
Xul			Òok Ni'	Pointed Stick
			Ok Tz'un	Planting Stick
			Tz'ikin	Bird
			Ch'oon	Vulture
			Ch'o'-Ni'	Mouse Nose
Kankin			Tzuw	Agouti
Pax			Pàax	Drum
Kayab			K'än Awa	Red-Yellow Macaw
			K'a'aw	Great-Tailed Grackle

TABLE 1.5

The Interim Set: innovative hieroglyphic names

Yukatek reference name Innovative hieroglyphic name	Language	Translation
Chen		
Ihk' Kum	Ch'olan	Black Stone
Yax		
Yäx Kum	Ch'olan	Green Stone
Yá'ax Kúun	Yukatekan	Green Thunder/Pot
Yá'ax Kun	Yukatekan	Green Seat
Yäx Tun	Ch'olan	Green Stone
Yá'ax Tùun	Yukatekan	Green Stone
Sak		
Säk Ha'	Ch'olan	White Water
Sak Sihoom	Yukatekan	White Soapberry
Ceh		
Chäk Tun	Ch'olan	Red Stone

TABLE 1.6
The Landa Set: manuscript names, dates, and pages

Names	Dates	Pages
Pop	16-VII	39r
Popp	None	38r–40r
Vo	5-VIII	39v, 40r
Zip	25-VIII	34v, 40r
Tzoz	14-IX	40v
Tzoz	none	41v
Tzec	4-X	41rv, 43r
Xul	24-X	37v, 42r
Yaxkin	13-XI	42v
Mol	3-XII	42v, 43r
Chen	23-XII	34r, 43v
Yax	12-I	34r
Zac	1-II	34v
Ceh	21-II	35r
Mac	13-III	35v, 37v
Kankin	2-IV	36r
Muan	22-IV	36v
Pax	12-V	37r
Kaiab	1-VI	37v
Cumhu	21-VI	38r
Uayeb	11-VII	Supplied

TABLE 1.7
The Landa set: reference names and Yukatek hieroglyphic names

Reference names	Yukatek names	Translations
Pop	Póop	Bat
Uo	Wooj	Glyph
Sip	Sip	Miss
Tzac	Tsek'	Skull
	Tse'	Lumpy Cornmeal
Tzoz	Tzòo's	Bat
Chen	Ch'éen	Cave
Yax	Yá'ax	Green
Ceh	Kéeh	Deer
Kankin	K'an K'ìin	Yellow Sun
Pax	Pàax	Drum
Kayab	K'ayab'	Drum
Kayab	K'àay-ab'	Horizontal Drum
Cumku	Kúum K'uh	Thunder God
Cumku	Kúumk'uh	Thunder of the God; Pot God
	Kùum K'uh	Cloud-Thickening God
	Kum K'uh	Throne God
	Kúum K'ú'	Thunder Cave; Potter's Kiln
	Kum K'ú'	Throne Cave

The standard form of Landa's Tzoz is Zotz.

TABLE 1.8

Yukatek definitions for Landa and the Books of Chilam Balam

Referent forms	Interpretive forms	Translations
Ceh	Kéeh	Deer
Chen	Ch'èen	Cave with Water
Cumku	Kúum K'uh	Thunder of the God; Pot God
	Kùum K'uh	Cloud-Thickening God
	Kum K'uh	Throne God
	Kúum K'ú'	Thunder Cave; Potter's Kiln
	Kum K'ú'	Throne Cave
Kankin	K'ank'ìn/K'an K'ìin	Days of Ripening; Harvest; Red Bead Days
	K'àan K'ìin	Cord Days; Field-Measuring or Field-Clearing Days
	Káan K'ìin	Net or Hammock Days
Kayab	K'àay-ab'	Horizontal Drum
	K'àay Yá'ab'	Song of Many
	K'ay Há'ab'	Announce the Year
Mac	Màak	Cover; Pit, Giant Sea Turtle; [Turtle] Shell; Granary
	Mak	Cornfield Measure
Mol	Mòol	Pile; Harvest
	Mó'ol	Paw
Muan	Muan	Hawk; Owl
Pax	Pàax	Upright Drum
Pop	Póop	Reed Mat
Sac	Sak	White; White Flower; Silversides Fish
Sec	Sek	Bat
Sip	Sìip	Offense; Ripe
	<Siip>	Miss; Deer Guardian; Shrub
Sotz'	Sòotz'	Bat
Tzec	<Tz'èk/Tzek>	Foundation
	Tzèek	Clearing
	Tzek'	Skull
	Tze'ek	Splinter
Tzos	Tzoz/Tz'òo's	Bat
Tz'e Yaxkin	<Tz'e'> Yáax K'ìin	Little First Sun; Little Dry Season
	<Tz'e'> Yá'ax K'ìin	Little Green Days; Little Reed Days; Little Dry Season; Little Green Grasshopper
Uayab	<Way-ab'>	Poison; Hex; Bed
	Wàay Há'ab'	Poison of the Year
	Wáay Há'ab'	Animal Companion of the Year
	<Way> Há'ab'	Bed of the Year
Uayeb	Wayib'	Bed
Uo	Wo'	Toad; Pitahaya Cactus

(continues)

17

TABLE 1.8   Yukatek definitions for Landa and the Books of Chilam Balam (*continued*)

Referent forms	Interpretive forms	Translations
	Wooj	Hieroglyph
Uol	Woʼol	Ball; Hole; Heap
Xuul	Xùul	End
	Xúul	Digging Stick
	Xúʼul	Lance Pod Tree
	\<Xuul\>	Striped Mullet, Goby
	\<Xul\>	Maimed
Yax	Yáʼax	Blue-Green; Reed
	Yáax	First; Great
Yaxkin	Yáax Kʼìin	First Sun; Dry Season
	Yáʼax Kʼìin	Green Days; Reed Days; Dry Season; Green Grasshopper
Zipʼ	Sìipʼ	Swelling, Ripe

The angle brackets (\<\>) indicate undetermined tone.

(fig. 14 Sip). On one (K6751: K3b) the normal, plainly different glyph for "red" surfaces nearby in the same text, suggesting that the substitution stems maybe not from a graphic similarity between the two signs attested elsewhere but rather from Chʼolan semantics. In Chʼol and Chontal, the two Western Chʼolan languages, *chäk* denotes "red" (Aulie and Aulie 1978: 51; Pérez González and Cruz 1998: 38); in Chʼoltiʼ and Chʼortiʼ, the two Eastern Chʼolan tongues, the cognate reads *chak* (Morán 1935: 12; Girard 1949: 1:110) or *chakchak* (Pérez Martínez et al. 1996: 32). In the first pair *yäx* means "green, blue, first" (Aulie and Aulie 1978: 145; Keller and Luciano G. 1997: 31, 296), while in the second pair *yax* has this meaning (Morán 1935: 2, 21, 30, 54, 67; Wisdom 1949: 766). The link between "red" and "green" comes first within Chʼol, where both *chäk* and *yäx* also mean "clear" (Aulie and Aulie 1978: 51, 145); and second between Chʼol, with its *chäk* "clean" (*sin hierba*) as well as "clear" (*limpio*) (Aulie and Aulie 1978: 51) and Chʼortiʼ, with its *yax* "clean, clear" (Wisdom 1949: 766). This synonymy might provide the mechanism for the substitution, but so far nothing supplies the motivation. Another sign for the month name, unregistered and undefined, replaces "red" on Monument 126 at Toniná, a stela (fig. 15 Sip).

Three suffixes displace the usual -**na** and -**ta**. The first, **ma** (fig. 16 Sip), indicates **táam*, as just discussed for **UO**. The second suffix, on a fragment from Copán (fig. 17 Sip), fuses signs read **pa** and **wa** to yield **PAW** "net bag"; this meaning parallels the main sign as **KʼAT**, the unexpected Kʼichean word for "large net for maize" (Ximénez 1985: 160). The third complement, -**ki** (fig. 18 Sip), seems to signal proto-Chʼolan/proto-Yukatek ****SIBʼIK/SABʼAK** "soot," from ****TAN/TÁʼAN-na** "ashes," itself from the original ****TAHN/ TÁAN-na** "chest, center." A rare superfix to the main sign likely indicates that it starts with **kʼä** or **ta** (fig. 19 Sip).

Sotz

A definite **la** suffixes the bat head on several ceramic vessels (fig. 21 Sotz), yielding **SUTZ'(i)-la** or **SUTZ'IL** "bat place or time," possibly a rebus for the Ch'olti' *sutz'il* "winter, rainy season," literally "cloud time." In the inscriptions, the usual name for this score of days is *suutz'* "bat" (Houston et al. 1998: 280). A superfix closely resembling a variant for **sak* "white" (figs. 58 and 59 Sac) accompanies the bat head at Resbalón (fig. 22 Sotz). As *sek* and *sik* function in Chontal (Becerra 1934: 33 [blanco *tsek*; huevo *siktok*]) as variants for *sak* "white" but also denote "bat" in Lakandon (Bruce 1979: 217, 285; Cassell 1974: 218), the name here seems to read *seek* "bat." The **ku** infix with the bat head (fig. 23 Sotz) prompts the same interpretation, as **SEEK-ku**, the complement vowel indicating the long root vowel.

Sec

Landa records this name as Tzec, but all other colonial sources offer **SEC**. The first apparently derives from the Yukatek *tsèek'* "skull" (Edmonson 1988: 248). It first appears about 714 CE on a ballcourt ring at Oxkintok, Yucatán. Two glyphs seem either to spell out the month name or record it as a logograph with a prefixed phonetic complement. The first represents **tze** or **se** (Bricker 1986: 7, 148), while the second, a cranium, presents either the syllable **kV** or **k'V** or the word *ts'eek* "skull" (fig. 24 Sec). This text remains the sole instance of the name for over 700 years. It resurfaces about 1500 CE, only in the Codex Dresden.

Normal versions of Casew, the Classic counterpart to Sec, invariably bear the suffix **-wa** to indicate the final consonant. Only two instances, both from the same text, represent **se** with a skull; the sign inside the skull is the usual glyph for this syllable of the name (figs. 25, 26 Sec). One of the two head variants and a few of the standard forms delete the suffix, leaving **KAAS** (figs. 26, 27 Sec). A noncalendrical *kaas* denotes the rattle or scepter that a royal dancer grasps (Schele and Friedel 1990, fig. 8.23). One instance omits the **ka-**, yielding **SEEW** (fig. 28 Sec), similar to the Landa variant (fig. 30 Tzec). Unrecorded in Yukatekan or Ch'olan, the term evades translation. Finally, Codex Dresden presents a hybrid (fig. 29 Sec), perhaps a rearranged **KASEW** (Bricker 2000: 96–97) or, alternatively, a normal **TSEC** with **-wa** cueing **SEEW** or **KASEW**. Epigraphers still debate whether the main sign yields **se**, **tse**, or both.

Xul

The compound for Xul consists of two parts, some sort of animal head followed by **ni**, which guarantees the final phoneme of the collocation as **n**. The unique example from Yaxchilán (fig. 32 Xul) offers a clue to the spoken name: The sign wedged between the head and **ni** is **ki**. The final lead is encoded in the heads. These arguably represent creatures like the coati (*chi'ik*), the dog (*tz'i'*), the rat or mouse (*tzuk*, *ch'o'*), the gopher or mole (*b'ah*), and the agouti or paca (*halaw*, *tzub'*); their names or the first parts of these terms might yield the initial syllable.

The Highland ethnographic counterpart to **XUL** is in most cases Tz'ikin Q'iij "Bird Days." Recalling **ki** and **ni**, this points to *tz'i'* and *tz'ijk*, putative Ch'olan predecessors to Ch'ol *tz'i'* "dog" and *tz'ik* "animal" (Attinasi 1973: 349). Together as **tz'i'-ki-ni** or **tz'ijk-[ki]-ni** they would virtually spell *tz'ikin* "bird" in proto-Mayan, proto-K'ichean, and proto-Q'anjob'alan. The Ch'olan equivalent, Chichin "Bird," might descend from *tz'ikin* through normalization (Fox and Justeson 1984: 41–42 n. 16). Other readings will likely derive from the values for the remaining heads. If "bird" is its basic sense, it remains odd that the compound never did develop a variant based on an avian image.

As some of the earlier examples of Xul do not sport the signature affix -**ni** (fig. 31 Xul; also Tikal Miscellaneous Text 27:A4), this form suggests an alternate value and perhaps even preserves the original hieroglyph. Some solitary heads resembling coatimundis might represent PC *chik* "coati." Homophony could bring this to *chik* "bird," a reflex reconstructed from the obsolete generic term in Ch'orti' (Wisdom 1940: 211 n. 13). Synonymy would then shift this to *tz'ikin*.

A parallel process might link *tz'ikin* with Xul. To the glyphic *tz'ikin* "Bird" it posits a Q'eqchi' equivalent *Xul "Animal, Bird" (Pinkerton 1976: 141). This entered Yukatek as a loan but slowly naturalized into *xúul* "digging stick."

Mol

Five distinct suffixes modify Mol. **Ma** and **la** indicate **mo-lo-ma** and **mo-lo-la** for *mol-om* and *mol-ol*, both "collector, harvester" (figs. 35, 36 Mol). An untranslatable *mol-ow*, *mol-aw*, or *mow* results from the addition of **wa** for **mo-lo-wa**, **mo-l (o)-aw**, or **mo[l]-wa** (fig. 37 Mol). If they are not parts of the **lo** component, a skull and an eye with volutes represent two more elaborations on the name (fig. 38, 39 Mol), currently without readings.

Chen

This name appears in six versions. With the first (fig. 40 Chen), the complement -**a** might cue to the long vowel of Sihoom "flower," a reading proposed, without explanation, for the main sign in each of the four consecutive color months Chen, Yax, Sac, and Ceh (Houston et al. 1998: 282). The -**a** might well indicate instead the vowel or glottal stop of **HA'** "soapberry" (*Sapindus saponaria*), the same and only Yukatek gloss for *sihoom*.

The second form of this month name (fig. 41 Chen) appends **la**, perhaps pointing to **HA'AL** "rain"; this assumes, first, that the many-stranded knot atop the stone (**hi**) derives from the ancestor of Mixe *hip* "scrape, sharpen, polish" and, second, that it also reads in proto-Yukatek as *HA' "scrape, sharpen, polish" as well as, by homophony, "soapberry" and "water," both *ha'*.

The third version of Chen (figs. 42, 43) pairs the consistent absence of knot variant prefixes with the presence of the suffix T74 -**ma**. This suggests a reading ending in **m** other than **SIHOOM**. The Yukatek *kúum* "thunder" and the putative Ch'orti' *ku'um* "stone" command consideration: first because the main sign depicts a stone, and second because the day **CAUAC** means "thunder, lightning." The variant with the "X" infix does render **ku** and so, arguably, *ku'um*. It develops during the Late Classic, first serving in Yax in 874 CE (fig. 53 Yax). It last surfaces in the Codex Dresden about 1250–1500 CE.

The fourth innovation on Chen remains unique (fig. 46 Chen). Dating to 736 CE, this compound suffixes **ni** to a knotless stone, giving **TUUN** "stone." The combination returns only some 800 years later, when Landa's informant draws the signs for Yax and Ceh (figs. 56 Yax, 62 Ceh).

A text at Palenque presents Chen as an open-palmed right hand over a left hand seen from the back (fig. 47 Chen), the month's fifth avatar. The last one graces a hieroglyphic staircase at Dos Pilas as a zoomorphic head, with its large eye abutting the index finger and thumb of the same left hand recorded at Palenque (fig. 48 Chen). Not enough details survive to generate a reading.

Yax

This name knows four variations. In one instance the compound sports the knot **PÁWOB'** "Net Bag" (fig. 51 Yax; Pacheco Cruz 1960: 140; McClaran Stefflre 1972: 237). With the main sign as **TUUN**, the compound **PAWOB'-TUUN** "Net-Bag Stone" might name the wizened wind deity God N, who in directional avatars upholds the earth or sky (Coe 1973: 14–15; Fox and Justeson 1984: 49). Several examples of the second version of Yax on pots display both the multistrand knot as well as **la**, signaling **HA'-la** or **HA'AL** "rain" for the combination (fig. 52 Yax). Rather late, in 874 CE, **ma** suffixes **KÚUM/ku**, rendering *kúum* "thunder," the third version (fig. 53 Yax). This same value seems to be recorded at Chichén Itzá (fig. 54 Yax). Finally, another date pairs the stone with a perhaps unusual **na** to produce the fourth alteration, **TUUN** "stone" or **KUN** "seat" (fig. 55 Yax). The first value derives from countless examples of "stone"-**ni** in Classic texts. The second finds its cue in the "X" infix of the sign for **cu** in Landa's "alphabet" and his compound for the month Cumku; it apparently depicts the diagnostic feature for the day sign "Flint" (fig. 56 Yax). This variant of the "stone" glyph, as **ku**, appears in a noncalendrical context at least some 800 years earlier (783 CE), at Palenque, on the Tablet of the 96 Glyphs (A8, C8, F6, H5).

The above variants all retain the stone as their main sign. Another version of **YAX**, however, replaces this with the glyph for *winik* "twenty days" or "person." Copán offers the earlier instance, on a fragment datable only to the fourth or fifth century CE (fig. 49 Yax). Ek' Balam preserves a text pinpointed to 29 July 794 CE (fig. 50 Yax).

Sac

The compound for this month alters the color glyph in two ways and complements the reading of the main sign in two ways. In one instance the color is written with a head variant (fig. 57 Sac), while in others an alternative initial logograph seems to proffer **SEK** as a dialect version of **SÄK** "white" (figs. 58, 59 Sac), the same prefix that prompts **SEK** "bat" for the **SOTZ** head (ig. 22 Sotz). Another instance of **SAC** suffixes **la-la** as a cue to **HA'AL-AL** "rainy season" (fig. 60 Sac). Only in the cave of Naj Tunich does **ka** appear at all, below the sign for "stone." It might complement the entire compound as **SAK** "white" (fig. 61 Sac). Like all the later color series except in Q'anjob'al, this implies that the main sign eventually fell mute, a process perhaps completed by about the time of this inscription, 744 CE. Very likely the texts at this cavern represent a form of Yukatekan or Ch'olan; this poses no difficulty, though, as the color months in each case offer no counterpart to the glyph for "stone."

Ceh

Two forms of Ceh, complete with their color prefix, delete the knot superfix and replace the usual **ma** complement with a different **ma**. In these two examples, as in six other color month signs, whenever one **ma** (T74) replaces the other (T142), the knot superfix, the cue to the reading **SIHOOM** vanishes. Based on the final **m** and on **ku** for the main sign, this combination seems a convention for the Ch'olan *kum* "stone" or the Yukatekan *kúum* "thunder." One of the two examples for **CEH** dates events on a ceramic cylinder of 600–800 CE (Schele and Miller 1986: 239, pl. 92A), while the other survives on a stela dating to 869 CE (Greene et al. 1972: pl. 101). The compound yields proto-Ch'olan *CHÄK KUM "Red Stone" or proto-Yukatek *CHÄK KÚUM "Red Thunder."

Mac

One of the four innovations on Mac (Houston et al. 1998: 283) is the glyph depicting a serpent segment **ma** fused with a turtle shell **MAJK** (fig. 63 Mac). This yields **MAM** "grandfather" via **ma-ma**. Another alteration replaces the usual -**ka** with another **ma** sign (fig. 64 Mac). If this version, however, represents instead the full form of **ma**, it records the ancestor of Ch'ol *ma* "grandson" or "great, old" (Attinasi 1975: 11 n. 2; Aulie and Aulie 1978: 78, 291; Schumann 1973: 26). Minus the subfix of three circles, the array represents **MA** (fig. 65 Mac). The last variant occurs on a throne from Palenque (Stuart 2005: 193). This omits the **ma** superfix found on all other known examples of this name based on the shell, producing **MAJK-ka** for *majk* "turtle carapace" (fig. 66 Mac). With the shell alone as **MAJK**, the **ma** might serve as a phonetic complement (Zender 2006). Its frequency, however, far surpasses that typical for other complements, so the **ma** perhaps represents a whole word integral to the name, perhaps **MA MAJK** "Elder Turtle Shell," a possible reference to some deity in a carapace.

Kankin

Two early significant examples of Kankin occur in the same text (figs. 67, 68 Kankin). The first appears without any complement. The second displays only **ni** for its suffix, pointing to **UN** "avocado." A later instance of the name sports a bifurcate complement, either a variant of **ni** or a fusion of this sign with **wi** or **wa** that produces the

-**niw** of Uniw (fig. 69 Kankin). Most examples of the name end with **ni-wi** or **ni-wa** or just **wi** or **wa** to indicate **UNIW**, spelled out on occasion as **u-ni-wa**. **UN**, the reading for the glyphs suffixed with just **ni**, suggests the more frequent form consists of *un* + *iw*, two terms for "avocado" still attested in Tzeltal, combined to describe a certain variety of the fruit. An alternative derivation casts Uniw as the Ch'olan "switched-consonants" form of the Tzeltalan *oven*, itself a Mayan modification of the Mixe-Zoquean *owi* "avocado."

K'an K'iin, the Yukatek name for this month, appears in one or two Classic texts. An unimpeachable example of the name presented as precisely those two words is recorded on Panel 2 from Xcalumkin, dated to 743 CE (fig. 70 Kankin). Another such instance seemingly emerges on a fractured block at Copán designated Stela 52 (fig. 71 Kankin). Style places this relic to somewhere in the span 400–600 CE.

Yet another value for this month name derives from an animal head nearly always suffixed with **wi** or **wa**, but never with **ni**, **ni-w**, or **u** (figs. 72, 73 Kankin). The sign (T753) first appears in the month name on Chinikiha's Throne 1 at 574 CE and last on Quiriguá's Structure 1 at 800 CE (Grube 1990: 117). Andrea Stone and Marc Zender (2011: 6, 180–81) identify the head with wrinkles and a notched trilobate ear as the portrait of a white-nosed coati (*Nasua narica*) then render it as *tz'utz'ih*, often spelled phonetically as **tz'u-tz'i-(hi)**, "coatimundi." Nikolai Grube and Werner Nahm (1994: 699) also tie images of coatis on ceramic vessels to phonetic spellings for *tz'utz'* "coati." Known in Spanish as the *tejón* and *pizote*, the coatimundi or *Nasua narica yucatanica* goes by *tz'utz'ub'* and *tz'utz'u'* in dialects of Ch'ol (Hopkins et al. 2011: 251–52; Stoll 1938: 62), Tzeltal (Stoll 1938: 62; Sapper 1897: 424, line 7), and Lakandon (McGee 1990: 32, table 3.4; Bruce 1979: 184).

The -**wa** and -**wi** suffixes to the coati head might indicate the -**u'** of *tz'utz'u'*; it seems that a final -**u** could serve this function, but it never appears in this context. Perhaps instead the -**w** represents a word-final variant of the -**b'** in *tz'utz'ub'*; a close parallel surfaces in Yukatek versions of "agouti, paca," namely *haaleb'* and *halew* (Barrera Vásquez et al. 1980: 176) as well as in Lakandon's cognates *hale'/halew* (Bruce 1979: 156, 171, 274, 283).

A present-day practice for defining types of avocados might preserve a rationale for the shift from the standard Uniw to the innovation Tz'utz'uw. The -**IW** of **UNIW** means "*aguacate del monte*/wild avocado" in some dialects of Tzeltal (Slocum 1953: Tzeltal-Español, 17; Slocum and Gerdel 1965: 12). Sometimes an animal name follows the term *un/on* "avocado" to denote a wild variety, as in Tzeltal's *yon chuch* "squirrel's avocado" (Berlin et al. 1974: 220), Ch'orti's *un ma'x* "monkey's avocado" (Wisdom 1949: 747), or Mopan's *ontzub* "paca's/agouti's avocado" (Ulrich and Ulrich 1976: 147). These suggest an unattested interim form *UNIW TZ'UTZ'UW, soon reinterpreted to **TZ'UTZ'UW**.

Pax

Two compositions determine interpretations of this month name. The first employs phonetic spellings: **pa-xa** (fig. 82 Pax) or **pa-xi** (figs. 76, 77 Pax) for **PAAX** "drum"

(Houston et al. 1998: 280), **pa-xi-la** for **PÀAX-IL** "music" (fig. 78 Pax). The second infixes **b'i** (fig. 84 Pax), or suffixes **-la** (fig. 85 Pax) or **-xa** (figs. 79, 80 Pax) to the standard two-part main sign; one example even infixes **pa** then suffixes **-xa** (fig. 81 Pax). The first of these two elements likely represents *pas* "open up, bloom" in Ch'olan, *pa'x* "break open" in Yukatek. The second part encodes proto-Ch'olan **HAB'*, proto-Yukatek **HA'AB'* "year, season." Together they yield proto-Ch'olan **PAS-HAB'* "sprout season," then proto-Yukatek **PA'X-HA'AB'* "break-open season." As *ha'ab'* abbreviates to *-ab'* in glyphic compounds, reinterpretation nudged the proto-Yukatek to **PAAX-AB'* "tap or strum" + "instrument," then **PAAX* "upright drum." This etymology indicates that written **PAAX** and **PAAX-IL** represent loans from Yukatek. The infix **b'i** cues to any form ending with **b**, while **-la** points to **PAAX-IL** or **TUNKUL** "horizontal slit-drum." The occasional **-ma** (fig. 84 Pax) hints at a dialect variant of **PAAX-AM** for **PAAX-AB'**, or **TUM** "drum" for **PAAX**.

Kayab

The normal version of this sign group corresponds to the Ch'ol **CANAZI**, interpretable as *k'an asij* "yellow cicada" or *k'an aj si'* "yellow woodchopper." The full-figure version of this month from Copán clearly depicts a parrot (fig. 91 Kayab). Almost everywhere the parrot head main sign (Justeson 1975; Stuart 1985) bears **K'AN** "yellow" in its eye. In renditions of a particular instance from Palenque one artist depicts this color infix (fig. 86 Kayab), while another records instead a lid and iris (fig. 87 Kayab). Normally **si** and **ya** accompany the head to form **-siy**, perhaps for **sii**, **si'**, or **sil**, with **si** absent rather more frequently than **ya**. This deletion pattern intimates that the main sign alone encodes the entire name. The candidate **ASIH** leads to Q'eqchi' *asih* "cicada" (Haeserijn 1979: 42; Sedat 1955: 25). The last innovative form, extant only on ceramics, deletes both **si** and **ya** (fig. 89 Kayab) representing by itself **K'AN ASIH**.

Cumku

The normal form of this name changes when four different glyphs each substitute for **OHL/OCH** "ball, sprout" and "food" as the superfix to **WAJ** "tamal, tortilla, food." The first displays a quincunx. The occasional presence of a suffix **la** here hints that this sign, too, reads **OHL** (figs. 93, 94 Cumku). The absence of any complement with the second alternate leaves its sound and sense unknown (fig. 95 Cumku). The third alternate might depict a bird wing with a blood scroll, but it does occasionally represent "nine" (fig. 96 Cumku). The name reads either Bolon Waj "Nine Tortillas" or Bolon Ol "Nine Balls." One evokes a stack of nine tortillas, apparently prescribed as an offering in the Codex Dresden (65b:A4). The other suggests heaped up balls of ground cooked maize, also indicated in that screenfold (31b:C3) as a meal for the rain god Chak, alongside fish and fresh tortillas.

On several artifacts the fourth substitute is **ha** (fig. 97 Cumku). As the logograph **HA**, the term denotes the large, spherical gourd called both *ha* and *hay* in Tzeltal, used to store tortillas (*waaj*) and catch the blood of recently slaughtered bulls (Berlin et al. 1974: 129, fig. 5.4a, 131; Slocum 1953: 18; Robles Urribe 1966: 32). This variant reads *HA-WAJ "gourd for tortillas." The roundness of the gourd and its function with blood recall the senses of the inscriptional **OHL** as "ball" and perhaps "pierce, let blood." They call up interpretations of **OHL** as the designation not only for a portal to the Otherworld opened during visions induced by bloodletting but also for the god pots used to collect and burn the paper fillets spattered with that blood (Freidel et al. 1993: 215). These associations recur in Kumk'u (Cumku) as "pot god." The phrase **u haay** "his bowl" often occurs on ceramic vessels, with *hay* spelled **ha-yi**, using for **ha** the same alternate superfix on **OL** (Tate 1985). The occasional -**la** suffix indicates that the compound represented not two words but just one, **OL** (fig. 98 Cumku).

Finally, one name for this month survives spelled out in syllables. On a block from the hieroglyphic stairway at Copán the text reads *tu jo'* **o-la** "on the fifth of Ohl" (fig. 99 Cumku).

Uayeb

Only four innovations of this name survive. The first prefixes **U** to the standard compound **WAY-HAB'/hab'** to render **U WAY HAB'** "poison/vision/co-essence/bed of the year"; the possessive *u* compels the interpretation of **HAB'** "year" as the possessor instead of the syllable **hab'** reduced to **ab'** for the instrumental -**AB'** (fig. 101 Uayeb).

The second form of **UAYEB** occurs on a plate commemorating a *k'atun* ending (Martin and Grube 2000: 39). Here the group consists of **AJAW ko-chu** for **KOCH AJAW** (fig. 102 Uayeb). In Yukatek *kóoch* denotes "inevitable, something that cannot fail to come to completion"; it is also the root "kick" (Barrera Vásquez et al. 1980: 324, 325). The name means "kick lord" or "lord of the inevitable completion." With no Ch'olan cognates, this designation derives from Yukatekan.

Stela 14 (A7) at Caracol preserves the third variant (fig. 103 Uayeb); **ko** comes first, for **ko-lo AJAW**. *Kol* means "loose" in Ch'ol (Aulie and Aulie 1978: 37), "left over" in Chontal (Keller and Luciano G. 1997: 59). *KOL AJAW "loose or left-over lord" refers to the period as an uncontrolled or extra demigod.

The fourth version for this brief period also surfaces in Caracol (fig. 104 Uayeb). It reverses the main signs, yielding **lo-ko**, apparently a Ch'olan root for "bend, fold" (Aulie and Aulie 1978: 73), a meaning appropriate for both the end of the year and the final page of a series in a codex.

THE CODEX DRESDEN SET

Dated to 1220–1320 CE by its astronomical content (Bricker and Bricker 2007: 114–16; 2011: 186–87), but more generally to about 1250 CE (Milbrath 1999: 5, 6; Thompson 1972: 15–16), the present manuscript of the Codex Dresden offers variations on eight month names. Three or four of these present major reinterpretations. The manuscript itself, like the other three surviving Maya books, consists of long strips of paper made from the bark of the fig tree; these are joined, then folded back and forth to produce a screenfold document. After coating the pages with a thin layer of plaster, scribes filled them with hieroglyphs and colorful pictures concerned with drought, rain, pests, and divination. The creators of the Codex Dresden, however, also focused intensely on solar and lunar eclipses as well as on the cycles of Venus and the motions of Mars. Table 1.1 displays the normal Dresden forms.

Uo

Of the two variants on this name, one adds a second, large "Black" and infixes **K'AT** there to fashion **ÉEK'-ÉEK' K'AT** "Black Black Crisscross" (fig. 12 Uo). The other discards the main sign for another glyph read **tan**, so the compound represents **ÉEK' TÁAN** "Black Center" (fig. 13 Uo). This second **TÁAN** translates well as "center, middle," with toponyms marking the travels of Cháak in this codex (pages 34–35, 40–41, 65–66, 68–69).

Tzec/Sec

This name remains troublesome to gloss. **TZEK** or **SEK** clearly corresponds to the usual compound **tze/se-ka**, extant only in this codex (fig. 25 Sec). The most straightforward solution views it as a reduced form of the Yukatek *tzek'* "skull," calling on the rare cranial variants for **tze/se** in the Classic counterpart Kasew (figs. 24, 25, 26 Sec). Word-final reduction of *k'* to *k* is attested but infrequent in modern Yukatek. Some examples are *k'olok'*: *k'olok* a plant (Roys 1931: 257), *nak'*: *nak* lean against (Bricker et al. 1998: 193, 194); and *tuk'*: *tuk* a plant (Barrera Marín et al. 1976: 148–49, #1186, #1188). The unique **TZEC** on Dresden 62:A10 (fig. 29 Sec) appends **wa** to the side of the normal **tse-ka**. Apparently the Yukatekan scribe, copying **KASEEW** from a Ch'olan manuscript, inadvertently set down the name he was most familiar with, **TZEK**, placing **ka** after **tze** instead of before it (Bricker 2000: 96–97); to address the lapse he could only add the **wa**. An alternative suggests that the -**wa** indicates the name can read as either **TSEK** or **KASEW**, just as the presence of -**wa** and -**b'u** both at the same time with the sign for **POP** allows **K'AN JALAB'** as well as **K'AN JALAW** (fig. 4 Pop).

Xul

The codex uses two versions of this month. The first, **OK-NI**, appears only twice (fig. 33 Xul). The major glyph depicts a dog and designates it with the Mixe term *ok* "dog," rebus for the Yukatek *òok* "enter; leg, staff." *Ni'* is glossed with "nose, tip, point" (Bricker et al. 1998: 196). These components together could produce **ÒOK NI'* "pointed stick," a virtual equivalent to Landa's **XUL** as *xúul* "planting stick."

The parts of the second form in Dresden read **ch'o** and **ni** (fig. 34 Xul), for **CH'ON**. In Yukatek this term would fit only as a dialect version of *ch'oom* "vulture," possible given the **-m/-n** variation attested in examples like *òom/òon* "avocado" (Bricker et al. 1998: 18).

Chen

An undeciphered sign replaces "Black" in each instance of this name (fig. 45 Chen), even though the usual **EEK'** does accompany **K'AT** in every example of Uo in the codex. As **ch'e-** or **CH'E'EN** "cave" the new glyph would document the designation recorded as **CHEN** by Landa in 1566 and date it to about a hundred years earlier, but its absence from any other context precludes verification.

Kankin

The sole form of this name always consists of two glyphs that both enjoy secure readings but together produce only **tzu-wa** or ***TZUW**, an unattested word (fig. 74 Kankin). This might, however, offer a version of the Yukatekan *tzuub'* "agouti."

The larger sign depicts a vertebral column with ribs, so it reads **tzu**, from **TZUL** "spine." The suffix normally renders **wa**, but just **w** word-finally. No sign representing **u** ever replaces it. This recalls the pattern of **-w** with the coati-head variant of the Classic Kankin. There the suffix signals *Tz'utz'uw for Tz'utz'u'/Tz'utz'ub' "coati," recording either the terminal glottal stop or the rare but attested shift of *b'* to *w* at the end of a word. In Yukatek *tz'uub'* and *tz'uu'* name the agouti or *Dasyprocta punctata* (Bricker et al. 1998: 45; Barrera Vásquez et al. 1980: 725, 865). The cognates are *tzub'* in Mopan (Ulrich and Ulrich 1976 :221), *tzu'* in Itzaj (Hofling 1997: 136) and *tzub'* in Lakandon (Baer and Merrifield 1971:2 38). The Dresden scribes apparently wrote ***TZUW** for **TZUUB'** or **TZUU'**. **UNIW** "wild avocado" transitions to ***TZUW** "agouti" via a proto-Ch'olan parallel to the Mopan term *ontzub'* "agouti's avocado" (Ulrich and Ulrich 1976: 147).

Pax

One of the two versions of this name appends **-xa** to the usual compound to spell **PÀAX** "drum" (fig. 80 Pax). The other both infixes **pa** and suffixes **-xa** to indicate the same term (fig. 81 Pax). A fully phonetic **pa-xa** spelling appears over 700 years earlier in a cave (fig. 82 Pax).

Kayab

For this name the Dresden scribe retains the main sign, a parrot head, with its infix **K'AN** "yellow," but discards **si-ya** to replace it every time but once with **wa** (fig. 90 Kayab). One interpretation casts this collocation as the Tzeltalan *kän awa* "red-yellow macaw," with *awa* an assimilated version of the colonial Tzeltal *ova* in *x-ova: can mut* "macaw: a red-yellow bird" (Ara 1986:4 18) and the colonial Tzotzil *ova* "parrot" (Laughlin 1988: 157). Another analysis yields **k'a-a-wa** for the Yukatek

K'A'AW "great-tailed grackle" (Bastarrachea Manzano et al. 1992: 99). The acrophony illustrated by the use of **K'AN** for **k'a** rarely affects color logograms, yet in this same document (Dresden 8c: C1, G1) **CHAK** "red" supplies the first syllable in the verb **U cha-k'a-ja** or *u chak'-aj* "he shut his eyelids."

Cumku

Dresden scribes substitute one version of a glyph for the usual allographs (fig. 92 Cumku). The absence of phonetic complements, however, precludes assigning it a reading.

THE INTERIM SET

Only three signs in Landa's records clearly represent values different from those of their counterparts in both Landa's list (discussed below) and the Classic period and Dresden corpora (table 1.1). Chen reads **i-kin-KUM-ma**, Ch'olan **IHK' KUM** "Black Stone," if the inverted stone represents the similar **ku** as on Dresden 17c:C1 (fig. 44 Chen). For Yax two features indicate many readings for the main sign, a stone (fig. 56 Yax). With the "X" or "flint" infix it reads Ch'olan **KUM** "stone"; the **-ni** indicates **TUN**, also "stone." Sometime after the debut of **KUM** in 874 (fig. 53 Yax), the name became **YÁ'AX TÙUN** (Yukatekan) or **YÄX TUN** (Ch'olan) "Green Stone." Or this same combination could instead spell out the referent as **ku-ni** or **KUN**, for the Yukatek *kúun* "thunder" and "pot" or *kun* "seat." With no suffix and perhaps no infix for the main sign, Zac seems to record the usual **SAK HA'/SIHOOM** "White Water/Soapberry." Finally, in Ceh the **-ni** and the absence of the "X" infix both select for **TUN** (fig. 62 Ceh), as they do in countless Classic inscriptions, though only once before with a color month (fig. 46 Chen).

THE LANDA SET

Diego de Landa arrived in Yucatán as a Franciscan missionary in 1549. He proved especially zealous in promoting and defending the faith. In 1566, in Spain, Landa penned or dictated his *Relación de las cosas de Yucatán*, perhaps as part of his defense against charges that he had overstepped his inquisitorial powers and maybe also as a reference and introduction for missionaries to the many aspects of the Maya culture. Landa returned, acquitted and consecrated as the new bishop, in October 1573, bearing either the original manuscript or a full copy. He died in 1579. In the section about the native calendar in his book, Landa appended to eleven of his eighteen months extra hieroglyphs that indicated Yukatek names distinct from those of the Classic and Dresden sets. To six he preposed syllabic spellings, namely Pop, Uo, Chen, Pax, Kayab, and Cumku. To one (Zip) he prefixed a phonetic complement, while for three others (Tzec, Yax, and Ceh) he suffixed one. Finally, in a single case (Kankin) he significantly altered or misunderstood the main sign. Table 1.1 presents his series.

Pop

To a usual version of this compound, **K'AN JAL-AW/AB'** "Yellow Loom," Landa prefixed **po-po** for **PÓOP** "Mat" (fig. 3 Pop). The glyph apparently depicts a small disc of copal resin incense molded around a flammable pith center, like those described among the Ch'orti' (Fought 1972: 140:5.125; Wisdom 1940: 183). As a logograph it denotes **PÒOM** "copal incense," source of phonetic **po**.

Uo

Landa's prefix (fig. 9 Uo) recalls an 880 CE text at Chichén Itzá (fig. 8 Uo), where **wo** follows the standard **ÉEK' TÁAN** "Black Center" to cue either **WO'** "Bullfrog" (Berlin 1987) or **WOOJ** "glyph," the more likely value because of the **wo-ji** spelling on a Late Classic vessel (fig. 7 Uo).

Sip

A perhaps stylized myriapod precedes the glyphic Ch'olan **CHÄK K'ÄT** "Red Clay" (fig. 20 Sip) as a likely **si**, cuing to **SIP** "slacken a snare or bow; miss, err." This names a Yukatek deer god who must be appeased lest he make the hunter miss. The change of the inscriptional form to that of Landa can derive from either of two shifts in meaning. The first reinterpreted *chäk* "red" as *chäk* "hunt, capture" (Ch'ol: Attinasi 1973: 250). *K'ät* "clay" probably also means, by analogy to Yukatek, "clay vessel, clay figure, idol." "Hunt Idol" could suggest to Yukatek borrowers their deer protector. The second alternative changes "red" to "deer" in Ch'olan. This is suggested first by dialect variants in Yokotan Chontal (Blom and LaFarge 1927: 2:477) and Ch'ol (Attinasi 1973: 253) that replace *chäk* "red" with *chik* and second by the Ch'olti' *chiik* "deer, stag" (Morán 1935: 11, 33). Yukatek could then transform *****CHIK K'ÄT** "Deer Idol" into **SIP**.

Tzec

Instead of the normal Dresden version **tze-ka** for *tzek* "foundation, house mound" or a reduced *tzek'* "skull," Landa presented **tse-wa** (fig. 30 Tzec). The simplest explanation attributes the error to his informant or copyist. As *tsew* and *sew* produce no valid lexemes, the mistake consists not in the substitution of **wa** for **ka** but in the omission of **ka**. The aberrant form does surface rarely in the Classic corpus (fig. 28 Sec).

Chen

Supplanting a putative **ÉEK'** "black," the scribe provides the spelling **i-kin** for **IK'** "black" in Ch'olan. Also unusual, he inverts the main sign then suffixes -**ma** (fig. 44 Chen). His reasons remain less than transparent. Resembling the **ku** on Dresden 17c:C1, the main sign here might record the Ch'olan **KUM** "stone." If it does, the original *k'* of **IK'** may have lost its glottalization by assimilation to its adjacent plain counterpart. One surmise for the change from **IHK' KUM** "Black Stone" to **CH'E'EN** "Cave" offers that **éek'* "black" slipped into **éek'* "spot." As *kax ek'* "forest pond" (Avendaño y Loyola 1996: 63) suggests, *éek'* can refer to a pond or puddle: by itself or with the main sign as **HA'** in **ÉEK' HA'* "spot or source of water," this term evokes *ch'e'en* "well, grotto, cistern."

Yax

By the time Landa penned his report, Yukatek scribes still wrote but no longer pronounced **TÙUN**, using just **YÁAX**. The practice of using solely color prefixes as entire names began to spread from northern Yukatekan centers following their deletion of the **SIHOOM** reading for the stone sign, after perhaps 700 CE. One hint of the new names appears on the walls of Naj Tunich, where in two instances the main sign for **ZAC** suffixes a large **ka**; this unique form dates to 744 CE (fig. 61 Sac). In 794 CE another clue debuts in a fresco at Ek' B'alam: **WINIK** replaces **SIHOOM** (fig. 50 Yax). Apparently **SIHOOM**, reinterpreted to *sih-om* "somebody born, a person," evolved into **WINIK** "person" then **WINIK** "month." Eventually regarded as superfluous, the second word faded from the spoken name. Color months with **WINIK** should predate those with **SIHOOM**; possible support survives on one lone battered stone at Copán, by style from about 400–600 CE, inscribed with **5+ YAX WINIK** (fig. 49 Yax).

Ceh

Landa's **CEH** "deer" also ends with **ni** and so probably took the same steps as his **YAX**; he omitted the "X" infix indicating **KÚUM** or **KUM**, but it would have marked all the color months (fig. 62 Ceh). At some point after 794 CE these muted their main sign and continued as only colors. Ch'olan subtraditions later shifted from *chäk* "red" to its variant *chik* then to *chik* "deer." Evidence lies in dialect variants of Yokotan Chontal (Blom and LaFarge 1927: 2:477) and Ch'ol (Attinasi 1973: 253) that replace *chäk* with *chik* and in the Ch'olti' *chiik* "deer, stag" (Morán 1935: 11, 33, 59). Bilingual Yukatekan scribes later translated *chik* into *kéeh* "deer."

Kankin

Landa's rendition of this compound (fig. 75 Kankin) preserves the -**wa** suffix present in the Dresden and Classic sets but also presents the logograms **K'AN** "yellow" and **K'ÌIN** "sun." These represent the month name in just one or two other texts, the first at Xcalumkin in 743 CE, complete with -**ni** (fig. 70 Kankin), and the second on a block at Copán from 400–600 CE (fig. 71 Kankin).

Pax

In front of the standard **PAX**-hab'-ma or **PÀAX**-AM' "tap-instrument, drum," Landa's informant drew **pa** atop an oval **xV**, to yield the Yukatek **PÀAX** "drum" (fig. 83 Pax). Most published versions of this prefix show the ovoid with a thumb-like appendage that evokes the hand representing *x* in Landa's "alphabet." The manuscript, though, clearly shows no appendage and no lines for fingers.

Kayab

In front of the normal Dresden version of **K'AN AWA** "Yellow Macaw" or **K'A'AW** "Great-Tailed Grackle" Landa's scribe supplied an abbreviated spelling of **KAYAB** as **k'a-b'a**, unless the uncommon eyebrow in his **k'a** represents **ya** (fig. 88 Kayab). James Allen Fox and John S. Justeson (1984: fig. 23q) analyze the Dresden variant as **KAYAB-AB**, interpreting the Landa prefix as **k'a-AB**.

Cumku

By Landa's day this month had transmuted for the last time. His prefixes clearly announce that, while retaining its written form **OHL/OCH K'AN/WAJ** "Ball or Food of Maize, Tamal," the compound now read as **KÚUM K'UH** "Thunder God" (fig. 100 Cumku).

The upper prefix depicts a stone, but throughout its history the sign suffixes **ni** to denote **TUN** "stone." This convention indicates a different original value. A hint lies in the extended senses of *tun* as "testicle" and "egg" in Ch'olan. These meanings also apply to the Ch'orti' *ku'um* (Aulie and Aulie 1978: 115; Fought 1967: 55; Girard 1949: 1:99, 105; Morán 1935: 33; Wisdom 1949: 604), which implies that this term once also meant "stone." The "stone" sign **ku**, then, might stand for **KU'UM** as well. Yukatekan speakers transformed *KU'UM into their *KÚUM "boom, explosion, thunder." Support for these readings, particularly for the usually unmarked presence of **m**, emerges from the optional presence of **ma** as a suffix on this sign first in the proper name of the captive Ah Ukum at Yaxchilán (L 54:D1–C2) and second in two instances of a title at Dos Pilas (HS 4, Step 3:H2: Houston 1993: 109, fig. 4.11; Panel 7:A3b–B3a: Houston 1992: 67, fig. 2.3). The absence of the knot superfix indicating **hi** and the **ni** subfix from all examples of the proper name and the title excludes **hi, HA', SIHOOM,** and **TUN,** while the presence of **ma** rules out **KAWAK.** To prompt the **KU'UM, KÚUM,** and **k'u** values, scribes devised the "X" infix sometime after 600 CE. This is the form that records the first word in Landa's **CUMKU.**

The drawing in the Landa manuscript (fig. 100 Cumku) clearly shows the second glyph in Cumku to be **K'UH/ k'u.** Published depictions of this compound, most of them copies of Brasseur de Bourbourg's 1864 edition, so distort the nose and mouth that the identity is not readily apparent. Several scholars independently read the second sign as **K'UH** (Barthel 1952; Ringle 1988; George Stuart, personal communication, 1988). Yuri Knorozov (1955, 1967) was the first to present Landa's sign as the glyph T1016. Only later did he read it, and then not as **k'u** but as **XIB** "youth, male" or *ŋom/nom (Knorozov 1982: 222, 428). Thomas Barthel (1952: 92–95) was the first to suggest in print that the glyph represents **K'U** "god." This value has proven valid, productive, and highly significant.

THE CHILAM BALAM OF CHUMAYEL SET

Munro Edmonson (1986: 1–2, 226) dates the final form of the Chilam Balam of Chumayel to 1824–37 and designates 1737 as the referent year for its calendar. Between approximately 1550 and 1750 the Maya of Yucatán preserved in a dozen or so manuscripts their history, myths, rituals, and calendars, recorded in their native tongue but with the Roman alphabet; scholars designate these documents as "the Books of Chilam Balam," and distinguish each by adding the name of the town of discovery. Only in the version from Chumayel do pictures accompany the month names, but they do not resemble their counterparts in any other set. As drawings (Brinton 1890: fig. 1; Roys 1967: fig. 5; Knorozov 1955: figs. 13a–b, 1967: pl. 4) and photos (Coe and Kerr 1998: fig.42; Miller 1986: 195, fig. 160) demonstrate, these pseudo-glyphs function as "little more than vestigial ornaments" (Miller 1986: 195, fig. 160) and mark "the end of the calligraphic tradition" (Coe and Kerr 1998: 219).

Conclusion

Table 1.9 tallies the Maya month name historical innovations by type, name, and set. For details, see "The Glyphic Calendars" in chapter 5.

TABLE 1.9

Types and totals of innovations by name and set

Type Name	Logographs C	D	L	Complements C	D	L	Both C	D	L	Neither C	D	L	Totals l	c	b	n	TL	%
Pop	0	0	0	0	0	1	0	0	0	0	0	0	0	1	0	0	1	.01
Uo	1	2	0	2	0	1	1	0	0	0	0	0	3	3	1	0	7	7.6
Sip	2	0	0	3	0	1	0	0	0	0	0	0	2	4	0	0	6	6.5
Sotz	0	0	0	3	0	0	0	0	0	0	0	1	0	3	0	1	4	4.3
Sec	3	1	10	0	0	0	0	1	0	0	0	1	5	0	1	1	7	7.6
Xul	0	2	0	2	0	0	0	0	0	0	0	1	2	2	0	1	5	5.4
Yaxkin	0	0	0	0	0	0	0	0	0	0	0	0	0	0	0	0	0	0
Mol	0	0	0	5	0	0	0	0	0	0	0	0	0	5	0	0	5	5.4
Chen	2	1	0	3	0	0	0	0	1	0	0	1	3	3	1	1	8	8.7
Yax	1	0	0	4	0	0	0	0	1	0	0	1	1	4	1	1	7	7.6
Sac	3	0	0	2	0	0	0	0	0	0	0	1	3	2	0	1	6	6.5
Ceh	0	0	0	1	0	1	0	0	0	0	0	1	0	2	0	1	3	3.3
Mac	1	0	0	1	0	0	0	0	0	0	0	0	1	1	0	0	2	2.2
Kankin	2	1	1	4	0	0	2	0	0	0	0	0	4	4	2	0	10	10.9
Muan	0	0	0	0	0	0	0	0	0	0	0	0	0	0	0	0	0	0
Pax	2	0	0	2	2	1	1	0	0	0	0	0	2	5	1	0	8	8.7
Kayab	1	0	0	2	1	1	0	0	0	0	0	0	1	4	0	0	5	5.4
Cumku	3	1	0	0	0	1	1	1	0	0	0	0	4	1	1	0	6	6.5
Uayeb	2	0	0	0	0	0	0	0	0	0	0	0	2	0	0	0	2	2.2
Totals	**23**	**8**	**2**	**34**	**3**	**7**	**5**	**1**	**2**	**0**	**0**	**7**	**33**	**44**	**8**	**7**	**92**	**100**
%	25	8.7	2.2	37	3.3	7.6	5.3	1.1	2.2	0	0	7.6	35.9	47.8	8.7	7.6	100	100
TL%: C	25			37			5.3			0							67.3	
TL%: D		8.7			3.3			1.1			0						13.1	
TL%: L			2.2			7.6			2.2			7.6					19.6	

Key: C: Classic (200–950 CE), D: Dresden (1250–1500 CE), L: Landa (1450–1550 CE, includes the Interim Set); l: logographs, c: complements, b: both, n: neither.

CHAPTER TWO

The Ethnographic Calendars

Many Maya traditions preserve a considerable calendrical corpus, but most of their sources are abbreviated or redundant. Remarks by original composers and later scholars usually seem irrelevant or repetitious, while translations prove either inaccurate or virtually identical to those presented here. This study examines only the more significant counts, definitions, dates, and comments. The next two chapters offer discussions about the calendrical names and dates organized by language.

Each calendar under discussion is named by its "referent year." This tag indicates the year of composition, reference, or publication for the source document. "Initial dates" denote the first day of a month. In the Maya area Europeans did not initiate the Gregorian calendar reform until 1584 (Edmonson 1988: 83, 178). Dates below are Julian before 1584, Gregorian after.

Calendrical Documents in Ch'ol

1589. John Eric Thompson (1932) assigns some if not all the names in the calendar from Lanquín, Alta Verapaz, to Ch'ol. Edmonson (1988: 154, 191) describes them as Q'eqchi' influenced by or mixed with Ch'ol. Lyle Campbell (1984b: 10 n. 2) derives them from Ch'olan, as do John Justeson and Campbell (1997: 47–49). The document omits the equivalents to the Yukatek Pop and Sotz, beginning with 1 January at 12 Zihora. The first roster (table 2.1) presents the original order of the calendar names in the 1589 document. The second (table 2.2) presents the order expected from comparison with the glyphic set of month names, preserving the sequence but starting with the first omitted name, probably K'an Jalib', and ending with Mahi. Dates here and in the 1631 source, as well as the translation "nauseated five days" for *holob cutan*, the gloss for Mahi, all seem to justify not only this numbering of the series but also the reversal in the lengths of the first and last units.

Nowhere does the original author of the 1589 document state its referent year. Suzanne Miles (1957: 743 n. 37, table 2 n. 2) places the manuscript in the late sixteenth or early seventeenth century based on its handwriting and style of spelling.

TABLE 2.1

Lanquín Ch'ol month names: original order

Month positions	Month names	Month dates	Month name variants	Sources
1	Zihora	21-XII		
2	Yax	10-I		
3	Zac	30-I		
4	Chac	18[19]-II		
5	Chantemac	11-III	Chantemat	G31, T450
6	Uniu	31-III	V Roziniu, Uiniu	G31, T450
7	Muhan	20-IV		
8	Ah Qui Xou	10-V	Ah Qui Ccu, Ahquicou	G31, T50
9	Ccanazi	30-V		
10	Olh	19-VI		
11	Mahi	9-VII		
12	[K'an Jalib']	28[29]-VII		
13	I Cat	3-VIII		
14	Chacât	23-VIII		
15	[Sutz']	12-IX		
16	Cazeu	2-X		
17	Chichin	22-X		
18	Ianguca	11-XI		
19	Mol	1-XII		

The 1631 calendar discussed below, clearly Gregorian, correlates 4 July to 0 *K'an Jalib', while the Lanquín almanac equates 14 July to the same *haab* date. The ten-day difference indicates 1589 as the referent year. The Gregorian intercalation means that any Western date after 29 February will coincide with its Maya equivalent only four times in a row, starting with the leap year; this indicates that 14 July coincided with the Ch'ol New Year's day in 1588, 1589, 1590, and 1591. If real rather than ideal, the document's capitalization of just the dominical letter *a* to indicate the first Sunday of the year as 1 January pinpoints the date of this calendar to 1589 (Bond 1889:3, 7, 42, 48, 54). The dominical *a* can apply to any day of the week, depending on the year (Bond 1889; Baaijens 1995: 51).

William Gates (1932a) and Thompson (1932) discuss the month names, but Gates transcribes the entire native text; this records planting or harvesting days and saints' feasts. In table 2.1, brackets enclose counting corrections and restored names. "G" denotes Gates and "T" Thompson. Translations are discussed in the next section.

1631. In his 1631 account Martín Alfonso Tovilla (Scholes and Adams 1960: 184–85) describes the Ch'ol year as composed of eighteen 20-day months plus a final

TABLE 2.2

Lanquín Ch'ol month names: revised order

Month positions	Month names	Month dates
1	[K'an Jalib']	14-VII
2	I Cat	3-VIII
3	Chacât	23-VIII
4	[Sutz']	12-IX
5	Cazeu	2-X
6	Chichin	22-X
7	Ianguca	11-XI
8	Mol	1-XII
9	Zihora	21-XII
10	Yax	10-I
11	Zac	30-I
12	Chac	18[19]-II
13	Chantemac	11-III
14	Uniu	31-III
15	Muhan	20-IV
16	Ah Qui Cou	10-V
17	Ccanazi	30-V
18	Olh	19-VI
19	Mahi	9-VII

period of five days, 29 June–3 July. The first of the year fell on 4 July. He names the dry season with a term derived from a glyphic compound and describes the five days with two phrases:

> *yazquin*: *el verano*/the summer. The Guatemalan "summer" is the dry season from November to February (Edmonson 1988: 90).

> last month: days that do not have a name.

> last month: [days] of great fasting This recalls apparent Ch'olti' equivalents.

The data indicate the reconstruction of the 1631 roster shown in table 2.3.

1971. Thompson offers sure or tentative translations for many of the calendrical names in this document. The numerals refer to pages in the text. The remarks are his, with the less likely ones omitted.

> [*kanhalib*]: no name in native document; no comment, 107

> *icat*: as *ikcat*, black jar or vase; as *ikkat*, black *kat* or black crossed-bands, 107–8, 453

> *chackat*: as *chaccat*, red jar or vase; as *chackat*, red crossed-bands, 108, 452; *chac*

TABLE 2.3

Ch'ol month names and dates given by Tovilla

Month positions	Month names	Month dates
1	no name	4-VII-1631
2	no name	24-VII
3	no name	13-VIII
4	no name	2-IX
5	no name	22-IX
6	no name	12-X
7	Yazquin	1-XI
8	no name	21-XI
9	no name	11-XII
10	no name	31-XI
11	no name	
12	no name	
13	no name	
14	no name	
15	no name	
16	no name	
17	no name	
18	no name	
19	no name	
Comment	[days] of great fasting	

[*zutz]: likely equivalent *zutz' bat*, 108

cazeu: no translation known, 108

chichin: barely possible form of *chichi*, nickname for "dog" in Mexico, 110

ianguca: no translation known, 110

mol: congregate, 110

zihora: not Ch'ol, maybe Spanish; no translation

yax: blue, green, 111, 452

zac: white, 111, 452

chac: red, 111, 452

chantemat: *mac*: cover, close, 113

uniu: no translation known, 113

muhan: sparrow hawk, kite, 114

pax: no comment, 116

> *kanazi: kan* yellow, 117
>
> *olh*: no translation, 117
>
> *mahi ikaba*: nameless days, 117

1988. In his comparison of the Tikal month glyphs with the names above and in the Q'eqchi' tradition, Edmonson (1988: 154, 191, 216–17) suggests the forms in table 2.4 as possible Ch'ol counterparts. Next to his forms are my versions with translations based on the Ch'ol and Ch'olti' calendars and the glyphs. Ma'y i K'aba' does not appear in the glyphs.

TABLE 2.4

Ch'ol month names given by Edmonson

Edmonson		Lamb	
Month names	Translations	Month names	Translations
Pop	Mat	K'an Jalib'	Yellow Shuttle
Ik Kat	1-Kat	Ihk' K'at	Black Wicker
Chac Kat	2-Kat	Chak K'at	Red Wicker
Zotz	Bat	Suts'	Bat
Zec	Skeleton		
Cazeu	none given	Kasew	Swamp Palm
Chichin	Birds	Ts'ikin	Bird
Yaxkin	Green Time	Yäx K'in	Green Days
Mol	Gather	Mol	Heap
Ik Zih	Black Flower	Ik'Sijom	Black Soapberry
Yax Zih	Blue Flower	Yäx Sijom	Green Soapberry
Zac Zih	White Flower	Sak Sijom	White Soapberry
Chac Zih	Red Flower	Chak Sijom	Red Soapberry
Mac	Cover	Majk	Granary
Oneu	none given	Un(iw)	Avocado
Muan	Owl	Muwan	Hawk
[Pax not given]		Paxab'	Upright Drum
Canaazi	Turtle	K'an Aj Si'	Yellow Woodchopper
		K'an Asij	Yellow Cicada
Ohl	none given	Ohl	Sprout
		Och Waj	Food Maize
		Ohl Waj	Ball of Maize
Mahi i Kaba	Nameless		
Mahi i K'aba'	There Is No Name		
		Jol Jab'	Leftovers of the Year
		Way Jab'	Co-essence of the Year
		Wayab'	Bed

Pop likely reads K'an Jalib' or K'an Jalam. Zotz' is a Yukatek form, with *suts'* as its Ch'ol cognate. Zec is also Yukatek, with its Ch'ol counterpart listed here as Cazeu. *Zih* means "flower" in K'ichean but not Ch'olan, so its presence here appears mistaken. The clear *zihora* of the manuscript represents either the misreading of a slovenly written *zihom* or a recasting of Yukatek *zihom* as *síih-<hom>* "well-up sinkhole" into Yukatek *síih-hóol-ha'* "well-up-hole-of water, wellspring" (see chapter 3 for apparent links among *zihom*, *zihora*, and *zih*). The manuscript has not *canaazi* but *ccanazi*, with *ccan* denoting *k'an* "yellow," the term found in the glyphic form (Justeson 1975).

Calendrical Documents in Ch'olti'

1700. The Ch'olti' Lakandon of Dolores/Zac Balam preserve the designations for two months in lists of personal names. These rosters date from the late seventeenth and early eighteenth centuries.

> *canhalib*: main component of the woman's name *nacanhalib*, with *na* ["mother, lady"] as the feminine prefix (Justeson, personal communication, 9 July 1991), or "first."

> *chichin*: a month name, used as a personal name (Justeson, personal communication, 9 July 1991)

Calendrical Documents in Ch'orti'

1949. Charles Wisdom (1949: 766) records a modern term for the dry season used in the glyphs for just a particular month. This is *yax k'in* "dry season" or *verano*.

1962. Rafael Girard (1962: 280–81, 1966: 230–31) records a term for the festival of the dead that recalls *tz'ikin* "bird," the synonym or cognate for the Ch'ol/Ch'olti' *chichin*. Girard dates and translates the term for the last five days of the year and describes the calendar.

> *tzi kin*: annual festival of the dead Tzi kin [*tz'ikin*]; connotes the idea of "bird" (1962: 281, 1966: 230).

> *días de duelo* 3–7 February; days of pain or mourning (1962: 3–4, 29–30, 1962: 29–30; 1966: 3, 20).

The Ch'orti' froze their calendar to the Gregorian system. This means that each of their days always correlates to its European counterpart; 29 February corresponds to a special extra day, so it does not disturb the pairings. They use eighteen months of twenty days and a final period of five days. The year starts on 8 February, while the last days run 3–7 February. They have forgotten the month names (Girard 1962: 3–4, 29–30, 325, 1966: 4, 20).

The count begins on 8 February because this marks the canonical but not possibly real day of the sun's first perceptible northward movement along the horizon out of its winter solstice extreme in the south (Girard 1962: 3–4, 29–30, 325). The assumed parallel of this system with the Ch'ol 1589 and 1631 systems implies for all three the equation O *K'an Jal-ib' = 14 July 1548. Although the Ch'ol did not freeze their calendar to the European count that year, the Ch'orti' did so at the next opportunity, sometime in 1552–55. To align them with Carnival, the start of the Five Days was shifted from 8 July to 3 February in 1598, 1693, 1701, or 1818, when the festival occurred on 3 February (Bond 1889: 138–42).

SUMMARY: With just three exceptions the names from Ch'ol, Ch'olti', and Ch'orti' correspond closely to those of the Classic glyphic tradition. The never-frozen Ch'ol calendars of 1589 and 1631 must be the same, given their correlations, while the Ch'orti' count becomes frozen early on, besides shifting the first month.

Calendrical Documents in Chuj

1930. The only original source for this count is the list from Santa Eulalia, a town in the mountainous northwest of Guatemala (Termer 1930: 391–94). In this settlement Chuj was a major language (Stadelman 1940: 95). Franz Termer preserves the original orthography and sequence of the list but is not sure of their correctness. The initial dates below for 1980–81 derive from the 5 Hoyeb' Ku = 1 March 1981 correlation reported by Judith Maxwell (Edmonson 1988: 95, 157). As their order is not certain, they are arranged here to end with the Five Days, because this does terminate the year in the town of San Mateo Ixtatán, source of the above equation. The calendar has been frozen only since the last daykeeper died in 1983.

The list below presents some glosses for the Five Days but names for only seventeen or eighteen months. It follows Edmonson (1988: 157), who for no stated reason inserts the empty slot between Xujim and Mol. The absentee might be Oyebin, because this can represent Hoyeb' K'iŋ "five 20-day periods." The former meaning of *k'iŋ* as "sun, day" allows the interpretation "Five Days" and thus the equation of Oyebin with Oja Kwal and Hoyeb' Ku. Oja Kwal is explicitly defined as a span of five days (Stadelman 1940: 96, 124). Hoyeb' Ku, also spelled H Oye' K'u "Five Days," clearly ends the year in one town at least (Edmonson 1988: 95, 157).

In nearby San Pedro Solomá the three-month sequence of Sivil–Tap–Oja Kwal begins in the referent year of 1937 sometime after Candlemas or 2 February (Stadelman 1940: 96, 99–100, table 3, 124). Using the Ixtatán correlation of 1 Siwil =

16 January 1981 (Judith Maxwell, personal communication, 1981), and assuming the recession rate of one day every four years in counts not correcting for leap years, 1 Siwil fell on 3 February, the first day after Candlemas, only in 1908 to 1911; in 1937 it began on 27 January. If the Solomá and Ixtatán calendars were once congruent and unfrozen, then for Raymond Stadelman's equation to be right the Chuj at Solomá must have fixed their calendar to the Gregorian count sometime in the period 1909–12. Termer's more likely remarks accompany the names in table 2.5.

1988. Edmonson (1988: 157) repeats Termer's list but leads off with Tap, equates Oyebin to Hoye' K'u, and offers a few translations (table 2.6). Stadelman (1940: 96, 124) and Termer position Tap immediately after Sivil.

Table 2.5

Chuj month names given by Termer, with modifications based on Edmonson and Stadelman

Month positions	Month names	Month dates	Name interpretations
1	Bex	2-III-80	perhaps from *bix* time, time span
2	Sacmay	22-III	white tobacco
3	Nabich	11-IV	
4	Mo	1-V	likely K'iche' *mo'*, sparrow-hawk
5	Bac	21-V	bone
6	Tam	10-VI	
7	Huatziquin	30-VI	
8	Kanal	20-VII	Ixil: yellow; Chuj: star; dance
9	Yaxaquil	9-VIII	
10	Yaxul	29-VIII	green or first *xul*; K'iche' *xul* flute; Q'eqchi' *xul* animal
11	Savul	18-IX	still not understood
12	Xujim	8-X	
13	Oyebin?	28-X	[position hypothetical]
14	Mol	17-XI	Yukatek, Poqomchi' *mol* collect; Q'eqchi' *mol* egg
15	Meak	7-XII	Meak like Yukatek *mac*; Tzeltal name: lid, fence
16	Oneu	27-XII	unclear
17	Sivil	16-I-81	Ixil *sibil*: steam, smoke
18	Tap	5-II	land crab in Highland languages
19	Oyebin	25-II	meaning questionable [13th name?]
	Hoyeb' Ku		San Mateo Ixtatán; Chuj; Judith Maxwell, personal communication, 1981]
	Oja Kwal		[San Pedro Solomá; Chuj; Stadelman 1940: 96, 124]

TABLE 2.6

Chuj month names given by Edmonson

Month name	Meaning
Tap	
Bex	
Sacmay	White Time
Nabich	
Mo	
Bac	
Tam	
Hua Tziquin	Bird
Kanal	Yellow
Yaxaquil	Green
Yaxul	
Savul	
Xujim	
Mol	Gather
Mak	Cover

SUMMARY: Chuj has sixteen of its nineteen names in common with other traditions, mostly Q'anjob'al. Thirteen of these find cognates exclusively with Q'anjob'al, specifically Bac, Bex, H Oye' K'u/Hoyeb' Ku, Oyebin, Kanal, Nabich, Oneu, Sacmay, Sivil/Siwil, Tap, Xujim, Yaxaquil, and Yaxul. The remaining three surface not only in Q'anjob'al but in several other systems, sometimes slightly modified. The first is Huatziquin, attested in Q'anjob'al, Ixil, K'iche', Kaqchikel, and Poqomchi'. Next comes Mo, known in Q'anjob'al and Ixil. The third, Mol, is recorded not just in Q'anjob'al but in Ixil, Poqomam, Q'eqchi', and Yukatek as well. The remaining three names, Meak, Savul, and Tam, occur only in Chuj. The large overlap between Chuj and Q'anjob'al might well derive from their mutual presence in Santa Eulalia.

Calendrical Documents in Ixil

1942. Jackson Lincoln (1942: 116–18, 1945: 110–11, 118) provides the most and earliest data (table 2.7). A Spanish manuscript from 1575 identifies the only villages that then spoke Ixil as Chajul, Nebaj, Cotzal, and part of Santa Eulalia (Lincoln 1942: 104), all villages in the mountains of northwest Guatemala. This allows Ixil loans into the Chuj and Q'anjob'al lists, as both were also used at Santa Eulalia.

Dates are assigned to just the first seven units of Chajul b and c (lists elicited from the calendar priest Sebastián C.), because only they follow the same order on both rosters. Note the unusually short list from Nebaj: its thirteen months span all

365 days, as displayed in table 2.8. From these Lincoln (n.d.: 111 n. 1, 120–21) separates with a line two more Nebaj names collected later from the old calendar priest Nicolás B.

As regards the Ixil calenders in the town of Chajul, Lincoln (1942: 115, 126 n. 27, 1945: 111, 117–18) correctly translates O'ki as "five days." The *clarinero* of Mol/Molchu glosses as "bugler," as well as "great-tailed grackle" or "melodious blackbird" (Schoenhals 1988: 381). Apparently relying just on visual inspection, he tentatively relates several names to counterparts in other systems but errs with about half of them. He correctly links Yax'ki "Green Days" to the Yukatek Yaxkin "Green Days" and the Tzeltal Yaxkin "Green Feast Days." He compares Mol Masat and Mol Tche, both "Herd of Deer," as well as Mol "Group" and Molchu "Flock of Grackles," to either Yukatek Mol "Group, Pile," Ch'ol and Chuj Mol "Group," or Yukatek Ceh "Deer." Chentemak he parallels to the Ch'ol Chantemac and the Poqomchi' Txantemac "Granary on Tall Poles" as well as the Yukatek Mac "Granary." Finally, he equates Tzikin'ki with the Chuj and K'iche' Tzikin Gih, "Bird Days." In all instances he is correct. Not so for the rest of the names. For Muen "Seedlings" or Muenchin "Seedling Bag," he finds the Yukatek Muan "Hawk, Owl," the Ch'ol Muhan "Hawk," and the Poqomchi' Muwan "Hawk." He moves from Pactzi to the Yukatek Pax "Upright Drum" and the K'iche' Pach "Incubation" or "Mistletoe"; from Tchotzcho/Tchohtcho/Tchoochcho "Small

TABLE 2.7

Ixil month names given by Lincoln

a. Chajul XII-1939 Diego M. or Tek Tus	b. Chajul 24-XI-1940 Sebastián C.	c. Chajul 27-XI-1940 Sebastián C.	Chajul b, c dates	d. Chel	Chel dates	e. Ilom
Mol Tche	Mol Tche	Mol Tche	6-XI	Tzu'ki	7-XI	Chentemak
Nol'ki	Och'ki	Och'ki	26-XI	Tzunun'ki	27-XI	Mol Masat
Xet'ki	Mek'aj	Mek'aj	16-XII	Chentemak		Nol
Talcho	Koj'ki	Koj'ki	5-I	Muenchin		Muenchin
Nimcho	Talcho	Talcho	25-I	Och'ki		Tchoch'ol
Metch'ki	Nimcho	Nimcho	14-II	Koj'ki		Talcho
Yax'ki	O'ki	O'ki	6/10-III	Talcho		Nimcho
Hui'ki	Chentemak	Nol'ki		Nimcho		Tchotzcho
Tzil'ki	Pactzi	Petzetz'ki		Tchoochcho		O'ki
O'ki	Nol'ki	Xukul		Avax'ki		
	Zil'ki	Zil'ki		Petzetz'ki		
	Zoj'ki	Zoj'ki		Yowal		
				Nol'ki		
				Mol Tche		
				O'ki		

Table 2.8

Ixil month names from Nebaj given by Lincoln

Month names	Month dates	Days	Interpretations or remarks
Muenchin	11-III	[34]	shadows from the straw; time of sowing
Mu	14-IV	[30]	
Mol/Molchu	14-V	[30]	when the *clarineros*/buglers assemble [1942: 111]
Tzanakbai	13-VI	[30]	little green animals
Tzikin'ki	13-VII	[30]	plantings that people used to make
Chen Temak	12-VIII	[30]	little is lacking before the harvest
Lajab'ki	11-IX	[30]	
Cajab'ki	11-X	[30]	
Onchil	10-XI	[30]	time when little animals come out of the earth
Pactzi	10-XII	[30]	kind of bird
Talcho	9-I	[30]	time when the corn is little
Nimcho	8-II	[26]	time of large animals
O'ki	6-III	[5]	five days
Tzijep			recorded later from Nicolás B., no dates assigned
Mama'ki			

The names and dates come from informants Diego B. and Nicolás B., the meanings from Vicente C. (Lincoln n.d.: 111 n. 1).

Rats," to Yukatek Zotz "Bat"; and from Ixil's Cajab'ki "Honey Days" to Yukatek's Kayab "Horizontal Drum." Here in every case he is mistaken.

The five days of O K'í always threaten misfortune and danger. Everyone fasts, and most remain indoors, engaging in virtually no work or activities. Shamans offer chickens, turkeys, and bulls at the mountain crosses and incense before all the crosses. Family members confess and reconcile. In the old days the men, who would always keep their heads covered with a white or colored cloth, went bare-headed; and the white-haired oldsters used to regain their black hair. People still believe that children born during these days will grow up impotent, sterile, and without molars and that boys will become effeminate. During these days of anxiety stones grow, snakes change their skin, and trees turn their foliage (Lincoln n.d.: 119–20; Nachtigall 1978: 305–6).

1973. In this year Horst Nachtigall (1978: 305–7) studied among the Ixil of Nebaj. The three most highly respected native Ixil calendar priests offered the month names in table 2.9, but they all had to ruminate quite a while and corrected the names several times days later. The year always starts between the first and the fifteenth of March. The calendar priests meet, or used to, each year at the beginning of March at the shrine of Huil near the town of Chajul and together set the start of the year. Nachtigall published the data here without translations. The names and interpretations for the last month all label it as a span of five days.

TABLE 2.9

Ixil month names from Nebaj given by Nachtigall, from three informants

#	Nebaj A: Don Jacinto de León	Nebaj B: Pap Tek	Nebaj C: Don Felipe	Month
1	Talchój	same	Talcho	March
2	Masnimchó	Moxnimchó	Nimchó	April
3	Ne'chó	same	none	May
4	Lem	none	none	May
5	Akmór	same	none	June
6	K'olk'óy	same	same	July
7	Tsanakváy	Tsanakbáy	Tsanakbáy	August
8	Maháb	same	none	August
9	Muentyín	same	Muanchím	September
10	Moch'ú	Molch'ú	none	October
11	Paksí	same	Paksî	November
12	Onchív	same	Onchíl	November
13	Mochó	Mochnimchó	Pekxitx	November
14	Sonchój	same	none	December
15	Chantemák	same	none	December
16	Kahabtsé	Kaháb	none	January
17	Sojnóy	same	same	February
18	Mamak'i Kejepk'í	none	none	March
[19]	Ok'í/Okok'í Ok'í/Oval k'í		five supernumerary days "five days" [last days of the year]	March

1988. In his survey of Lincoln's lists, Edmonson (1988: 185–86; 216–17) counts thirty-nine different names, seven cognate with terms in other systems and seventeen mutually exclusive (table 2.10). By placing the cognates first and then making a best fit of the rank orders of the mutually exclusive names, he comes to the "solution" in the first column in table 2.10, a tentative reconstruction of the original names and their positions. His sequence letters and rank orders indicate the possible synonyms for some of these in the second column. Throughout he divides the names into grammatical words and slightly alters the orthography.

1999. A recent dictionary explicitly identifies as Ixil one name otherwise known only from the Chuj and Q'anjob'al calendars.

> *Mo* name of a month in the Ixil calendar (Chel and Ramírez 1999: 162).

Between them Lincoln and Nachtigall preserve nine rosters, reducible to six different calendars. Altogether these present some fifty distinct names. Only thirteen names find cognates in other systems, just three come in as loans, and only two go

TABLE 2.10

Order and comparison of Ixil month names given by Edmonson

Name meanings	Possible synonyms
Cajab Ki	Och Ki, Onchil
Pac Tzi	
Koj Ki	Chochol
Tal Cho	
Nim Cho	
Chotz Cho	Mech Ki, Tzijep
A Ki	Awax Ki, Mam A Ki
Tz'ikin Ki (Bird)	Kucham
Yax Ki (Green)	Mekaj, Mu
Yax (Green)	
Mol (Gather)	
Mol Che, Masat	
Tzil Ki	
Petzetz Ki (Flower)	Tzu Ki, Tzanakbal [*sic*]
Xukul Ki	Hui Ki
Yowal	Tzunun Ki
Chente Mac (Cover)	
Nol Ki	
Muen Chin	
Muen (owl)	
Xet Ki	Zoj Ki, Lajab Ki
O Ki ("the five days")	

The root meanings appear in Edmonson (1988: 216–17, fig. 16D).

to distinct traditions. As table 2.11 demonstrates, almost every source offers at least one unique name.

Calendrical Documents in K'iche'

1554. An unknown author penned the original *Título de Totonicapán* (Carmack and Mondloch 1983) as a title to territory and to cacique privileges for his ruling princes in 1554. Sometime between 1650 and 1725 two or three scribes composed the present copy of the original manuscript. The work preserves one unique term for the period of five days:

> *pubaix* "the unlucky days," folio 21r (ibid.: 122–23, 188).

1698. On folio 160v of his *Vocabulario de la lengua Quiché*, finished in 1698, Fray Domingo de Basseta lists the calendar names and their initial dates. He begins the

TABLE 2.11

Unique Ixil month names by calendars (as listed in tables 2.7 and 2.9)

Lincoln	Lincoln	Lincoln	Lincoln	Lincoln	Chel and Ramírez	Nachtigall	Nachtigall	Nachtigall	Nachtigall	Nachtigall	Nachtigall
Chajul a	Chajul b, c	Chel	Ilom	Nebaj	Nebaj	Nebaj a	Nebaj b	Nebaj c	Nebaj abc	Nebaj ab	Nebajd
Hui'ki	A'ki	Avax'ki	Nol	Cajab'ki	Mo	Kejep'ki	Mochnimchó	Muanchim	K'olkôy	Akmór	Okok'i
Metch'ki	Kucham	Tchoochcho	Tchochôl	Lajab'ki		Lem	Moxnimchó	Pekxitx	Sojnóy	Kaháb (Tsé)	Oval K'i
Tzil'ki	Mek'aj	Tzu'ki	Tchotzcho	Mu		Masnimchó				Kaháb	
Xet'ki	Tchohtcho	Tzunun'ki		Tzijep		Mochó				Moch'ú	
Yax'ki	Xukul (Ki)			Tzikin'ki						Muentyin	
	Zil'ki									Nechó	
	Zoj'ki									Sonchój	

Lincoln attributes his Chajul a to Diego M. or Tek Tus, Chajul b and c both to Sebastián C., and Nebaj to Diego B., Nicolás B., and Vicente C. He mentions no informants for Chel and Ilom. Nachtigall assigns his own Nebaj a to Don Jacinto de León, Nebaj b to Pap Tek, and Nebaj c to Don Felipe. Chel and Ramírez (1999) mention nobody specifically for their one name.

series not with Nabe Mam or Tacaxepual, as others do, but Nabe Tzih, explicitly labeling it the natives' first month. He gives nineteen months, with Hunobixqih lasting five days and the rest twenty (table 2.12). He indicates no referent year, but correlations in the 1722 and 1976 calendars point to 1588–91.

In his edition of this work, René Acuña (Basseta 2005: 327 n. 4) proposes that the clearly written Nabe Liquin Ca could represent Nabe Liquim Ha, "the first collection of water in vessels," and that *pubaix*, the name for the five-day span in K'iché, translates as "five lunar days" (327 n. 3). K'iché dictionary entries indicate instead "sudden misfortune" or "blowgun death" (see chapter 3).

1722. *Ch'ol Poal Σih*, *Maceval Σih*, and *Ahilabal Σih*, native sources from the Quezaltenango area copied by the scholar C. H. Berendt in 1877 from a manuscript, constitute his unpublished *Calendario de los indios de Guatemala* (Carmack 1973: 165–67) (table 2.13). The full 365-day calendar on pages 1–17 presents the month names, while a list of them appears on pages 18–22 and 50. Unmarked forms are from the roll on page 22, with variants as noted. Nabe Mam heads this list. Edmonson (1988: 92) equates 9 Keh 1 Nabe Mam to 3 May 1722.

TABLE 2.12

K'iche' month names given by Basseta

Month names	Month dates
Nabe Tzih	21-XII-[1590]
V Cab Tzih	10-I-[1591]
Roxtzih	30-I
Che	19-II
Tecoxepual	11-III
Tzibapop	31-III
Zac	20-IV
Chab	10-V
Hunobixquih	30-V
Nabe Mam	4-VI
Vcabmam	24-VI
Nabe Liquinca	14-VII
V Cab Liquinca	3-VIII
Nabepach	23-VIII
V Cab Pach	12-IX
Tziquin Quih	2-X
TzizilaΣan	22-X
Cacam	11-XI
Botan	1-XII

TABLE 2.13

K'iche' month names of 1722 copied by Berendt

Month names	Month dates	Month name variants and pages in 1722 MS
Nabe Mam	3-V-1722	Mam 50
Ucab Mam	23-V	
Nabe Liquin Ca	12-VI	Liquin Ca 50
Ucab Liquin Ca	2-VII	
Nabe Pach	22-VII	Pach 50
Ucab Pach	11-VIII	
4,içi Lakam	31-VIII	4,içilakan 50; 4,içilakam 19–21
4,iquin Σih	20-IX	
Cakam	10-X	
Botam	30-X	Batam 18
Nabe Çih	19-XI	Çih 50
Ucab Çih	9-XII	
Urox Çih 50	29-XII/17-I-1723	Rox Çih
Chee	18-I	
Tequexepual	7-II	
Tequexe Pual 14	"This is when they dress up, all the people"	Edmonson 1997: 118
4,ibapopp 14	27-II	4,ibapopp 19, 20; 4,ibba popp 21; 4,ibapop 50
Çac	19-III	Cac 15, 21?
4,hab	8-IV	"Great celebration to be given" Edmonson 1997: 118
4,api Σih 50	28-IV/2-V	

1722. Daniel Brinton (1893: 300–302) bases this list (table 2.14) on the copy of the 1722 native calendar in his possession and on the list published by Juan Gavarrete (1868: 82), itself compiled from seventeenth-century indigenous documents. A few names are spelled differently than their counterparts. The sign *4,* stands for tz'. The etymologies are Brinton's. He omits Tz'api Q'ih. This list appears again in Adrián Recinos (1947: 121 n. 82) as well as Recinos et al. (1950: 108 n. 2), but these authors spell and interpret several names differently. Entries from these sources follow below Brinton's, prefixed with "RGM." Asterisks mark my interpretations of the Recinos glosses that differ from those by Recinos et al.; in these cases the Recinos counterparts appear.

1935. During the 1930s Juan de León gathered astronomical and calendrical data from aged native inhabitants of Santa Cruz del K'iche' (Carmack 1981: 229) and published most of the corpus in his *Mundo quiche miscelánea* (1945). He included much of this information in his *Diccionario quiché–español* (1954), a collection of

Table 2.14

Table 2.14

K'iche' month names of 1722 given by Brinton

Month names	Translations, comments
Tequexepual	corruption of Nahuatl Tlaca Xipeualiztli
	RGM: time to plant the cornfields
Tziba Pop	painted mat
Zac	white, referring to white flowers of the season
Ch'ab	from *ch'aban* mud, mire
	RGM: muddy ground
Nabey Mam	first old man
Ucab Mam	second old man
	RGM: both are months of bad augury
Nabey Lik'in K'a	first soft to the hand
	RGM: soft and slippery soil
Ucab Lik'in K'a	second soft to the hand
	RGM: second month like the one before
Nabey Pach	first hen hatching
	RGM: *primera echada, tiempo de empollar* first time of hatching/ *first casting [of seed] or first covering [of eggs]; time of hatching
Ucab Pach	second hen hatching
	RGM: *segunda echada* second time of hatching/*second casting or covering
Tzizi Lakam	sprouts or shoots like flags appear
	RGM: the sprouts show
Tziquín Kih	season of birds
Cakam, Cakan	derived from the season's red clouds or more probably from a species of red flowers blossoming at this time
Botam	roll of mats
	RGM: *esteras enrolladas* tangled mats/*rolled-up mats
Nabey Zih	first tree with abundant white flowers
	RGM: *árbol frutal de flores blancas* first month of white flowers/ *fruit tree of white flowers
Ucab Zih	second tree with abundant white flowers
	RGM: second month of white flowers
Rox Zih	third tree with abundant white flowers
	RGM: third month of white flowers
Chee	trees or wood

RGM = Recinos et al. (1950).

lexemes and lore compiled over perhaps thirty years in Chichicastenango, Chiché, and Sacapulas. In this second book he attributes the calendrical data to Juan Zapeta and the astronomical data to Santos Reynoso, describing both as *quicheístas* (León 1954: 10). He seems to have recovered highly esoteric elite traditions lost to other communities (Carmack 1981: 357). The innovative names appear in table 2.15 in publication order, with identical spellings or definitions omitted (León 1945: 66–67, 1954: 79). The last two names might be equivalents of Tz'api Q'ih, the five intercalary days, not listed here. Nabé Makuk and U Cab Makuk likely correspond to the standard Nabe Mam and U Kab Mam, months beginning on 2 and 22 March in 1976 (see the 1976 almanac below). In 1979 an excellent native Ch'ol informant from Palenque in Chiapas, Mexico, offered the Ch'ol term *man k'uk'* in a vocabulary survey, glossing it not only as "quetzal" but also as *primavera* or "spring" (Hopkins et al. 2011: 142). This supplies perhaps a vestigial link between the Ch'ol and K'iché counts.

TABLE 2.15

K'iche' month names given by León

Month names	Meanings, comments
Nabé Tzij; Nabé Tsij	first counsel; first counsel of gods (1954: 79); to create air (1945: 67)
U Cab Tzij; U Cab Tsij	second counsel; to create water (1945: 67)
U Rox Tzij; U Rox Tsij	third counsel; to create fire (1945: 67)
Cak Che	red tree (1945: 66, 1954: 19); tree of fire (1954: 79); name of the Pleiades (1954: 91)
Cox Xe Puatl [*sic*]	sowing in islands or chinampas
Tzibé Pop; Tsibé Pop	painting of mats
Zac Imuj; Sac Imuj	white cloud; shadowless days [sun directly overhead?] (1954: 56)
Ajau Chap	god of the arrows
Jun Bix Kij	one song to the sun
Nabé Makuk [*sic*]	first quetzal; refers to Venus (1945: 67); poetic name of Venus (1954: 91)
U Cab Makuk [*sic*]	second quetzal; refers to Mars (1945: 67)
Nabé Li Quin Cab	first twins; Catholics' "Eyes of St. Lucy" (Castor and Pollux) (1945: 67)
U Cab Li Quin Cab	second twins; other paired bright stars (1945: 67)
Nabé Pach	*primer armado*/first armadillo; a bright star (1945: 67)
U Cab Pach	second armadillo; a bright star (1945: 67)
Iquim Kij	sun of the south
Chi Il Lacan	banners of the sky
Caam Kij	sun that measures
Chacan	week of five days (1954: 26)
Jop Bix	week (1954: 38)

1965. In his dictionary Edmonson (1965) presents the calendar names, cites Recinos (1947) for most of the etymologies, and omits Tz'api Q'ih. Translations deduced from Edmonson's own definitions are enclosed in brackets. Page numbers end each line (table 2.16).

1976. In an article linking astronomy with hawks Barbara Tedlock (1985: 84–85) presents the Momostenango agricultural almanac for 1976. Here she labels 26 February–1 March as the last five days of the year and 2 March as New Year's; in 1990 this day (1 Nab'e Mam "First Grandfather") was 27 February (Tedlock 1992b: 26, 36). Barbara Tedlock (1982: 103) records the only two *winaq* names retained in Momostenango, those for the first two months. This calendar leads with the same score of days as does the calendar of 1722. Dennis Tedlock (1993) gives the names and positions of two months and the translations for three. The positions here also indicate a roster commencing with Nabe Mam (table 2.17).

1979/1981. Both in the Spanish version of his work and in the later English edition Robert Carmack (1979: 48–50, 1981: 87–89) offers a reconstructed calendar for 1979–80 (table 2.18). For each unit he not only notes native associations and his own but also reconstructs approximate dates, given here as initial. Identical spellings and

TABLE 2.16
K'iche' month names given by Edmonson

Month names	Definitions or comments
Tekexepual	corn sowing time (120)
U Kab Pach	[second throwing] (84)
Tz'iba Pop	painted mat (91, 130)
Tzitzil Lakam	shoots come (135); shoots coming (65)
Zaq	[white] (158)
Tz'ikin Q'ih	bird season; bird days (130)
Ch'ab	muddy soil (19)
Kaqam	red (clouds) (56)
Nabe Mam	first elder; bad luck (70)
Botam	rolled (mats) (16)
U Kab Mam	second elder; bad luck (70)
Nabe Zih	first month of white flowers (161)
[U Kab Zih]	[second month of white flowers] (161)
Nabe Likin Ka	[first] soft, slippery earth (66)
U Kab Likin Ka	[second soft, slippery earth] (66)
R Ox Zih	[third month of white flowers] (161)
Nabe Pach	first throwing (84)
Chee	[tree] (25–26)

Table 2.17

Several K'iche' month names given by the Tedlocks

Month names	Dates	Translation, comment
Nabe Mam	2-III-76	First month (B. Tedlock 1982: 103); begin solar year
Ucab Mam	22-III	Second month (B. Tedlock 1982: 103)
[Rox Tzij]	28-X	Third stick of firewood (D. Tedlock 1993: 104, 245)
[Che']	17-XI	Trees, logs, poles, and stakes (D. Tedlock 1993: 108)
Full Measure		Counterpart to Mac [K'iche' not given]; D. Tedlock 1993: 36

Table 2.18

Reconstruction of K'iche' month names by Carmack

Month names	Month dates	Definitions or comments
Tequexepual	*10[11]-III-79	plant milpas, begin year; planting minor; collect tribute
Q'uibapop	30[31]-III	forty days' rain, insects
Sak	19[20]-IV	White Flowers, rain, insects
Ch'ab	9[10]-V	planting; muddy soil; planting major
Hun Bix K'ij	29[30]-V	First Song to the Sun
Nabe Mam	18[19]-VI	First Old Man, bad to plant
Ucab Pach [*sic*] Ucab Mam	8[9]-VII	Second Old Man; same [rain, insects]
Nabe Liquincá	28[29]-VII	muddy and soft; mud, water from rains
Ucab Liquincá	17[18]-VIII	same [muddy and soft]; first cutting; clear weeds
Nabe Pach	*7-IX	casting [seeds]; Hatching of Birds
Ucab Pach	27-IX	same [casting [seeds]; Hatching of Birds]
Tz'ici Lakam	17-X	Sprouts; sprouting corn tassels
Tz'iquin K'ij	6-XI	Birds; flying south; major preconquest ritual
Cakam	26-XI	heat, Red Clouds; heat around harvest time
Nabé Sij	16-XII	dry, burning; White Flowers grow in maize fields; dried weeds, stalks from past crops burned
Ucab Sij	5-I-1980	same [dry, burning]; plant in mountains
Rox Sij	25-I	same [dry, burning]
Che'	4-II	Trees; cold
Batam	6-III	five unlucky days
Tz'ap K'ij	6-III	"Closing Days"; associated with underworld god Batam

Initial capital letters represent translations possible with Carmack's own words, though he does not expressly offer any and lists all texts after dates under "Ecological Association."

dates are not repeated. My changes are enclosed in brackets. Asterisks mark counting errors discussed below. The names appear in Carmack's spelling.

In an earlier work Carmack (1973: 324 n. 103) began the count with Nabe Mam, following the 1722 calendar. Here he starts with Tequexepual, apparently relying on descriptions of this as the beginning of the year; the sources, however, may deal with only the Kaqchikel calendar. He ends with Tz'ap K'ij, explicitly attested as the end of the year in both K'iche' and Kaqchikel. Without explanation, he reverses the number of days assigned to Hunobixquih (5) and Botan (20), then moves Botan several slots to the end of the roster as an erroneous equivalent to Tz'api K'ij. The asterisks in table 2.18 highlight two further computing errors: Tequexepual cannot begin on 10-III because that is the last day of Batam/Tz'api K'ij; and Ucab Liquinca cannot start on 17-VIII and end on 6-IX, as this would make it twenty-one days long. The brackets present the solution: add one day to the initial dates for the first nine months.

1988. Edmonson (1988: 216–17, 237) presents two rosters. The first he bases on the 1722 calendar with his own definitions. The second he renders with his interpretations limited by space constraints to one word each. Every entry in table 2.19 records the first list's version. Edmonson (1988: 237) derives Tequexepual/Tacaxepual, Mam, and Pach from the Nahuatl dialect Pipil. For ethnohistorical and phonological reasons Lyle Campbell (1970: 4–7, 1977: 106–9, 1978: 38–40) attributes only Tequexepeual/Tacaxepual and Pach to the post-1200 CE Gulf Coast Nahua of Tabasco and southern Veracruz instead. Edmonson's Balam and Kak/Zaq "Fire/White" (also Edmonson 1988: 237-J "Kak") are likely errors.

1997. In this study of the 1722 calendar, Edmonson (1997: 117–19, 147 nn. 27–40) presents slightly variant translations for some names, along with native comments on activities (table 2.20). The unusual spellings are his own; "[*sic*]" marks the most egregious. He reads 4,içila kan or Tz'izil Akam as Tz'izil Aqan (147 n. 32) and notes the occurrence of the variant Tz'izil Akan on page 50 of the manuscript. Asterisks mark glosses different from those in his 1988 work.

Summary: Perhaps the most noteworthy feature of the K'iche' calendars is the variety in the names for their first and last months and the number of loans from Nahuatl. Two or three months begin the year. One is Nabe Çih, while another is Nabe Mam. The third, Tequexepual, might start the year only among the Kaqchikel. Of the two names for the last five days, several sources explicitly describe 4,api Σih as the end of the year. The other, Hunobixquih ("five days pass"), appears once only, in the 1698 list. K'iche' has incorporated a string of names from Nahuatl; Tequexepual is certain, while Pach seems likely, but Mam just possible.

TABLE 2.19

Translations of K'iche' month names in Edmonson (1988)

Month names	Translations
Nabe Mam	1 Lord
U Cab Mam	2 Lord
Liquin Ca	Soft Earth
U Cab Liquin Ca	2 Soft Earth
Nabe Pach	1 Moss
U Cab Pach	2 Moss
Tz'izil Lacam	Shoots
Tz'iquin Q'ih	Bird Time
Cakam	Red Clouds
Balam	Jaguar
Nabe Zih	1 Flower
U Cab Zih	2 Flower
R Ox Zih	3 Flower
Chee	Trees
Tequexepual	Flaying
Tz'iba Pop	Painted Mat
Kak/Zaq	Fire/White
Ch'ab	Arrow
Tz'api Q'ih	Extra Days

TABLE 2.20

Translations of K'iche' month names in Edmonson (1997)

Month names	Translations
Nabe Mam	*First Elder
U Kab Mam	*Second Elder
Liqin Ka [sic]	Soft Earth
U Kab Likin Ka [sic]	Second Soft Earth
Nabe Pach	First Moss
U Kab Pach	Second Moss
Tz'izil Akam	*Sprouts Rising
Tz'ikin Q'ih	Bird Time
K'aqam [sic]	Red Clouds
Botam	*Rolled Mats
Balam [sic]	Jaguar [batam in manuscript]
Nabe Zih	First Flower
U Kab Zih	Second Flower
R Ox Zih	Third flower
Chee	Tree
Tekexepual [sic]	*Sowing; They Dress Up
Tz'iba Pop	Painted Mat
Zaq	White
Ch'ab	Arrow; Great Celebration
Tz'api Q'ih	Extra Days

Calendrical Documents in Kaqchikel

1550. When he composed his lexicon Fray Domingo de Vico (1555) included not only the month names of the K'iche' but those of the Kaqchikel as well (table 2.21).

The Newberry Library houses the photostatic copy of another version of the lexicon. This work glosses *obota/ibota* as "name of a month of the Indians; five days," an entry that misled scholars to apply the name to the period of five days. The copyist omitted several words found in the Bibliothèque nationale version of the dictionary (here in boldface): it reads "*Obota/ybota*. name of a month of the Indians. ***oo* five. *obis* it has been five days**" (Vico 1555: 142v). This complete entry confirms that *obota/ibota* lasts not five days but twenty.

1600. The "Calepino en lengua kaqchikel" is a copy made in 1699 of a lexicon written about 1600 by Fray Francisco Varea, itself based on lost works from some fifty years earlier (Carmack 1973: 116–17). It contains the following references to the months, with numerals denoting pages in the "Calepino."

> *tacaxepual*: beginning of the year or time of sowing the first cornfields, 283; while *tacaxepual* is seated, cornfield days, 283

> *nabeytumuzuz*, 311

> *ru cam tumuzuz*, 311

> *tumuzuz*: two months [of the same name] come one after the other, 311; first downpours of the winter [rainy season] begin. There mate and go off flying some little worms that they also call *tumuzuz*, 311

> *tumuzmuz*: sowing days *tumuzmuz* and *çibixic*, 60

> *çibixic*: a time of the year in which the Indians sow, 60

> *vchum*, 60

> *nabey mam*, 174

> *ru cam mam*, 174

> *mam*: they hold it as a bad sign and it goes with the name of him who is born in it, 174; evil days the two months of the days of *mam*, 174;; they do not grow, they stay small, 174; one should not plant in Mam; plants do not grow when Mam is seated, 174

> *liquinca*: sign, 161; Liquinca the month in which the Indians sow, 161

> *nabey to∑ic*, 305

> *ru cam to∑ic*, 305

toΣic, 305

nabey pach, 239

ru cam pach, 239

pach, 239

4,iquin Σih, 439

katic: possibly referred to under the entry *vchum*, 339

ytzcal: when they sow in the Highlands, 149

payriche [*sic*], 242

4,apiΣih, seven [*sic*: five] days that usually fall in Lent in which, say the Indians, all the wild animals take shelter, 436; they shut their animals inside now in *4,apiΣih*, iguanas, crabs, armadillos, pigs, or peccaries, 436

TABLE 2.21
Kaqchikel month names given by Domingo de Vico

Month names	Pages	Comments or variants
Tacaxepual	177	beginning of the year among them
Tumuzuz	195v	
Çibixic	42v	
Vchum	210v	
Mam	124r	
Liquinca	118r	
ToΣic	192r	
Pach	152r	
4,iquin Σih	270v	
Cakan	33v	
Obota	142v	*ybota* (under *obota*)
Ytzcal	106r	
Pariche	150v	
4,apiΣih		

1650. The dictionary of Fray Thomás de Coto reads only from Spanish to Kaqchikel, but the 1983 edition provides a Kaqchikel–Spanish index. Scholars date it to 1643–46 (Tedlock 1992b: 28) or 1647–56 (Acuña in Coto 1983: xliv). As it repeats most of the glosses in the 1600 lexicon, only the new remarks appear below.

vchun: good for making seedbeds and sowing vegetables, 522, under *signo*

mam: a bad sign because they say that he who is born under it remains disproportionately short [*desmedrado*], 522, under *signo toΣic*: among the summer or dry season months, 385, under *otoño*

pach: Moh, Pay, Pach, they are other names of months, 522, under *signo*

pay: Moh, Pay, Pach, they are other month names, 522, under *signo*; this recalls the name *payriche*

4,api Σih: name of the five unlucky days, 522, under *signo*

moh: Moh, Pay, Pach, they are other month names, 522, under *signo*; this might correspond to the Mol, Mu, Mux, or Muy of other calendars; Acuña (Coto 1983: 522 n. 437r) doubts its reading

1685. The anonymous priest who penned the "Crónica franciscana" quizzed a knowledgeable native informant for the data of "Calendario de los indios de Guatemala 1685 Kaqchikel." In 1878 Berendt copied the calendar in Guatemala City. In his preface Berendt locates this calendar in chapter 7, folios 21–25, of the "Chronica de la St. Provincia del Santissimo Nombre de Jesus de Guattemala." Known as the "Crónica Franciscana" (Calderón 1957: 17), it is an undated and since-lost manuscript with 283 folios. Berendt's copy served as the source for several photostats (Carmack 1973: 165). One records the information presented here.

All the month names appear on page 5 of the manuscript; page numbers below locate other occurrences. Some glosses translate the names, but most relate them to seasonal activities. Spanish is retained only when the vocabulary might allow differences in the translation. Though the manuscript lists all the names, only those with novel comments are cited below.

Tecaxepual [*sic*]: time of sowing the first cornfields, 4; a footnote on page 9 suggests the source of this *winaq* name: "The second Mexican month, according to Torquemada, was called *Tlacaxipehualiztli*."

CibixiΣ: 12; smoke, burning of underbrush, 5; time of smoke in which one sows, or because there usually were the burnings of the underbrush for sowing, or, as a metaphor, overcast or opaque due to the big clouds [smoke] that are usually made; 1-IV 12

Vchum: 13; time of resowing, 5, 13; 21-IV 13

Nabeimam: 14; first time of those prematurely old [*revejecidos*], 5; first time of those prematurely old, because the cornfield that was sown at this time did not grow very tall, and even the babies that were born [did not grow very tall], 14; 11-V 14

LiΣinΣá: 16; time of soft earth, 5; time in which the earth is soft and slippery because of the many rains, 16; 20-VI 16

NabeitoΣiΣ: 17; first harvest of cacao, time of cutting off [pods?], 5, 17; 10-VII 17

Nabeipach: *primer tiempo de empollar la clueca*/first time for the hen to sit on the eggs, 5; 19-VII 19

Cakan: 22; time of red clouds, 5; time of red clouds and yellow flowers, 2; 18-X 22

Ybotâ: time of various colors or of rolling up mats, 23; 7-XI 23

Katic: 23; *pasante o siembra común*/the walking or general sowing, 5, 24; 27-XI 24 [broadcasting, no digging sticks]

Yzcal: 25; time of sprouts or shoots, 5; which is to sprout or send out shoots, 25; 17-XII 25. A footnote on page 25 suggests that this is a loan: "The 18th Mexican month, according to Torquemada, was called *Izcalli*."

Pariche: 26; time of covers, of cold, 5; time of covers, to guard oneself from the cold, 26; 6-I-1686, 26

Tzapi Σih: 6, 27; door that closes the year, days, and seasons, 6; 26–30-I-1686, 27

1703. Between 1701 and 1703 Fray Francisco Ximénez produced a large dictionary of K'iche' (Carmack 1973: 162). It presents many Kaqchikel month names. Only those with notable glosses or variants are shown here.

tacaxepual: beginning of the year, 516

pariche, pairiche: 444; Payiriche [*sic*], 442, 443

tzapiquih: the five intercalary days that they added to their year to adjust it to ours, 558

Seven sources start the year with Tacaxepual. Ximénez specifies one New Year's Day as 21 February. Five sources (the calendars from 1550, 1675, 1685, 1693, and 1750, not discussed here) describe 4,api Σih explicitly as the end of the year. 1 Tacaxegual in this count falls on 21-II, but on 31-I in that from 1685; this means that Ximénez recorded a calendar frozen to the Gregorian in 1601–1604.

18th century. Sometime in the eighteenth century Fray Ángel penned his "Vocabulario de la lengua kaqchikel," recording most of the names as lexical items (Ángel 18th century b; Smailus 1989a: 13). Two with perhaps significant spelling variants appear here:

obotha, ybota, 105v

parichee, pairichee [*sic*], 112v

1813. The anonymous "Vocabulario de la lengua kaqchikel y española, con un arte de la misma lengua" preserves most of the names with brief remarks. Here two variants are of interest:

taΣaxepual [*sic*]: beginning of the year, 79v

payi ri che [*sic*], 63v

1885. Brinton (1885: 28–30) follows the 1685 calendar closely but does offer novel interpretations for a few:

Çibix [CibixiΣ]: smoky, or clouds

Cakan: red clouds

Katic: drying up

Itzcal Σih [Yzcal]: bad road days

Pariche: in the woods

Tzapi Σih: days of evil or days at fault

1893. Brinton (1893: 297–300) mentions several extant native Kaqchikel calendars and that he possesses a careful copy of one, dated 1685, probably a version of the work above. Some of his spellings vary even from those in that calendar, but many of his translations derive from notes identical to remarks following some of the names there. His own etymologies and some comments appear below:

Cibixic: "From *cib*, smoke, mist, or vapor. . . . It derived its name from the smoky appearance of the atmosphere at this season, or from the custom of burning brush in clearing the ground." 298

Lik'in k'a: Soft to the hand, "refers to the soil which was then soft owing to the rains." 299

Cakan: "Derived from *k'ak* [*sic*], red; according to the Cakchiquel Calendar, from the reddish clouds (*celages rojas*) often seen at this season; according to others, and more probably, from a species of red flowers which blossom at this time." 299

Ibota: The season of various colors or of mats rolled up (from a note to the calendar), 299

Katic: *pasante o siembra común* [walking or usual sowing] (in a note to the calendar), 300

Izcal: The season of sprouts or of throwing out shoots (translated in the calendar), 300

Itzcal: "The word is undoubtedly the Nahuatl *Itzcalli*, the name of the eighteenth month in the Mexican year. Its signification is 'renewal,' or 'resurrection,' or 'growth.'" 300 [The Nahuatl name is really spelled Izcalli, without the *t*, but its meaning matches the one given here. See chapter 3.]

Pariche, Payriché [*sic*]: The season for covering, in order to protect one's self from the cold (a note to the calendar), 300

1950. Recinos et al. (1950: 33–34) offer much the same spellings and translations as does Brinton 1893. His few original interpretations are included below:

Liquin ka: Time when the earth is furrowed (*labrado*) and slippery because of the many rains

Nabey tokic: The first cut, the first wound or bleeding. Recinos remarks, "Possibly an allusion to pruning or to incisions made in certain trees to extract the sap."

Cakán: Hot season

Katic: Burning or clearing, or period of drought

1988. Edmonson (1988: 145) presents his list based on the 1685 calendar, with his own definitions (table 2.22). He derives Pach and Yzcal from the Nahuatl dialect Pipil. Curiously enough, he does not ascribe Tacaxepual, Nabei Mam, and Ru Cab Mam to this language as well, though he does designate their K'iche' cognates as such (1988: 237). Campbell (1970: 4–7, 1977: 106–9, 1978: 38–40) attributes Tacaxepual, Pach, and Tamuzuz not to Pipil but the post-1200 CE Gulf Coast Nahua of Tabasco and southern Veracruz. Edmonson (1965: 118) earlier also assigned Tamuzuz to Nahuatl but here no longer does. Misread for Tumuzuz, Tamuzuz appears in no original source.

Table 2.22 Kaqchikel month names given by Edmonson

Month names	Translations
Tacaxepual	Flaying, Flay
Nabei Tamuzuz	1 Termite
Ru Can Tamuzuz	2 Termite
Cibixiq	Smoke
Uchum	Reseed
Nabei Mam	1 Elder
Ru Cab Mam	2 Elder
Liquin Qa	Soft Earth
Nabei Toqiq	1 Damp
Ru Cac Toqiq	2 Damp
Nabey Pach	1 Moss
Ru Can Pach	2 Moss
Tziquin Qih	Bird Time
Cakan	Red Cloud
Ybota	Rolled Mat
Katic	Burning
Yzcal	Rebirth
Pa Ri Che	In the Trees
Tzapi Qih	[extra days]

Summary: Like its sibling K'iche', Kaqchikel includes a number of loans. From Nahuatl come Tacaxepual and Izcal, probably Pach, and perhaps Mam. Another surprise is Uchum, from Q'anjob'al, Chuj, or Tojolab'al. The many names that Kaqchikel innovates and borrows as counterparts to K'iche' terms also seem remarkable, specifically Tumusus, Sibixic, Uchum, Toqic, Ybota, Katic, and Izcal. Unlike its sister, Kaqchikel designates just one month, not several, for its initial unit.

Calendrical Documents in Poqom

The term "Poqom" includes both Poqomam and Poqomchi'.

Calendrical Documents in Poqomam

1720. The dictionary by Fray Pedro Morán preserves two names.

> Mol: December 1–20 (Miles 1957: 744, table 2) This would be the sole instance of Mol in a Poqom source.

> Muan: the time from the twentieth of April to the ninth of May

Calendrical Documents in Poqomchi'

1725. This dictionary of Fray Pedro Morán dates to about 1725 and represents the dialect of San Cristóbal Cajcoj, now called San Cristóbal Verapaz, a town in the central mountains of Guatemala. It appears to derive ultimately from a pre-1574 work of Fray Francisco de Viana (Carmack 1973: 120–21).

> Canazi: the time that there is from the 4th of June until the 23rd of the same [no page]

> Petcat: the time from August 3 until its 22nd inclusive (355v)

1788. León Fernández (1949: 112) preserves and loosely glosses two names: Chab "spring" and Muan "autumn."

1897–1914. A schoolteacher in Cajcoj, Víctor Narciso drew up a month list in 1897, 1906, and 1914, each published in turn by a different scholar (Sapper 1906; Termer 1930; Gates 1932a). The version here combines them (table 2.23). All lead with Yax, either because this includes 1 January or because it immediately follows the Five Days.

At best the sources indicate that the Poqomam count started with Kanjalam on 14 July, like the 1631 Ch'ol calendar, while the Poqomchi' began on 16 July, like the Yukatek calendar. If so, the earliest dates that the Maya froze their year to the European round would be 1548 for the Poqomchi' and 1556 for the Poqom. The data at hand cannot demonstrate whether, why, or when the Five Days shifted position. Two later accounts (Gates 1932a; Termer 1930) record it as spanning 23–27 December; perhaps the Poqomchi' used it to mark the winter solstice (20–23 December).

TABLE 2.23

Poqomchi' month names given by Narciso

Month names	Month dates	Translations or comments
Yax	28-XII	time of the clearing of the first cornfield; on the tenth day of the winal ends the time for the sowing of the summer corn field; green
Sak	17-I	sowing of small corn in forest areas; snow; white
Tzi	6-II	Dog
Kchip	26-II	the same [sowing of small corn in forest areas]; the last thunder
Chantemak	18-III	time of the first corn harvest [of summer corn sown in Yax?]; first sowing of *kanjal* or large corn
Uniw	7-IV	second sowing of large corn
Muwan	27-IV	sowing of corn in the village
Kcham	17-V	first cleaning; last sowing of corn in Cristóbal
Sak-kojk	6-VI	White Chilacayote, Gourd
Ojl	26-VI	entrance of rainy season; readying of the granaries; setting out of bananas; avocado tree; a tree with light-weight wood
Kanjalam	16-VII	bean harvest
Makux	5-VIII	cleaning for another sowing of lands already made arable
Kasew	25-VIII	peak of rainy season; slender palm tree of cold region
Kanasi	14-IX	cleaning of the *kanjal* corn
Kanajal	4-X	Ripening Time of Corn
Tzikin-kij	24-X	corn harvest; time of birds, swallows or martins [*Schwalben*]
Mox-kij	13-XI	Time of the Insects
Tik-cheik	3-XII	Stem of the Tree
Kaxik-laj-kij	23-XII	days of trials, *días penosos*

1988. Edmonson (1988: 235) offers his version of the 1906 and 1914 lists (with several interpretations) of paired names and alternative spellings (table 2.24). He starts the year with Tam or Sac Cohk, but here Kan Halam leads, to parallel the glyphic roster.

TABLE 2.24

Poqomchi' month names given by Edmonson

Month names	Translations
Kan Halam	
Makux, Pet Cat	
Kazeu	
Kanazi	
Kanahal	Yellowing

(*continued*)

TABLE 2.24

Poqomchi' month names given by Edmonson (*continued*)

Month names	Translations
Tz'ikin Kih	Bird Time
Mox Kih	
Tik Cheik	Standing Trees
Yax	Green
Sac	White
Tsi	
Chip, Tzip	
Chante Mac	Cover
Uniu	
Muuan	
Tam	Crab
Sac Cohk	White Squash
Ohl	

SUMMARY: Two features of the Poqom names stand out. First, over half of them (twelve of twenty-one) represent pure or modified versions of their glyphic antecedents. This seems due to lending from Ch'olan. Second, it is striking how differently these names line up with each other when compared to the glyphic roster. The changes belie either the informants' uncertainty or considerable vicissitudes over the centuries.

Calendrical Documents in Q'anjob'al

1932. The sole source of calendrical documents in Q'anjob'al is Oliver LaFarge's (1947: 168, table 2) lists of Q'anjob'al months as given by several native informants in Santa Eulalia, a town in the mountains of the extreme northwest of Guatemala (table 2.25). The assistants disagreed on the positions of many months and even on the names of a few. As the referent year 1932 is a leap year, the fact that the days between the start of Tap on 17 February and the first day of Oyeb K'u on 9 March total twenty-one instead of twenty indicates that the Q'anjob'al at some time froze their calendar and adjusted for leap years. The names and sequence of the first eight months enjoy almost unanimous agreement among informants. Brackets enclose names supplied in the original footnotes, but the footnotes do not relate the names to dates. According to a comment for "list f," some native informants have redefined K'eq, Sax, Yac, and K'aq Sihom each as a period of eighty days.

LaFarge (1947: 38, 79, 168–69) offers a few remarks about the months. The seven from the beginning of January to June, namely Onew (starts 8-I) through Mo' (ends

TABLE 2.25

Q'anjob'al month names given by LaFarge

a Yego	b Xuarís	c Simón	d Cuxil	Dates a–d	e Xwarís	f Juárez
Onev	Oneu	Oneu	Onev	8-I	Oneu	Oneu
Sivil	Sivil	Sivil	Sivil	28-I	Sivil	Sivil
Tap	Tap	Tap	Etap	17-II	Tap	Tap
Oyeb K'u	Oyeb K'u	Oyeb K'u	Oyeb K'u	9-III	Oyeb K'u	Oyeb K'u
Saqmai	Wec	Wec	Wec	14-III	Wec	Wec
Wec	Saqmai	Saqmai	Saxmaih	3-IV	[Saqmaih]	Saqmaih
[Nabitc]	Nabitc	Nabitc	Nabitc	23-IV	Moo	Nabitc
Moo	Moo	Moo	Mo	13-V	Nabitc	Moo
Vak	Cuxem	Cuxem	Bak'	2-VI	Bak/Mak	Bak'
Bak'	Bak'		[K'aq/K'an Sihom]			
K'aq Sihom	K'aq Sihom		[K'eq Sihom]			
K'eq sihom	K'an Sihom	K'an Sihom	Kanal	22-VI	Kanul	Yacakil
Mol	Mol		[Sax Sihom]			
			Yacul	12-VII	Yacul	Cuxem
			Yacakil	1-VIII	Watsikin	Gwatsikin
			Cubihl	21-VIII	Cuxem	Baktan
			K'aq Cuxem	10-IX	K'eq Sihom	Kanal
			Mol	30-IX	K'an Sihom	Mol
						[K'eq Sihom]
						[Sax Sihom]
						[Yac Sihom]
						[K'aq Sihom]
					Mol	Mak

1-VI), cover the planting period and the advent of the rains. Those that include the periods of planting and the arrival of the rains and the high winds total 185 days [= 9 × 20, +5; day 185 is 11 July]. In June or Bak' high winds threaten the milpas. During July or Cuxem, as the tender green ears used for roasting mature, the clay flute is played in fields to protect the maize from animals and promote its growth. Several names signal outstanding events, such the bird plague (Wah Tz'ikin "Birds That Eat"?), the flute ceremony (Xuub'-b'il "Flute"?), and the harvest (Mol "Group; Pile"). The names identify periods of agricultural activity.

1988. Edmonson (1988: 190) derives his interpretation from the sequence in cognate systems, apparently the Chuj and the glyphs (table 2.26). He omits Baktan, Cubihl, and Yacul.

Table 2.26
Q'anjob'al month names as translated by Edmonson

Month names	Translations
Wex	Time
Saqmay	White 20
Nab Ich	1 Moon
Moo	Parrot
Bak'	Bone
Xuhem	—
Kanal	Yellow
Wa Tsikin	1 Bird
Yaxakil	Green Time
Oyeb Ku	5 Days [no position assigned]
Mol	Gather
K'eq Sihom	Black Flower
Yax Sihom	Green Flower
Zaq Sihom	White Flower
K'aq Sihom	Red Flower
Mak	Cover
Oneu	—
Sivil	Vapor
Tap	Crab

1996. Without referencing any source, authors of a dictionary (Diego Antonio et al. 1996: 17, 20–21, 265–66) gloss the following names, also recorded in 1932.

B'ak: name of a month of the Q'anjob'al Maya agricultural calendar, which contains twenty days

B'aktan: month of the Maya agricultural calendar which corresponds to August. "The first twenty days of August . . . in these days one can't sow."

Tap: a month of the agricultural calendar corresponding to February, containing twenty days; in Tap the corn field is sown.

SUMMARY: Only the Q'anjob'al tradition preserves the *sihom* names, each prefixed with a different color, probable relics of the four glyphic units. Each of these consists of a color prefix plus the "stone" logograph. Often this sign reads **SIHOOM** "Flower" or "Soapberry." The Q'anjob'al names also keep the Classic colors black, green, white, and red, but not in the Classic order; "yellow sihom" constitutes an innovation. Also unique to Q'anjob'al are Baktan and Cubihl.

Calendrical Documents in Q'eqchi'

1788. In a routine vocabulary list León Fernández (1949: 107–15) preserves and glosses two month names: Chantemac "autumn" and Muan "spring."

1930. J. Eric S. Thompson (1971: 118) points out that the Q'eqchi' transferred their five unlucky days to Easter and gives the name *rail cutan* but no definition. It means "grievous days" (Wirsing n.d.).

1979. Don Francisco Curley García of Cahabón in Estéban Haeserijn (1979: 391–92) presents the thirteen month names in table 2.27; Haeserijn's transliterations follow in parentheses. This set constitutes a 365-day maize agriculture calendar in the peculiar format of thirteen months of 28 days plus one day, called Mayejick "Offerings Time." Every four years, in a cycle known as Gkoban [?], when the yearbearer is the day Camnck ["Death"?], a further day is added. The Q'eqchi' title it Utzumayejic "Flower Offerings Time" and equate it to 1 January. Clearly this represents a leap-year correction. Using the explicit equation 31 December = Mayejick produces the correlation in table 2.27 (Haeserijn 1979: 392). To serve as a bissextile or leap year day and fall on New Year's day, Utzumayejic must be the first of a pair of days designated 1 January.

Tap: the last month; Haeserijn (1979: 317) offers this name with no source.

TABLE 2.27
Q'eqchi' month names given by Curley García

Month names	Month dates	Translations
Pop	16-VII	Mat
Rakol	13-VIII	Mark Off [fields]
Gkalec (C'alec)	10-IX	Clear and Weed Fields
Gkatoc (C'atoc)	8-X	Set on Fire
Auuck (Aauc)	5-XI	Sow
Cheen	3-XII	Mosquito; Well
Mayejick	31-XII	Offering
Tzaack (Tzaak)	1-I	Hunted Game; Walls
Raxgkim (Raxk'im)	29-I	Green Leaves [of maize]
Gkan (K'an)	26-II	Yellow, Ripe
Gkoloc (K'oloc)	26-III	Harvest Maize
Moloc	23-IV	Collect [into granaries]
Yulic	21-V	Anoint
Oxlajuel	18-VI	Thirteen(th) Group

1988. Edmonson (1988: 191) presents the names from the Lanquín list, treated in the present study as a Ch'ol roster. He defines nearly all of them and matches some to those recorded by Curley García (in Haeserijn 1979), though he cannot link any to Rakol, Gkalec, or Gkoloc. Curley García's names follow slashes in the list below.

Pop: Mat

Icat/Gkatoc: 1 Cat

Chacc'at: 2 Cat

Cazeu/Yulic

Chichin: Birds

Ianguca/Raxkim: Green Time

Mol: Gather

Zihora/Cheen: Well

Yax: Green

Zac: White

Chac: Red

Chantemac: 1 Cover

Uniu/Kan: Yellow

Muhan: Owl

Ahquicou

Ccanazi: Turtle

Olh

Mahi y Ccaba: Nameless

As a counterpart to the last month Edmonson (1988: 191) suggests Gkimuch, the primary source's (Haeserijn 1979: 391) name for Saturday, the last day of the week.

SUMMARY: The Q'eqchi' tradition borrows much from the Ch'olan system, likely a nonfrozen count, but it also recently innovated another calendar, this one fixed. Remarkably, nine of the modern names represent innovations. Nine names in all, not all of them innovations, clearly refer to agriculture, which are useful in a frozen count. Thirteen, the total tally of months, is an ancient sacred number.

Calendrical Documents in Tzeltal

1972: proto-Tzeltalan. For most of the Tzeltal names Terrence Kaufman (1972) reconstructs the proto-Tzeltalan counterparts but gives no translations (table 2.28). Brackets enclose reconstructions that are not presented by Kaufman but are based on

TABLE 2.28

Tzeltal month names in proto-Tzeltalan

Position	Month names	Page in Kaufman 1972
1	B'ats'ul	95
[2]	[Sakil Hä']	115, 103
[3]	['Ajil Ch'ak]	94, 101
4	Mäk	109
5	Olal/Ohlal Ti'	113
6	Hulol	103
7	J Ok'-i/E-n Äjäw	123
[8]	[Ch'in J'uch]	101, 120
[9]	[Muk' J'uch]	110, 120
[10]	[Huk Winkil]	103, 122
[11]	[Wak Winkil]	121, 122
[12]	[Ho' Winkil]	103, 122
[13]	[Chan Winkil]	99, 122
[14]	[Ox Winkil]	113, 122
15	Pom	114
16	Yax-K'in	124
17	Mux	111
18	Ts'un or *Ts'uhn	99
[19]	[Ch'ay K'in]	107

his data. Numerals to the far left indicate position in the sequence, while page numbers end each line. The calendar in proto-Tzeltalan times might well have displayed different names. For example, the series that prefixes 7, 6, 5, 4, and 3 to Winikil likely once had 2 Winikil and 1 Winikil as well. All subsequent calendars are Tzeltal.

1600. The earliest reference to a Tzeltal name occurs in Domingo de Ara's (1986: 433) dictionary as *chayquim hohel cacal los días p[er]didos*. Analysis parses this to *ch'ay k'in, ho-hel k'ak'-al* "lost days, five successive days."

1788. León Fernández (1892: 60) records one month name that lacks its numeric prefix (this should be between seven and three) and a second under the term for a season (table 2.29). Other Tzeltal dialects begin the first name with *b* or *v* (Robles Urribe 1966: 15–16).

1887. Vicente Pineda (1887: 130–31, 177, 1888) preserves the first complete set of names, presented here with his orthography, sequence, and translations (table 2.30). Asterisks mark variants in his second list, while page numbers in parentheses accompany names from his 1888 dictionary. In nearly every instance his *h* represents ' (glottal stop). He offers no dates, but Edmonson (1988: 258) attributes to him the base for the correlation 15-IX-1888 G = 1 Batz'ul. While Pineda (1887: 177,

TABLE 2.29

Two Tzeltal month names from 1788

Month name, as written	Month name, in modern spelling	Gloss/translation
<*guincil*>	Winkil	*otoño*/autumn
<*jul el*>	Hulel	*primavera*/spring

TABLE 2.30

Tzeltal month names given by Pineda

Month names	Place	Month dates	Meanings
Batzul	1	15-IX-1888	first amaranths
Saquiljá	2	5-X	clear water
Agelchac	3	25-X	abundance of fleas
Mac	4	14-XI	cover, fence; cover, enclose
*Olaltí; also Alal-ti,*Olal-ti, Alalti; Halalti (88)	5	4-XII	mouth of a child
Tzun	6	24-XII	planting
Jul-hol; also Jul-hal; *Julol, *Julal	7	13-I-1889	the baby came
Hoquén-hajab	8	2-II	the big hill wept
Yal-uch; also Yahal-uch (139)	9	22-II	child of the opossum
Muc-uch	10	14-III	large opossum
Juc-binquil; also *Juc-vinquil	11	3-IV	seventh birth
Guac-binquil; also *Guac-vinquil	12	23-IV	sixth birth
Jo-binquil; *Jo-vinquil	13	13-V	fifth birth
Chan-binquil; *Chan-vinq[u]il	14	2-VI	fourth birth
Osh-binquil; *Osh-vinquil	15	22-VI	third birth
Mush	16	12-VII	general softening of the earth due to excessive rain
Yash-quin	17	1-VIII	moist feast day
Pom	18	21-VIII	incense
Chay-quin; *Chai Quin	19	10-IX	the feast day was lost

1888) assigns the ordinal positions twice, Edmonson doubts their accuracy, favoring instead the slots and sequence of the 1979 Tzeltal calendar from San Cristóbal de las Casas; they hold for nearly all the others. The unusual placement of Tzun, Mush, Yash-quin, Pom, and Chay-quin marks this calendar as eccentric. Its initial dates make clear that it cannot be the original, nonfrozen calendar fixed at either 17-I-1585 G (1 Batzul) or 22-V-1585 G (1 Hoquen-hajab) (Edmonson 1988: 258).

1898. Below appears Eduard Seler's (1902: 1:706–11) version of Vicente Pineda's roster. As Seler quotes Pineda's translation for each name, this list records Seler's glosses only when they differ from Pineda's.

Tzun: Sowing

Batzul: First Green

Saquil Já

Ajel Chac

Mac: Lid, Fence

Olaltí

Julol

Oquén Ajab

Yal-Uch: Little Opossum, 710; Little Louse

Muc-Uch: Big Opossum, Big Louse

Juc Vinquil: Seven Score

Wac Vinquil: Six Score [also Guac Vinquil]

Jo Vinquil: Five Score

Chan Vinquil: Four Score

Ox Vinquil: Three Score: [also Osh Vinquil]

Pom

Mux [Mush]

Yax-Quin: First Feast of the Year

Chay-Quin: The Omitted, Lost Days

1915–20. As a schoolteacher during these years Marcos Becerra collected month rosters in several towns among the mountains of Chiapas (the state in the southeast corner of Mexico) then published them (Becerra 1933: 45–46, table 8). This condensed list presents all his data (table 2.31). The towns all preserve the same sequence. Letters mark variants as from Santo Tomás Oxchuc (O), Presentación Cancuc (C), San Ildefonso Tenejapa (T), and San Pedro Chanal (CH); unmarked forms appear in all these pueblos. The initial dates for Santo Tomás Oxchuc appear in the second column, with those from Presentación Cancuc starting one day later and those of San Ildefonso Tenejapa two days later. Becerra does not state the referent year, but comparisons with other lists indicate 1916–19. The interpretations are Becerra's (1933: 51, table 9, 56–70).

1917. Ramón P. C. Schulz (1953: 114–15, table IIb) reconstructs the Cancuc calendar for this year. He recounts that in about 1917 the native calendar in Cancuc, Oxchuc, and Tenejapa was freed from the European count, due to social, political, and economic disturbances; the change of authorities; and the dissolution of relations with the church and civil authorities. Treating bissextiles as common days, the

Table 2.31

Tzeltal month names given by Becerra

Month names	Month dates	Meanings and pages
Jokinajau, Jokenajau O	6-V	five day lord 62
Yajuch CH, T; Yalajuch O	26-V	small louse, small opossum 62
Alajuch CH, C, T; Alaluch C; Bikituch T; Chinuch O	26-V	small louse, small opossum 62
Mukuch, Mukuluch T	15-VI	big louse, big opossum 62
Jukbinkil	5-VII	seventh companion 65
Guakbinkil	25-VII	sixth companion 66
Jobinkil	14-VIII	fifth companion 66
Chanbinkil	3-IX	fourth companion 66
Oshbinkil	23-IX	third companion 66
Pom	13-X	incense 66
Yashkín	2-XI	humid time 67
Mush	22-XI	variant of Mosh, for Imosh, a contraction of *im + mush* "nipple + navel," a kenning for the pointed fire drill and its hearth stick with holes 16, 21; table 6, under "chiapaneco" 67
Tsun	12-XII	to plant; hummingbird 6)
Batsul	1-I	first amaranth 67–68
Sakilja'	21-I	clear water; white ear of corn 68
Ajilchak	10-II	red lord, a deer 69
Mak	2-III	cover, close; end of year at spring equinox 69
Olalti'	22-III	half a small meal, fasting 61
Julol	11-IV	pierce; eat half; bloodletting, fasting 61
Chaikín [*sic*]	1-V	lost time, lost days 70

native round receded from the Western calendar one day every four years. The first month, Batsul, begins on 1 January.

1938. In February and March of this year Alfonso Villa Rojas visited fifteen Tzeltal communities and recorded lists from at least Chanal, Cancuc, and Oxchuc (Redfield and Villa Rojas 1939: 107, 117). He presents no dates, noting only that Yax-kin "ended in November so that the new year would begin in the second week of January," a correlation unattested in any calendar for or near this year. He notes one activity for each of twelve months. Bracketed numerals mark members of the countdown series. Pom does not mean "second" but falls in that slot; the *yax* of *yax-kin* sometimes by extension denotes "first."

Batzul: time of harvest

Zakilab: clearing the milpa begins

Agelchac: the clearing continues

Mac: time of sowing

Olalti: time of sowing

Jul-Ol: time of sowing

Uec-Cachaquin: no work ["unlucky five days"]

Joquen-ahau: time of sowing in tierra caliente

Chin-Uch: time of weeding

Mucul-Uch: time of weeding

Juc-Uinquil: time of resting [7]

Uac-Uinquil: [6]

Jo-Uinquil: [5]

Chan-Uinquil: [4]

Ox-Uinquil: [3]

Pom: new ear forms; October [2]

Yax-Kin: ends with November [1]

Mux

Tzun

1942. In this year Alfonso Villa Rojas (1946: 573–75) elicited near Oxchuc in highland Chiapas virtually all the data for his calendar (table 2.32). His comments are slightly abbreviated.

1944. Fernando Cámara Barbachano (1945b: 116) recorded two lists from Tenejapa for 1944. Both represent a calendar not frozen or fixed to the Gregorian dates because they count February 29 as one of the twenty days in Mák. Schulz (1953: 114, table 1) published a calendar for 1952, another leap year, from not only Tenejapa but Cancuc and Oxchuc as well. In all three the native informants explicitly counted February 29 as a day in Mac.

1953. At the end of the Spanish–Tzeltal section in her dictionary, Marianna Slocum (1953: 87) offers the 1953 Oxchuc calendar (table 2.33). The orthography for the first time consistently distinguishes glottalized from plain consonants.

TABLE 2.32
Tzeltal month names given by Villa Rojas, with abbreviated comments

Month names	Month dates	Comments
Batzul	26-XII-1941	change of office; chilis sown; a little brush cleared or cut away; not much work yet
Zakilab	15-I-1942	same as in the previous month
Agilchac	4-II	Carnival; start of clearing brush
Mac	24-II	clearing brush; start of sowing in cold upland
Alalti	16-III	clearing; best time to sow
Jul-ol	5-IV	same as in previous month
Chaykin [Uec-cachaquin]	25-IV	no work; Chaykin ended 29 April [1942], principal day of San Juan, saint of Cancuc
Joken-ajau	30-IV	start of weeding in cold upland; still not much work here; some scrape a little agave fiber for bag nets and cords to be sold
Chin-uch	20-V	start of weeding corn fields around here
Muc-uch	9-VI	weeding and doing other small tasks
Juc-uinkil	29-VI	little work in corn fields, clearing them of weeds; houses built; in cold upland some brush cleared
Uac-uinkil	19-VII	weeding the corn fields continues, as here they weed twice
Jo-uinkil	8-VIII	some rest here; in cold upland beginning of clearing brush for next year's cornfield
Chan-uinkil	28-VIII	inactivity; men go to work; cold upland cornfields, coffee fields
Ox-uinkil	17-IX	as in the previous month; 100 days of just *uinkiles* [all 5 *uinkil* months total 100 days]
Pom	7-X	making of ropes and agave fiber bag nets
Yaxkin	27-X	making houses, chicken coops, bag nets, etc.
Mux	16-XI	they start to work a bit in the cornfields
Tzun	6-XII	10th of this month perhaps real feast day of St. Thomas of Oxchuc

1966. Brent Berlin, Dennis E. Breedlove, and Peter H. Raven (1974: 119: tables 5.16, 5.17. 120–24) present four sets for 1966 from Tenejapa. The first appears in table 2.34, with variants in brackets. All the informants begin the list of months with Bats'ul. The one clear correlation equates 13 Bats'ul to 1 January 1966. The identity of the dates with those of the 1979 roster from San Cristóbal de las Casas indicates that this calendar was frozen in 1964 or 1965. Berlin et al. (1974: 120–24) describe in detail the agricultural activities for each month in both the cold country and the hot country.

1979. The *Calendario Sna Holobil* from San Cristóbal de las Casas presents a recent list of month names (table 2.35). The initial dates derive from the correlation of

TABLE 2.33

Tzeltal month names given by Slocum

Month names	Month dates
Bats'ul	23-XII-1952
Sakilab	12-I-1953
Ajilch'ak	1-II
Mak	21-II
Jul'ol	2-IV
Ch'ayk'in	22-IV
Jok'en'ajaw	27-IV
Ch'inuch	17-V
Muk'uch	6-VI
Jukwinkil	26-VI
Wakwinkil	16-VII
Jo'winkil	5-VIII
Chanwinkil	25-VIII
Oxwinkil	14-IX
Pom	4-X
Yaxk'in	24-X
Mux	13-XI
Ts'um	3-XII

TABLE 2.34

Tzeltal month names given by Berlin et al.

Month names	Month dates	Name variants
Bats'ul	20-XII-1965	
Sakil Ha'	9-I-1966	
Ahil Ch'ak	29-I	[Ch'ak, 119]
Mak	18-II	
Olal Ti'	10-III	
Hul Ol	30-III	
Ch'ay K'in,	19-IV	Ho'eb Sóre Ahtal
Hok'en Ahaw	24-IV	[Hok'in Ahaw, 121, 122]
Ch'in H'uch	14-V	
Muk' H'uch	3-VI	[Muk'ul H Uch, 119]
Huk Winkil	23-VI	
Wak Winkil	13-VII	
Ho' Winkil	2-VIII	
Chan Winkil	22-VIII	
Ox Winkil	11-IX	
Pom	1-X	
Yax K'in	21-X	
Mux	10-XI	
Ts'un	30-XI	

TABLE 2.35

Tzeltal month names of the *Calendario Sna Holobil*

Month names	Translations	Month dates
Jok'en Ajaw	5 Day Lord	24-IV-1978
Ch'in J'uch	1 Opossum	14-V
Muk' J'uch	2 Opossum	3-VI
Juk Winkil	7 Score	23-VI
Wak Winkil	6 Score	13-VII
Jo' Winkil	5 Score	2-VIII
Chan Winkil	4 Score	22-VIII
Ox Winkil	3 Score	11-IX
Pom	incense	1-X
Yax K'in	Green Time	21-X
Mux	Mud	10-XI
Tz'un	Plant	30-XI
Batzul/Batz'ul	1 Amaranth	20-XII
Sakil Ja'	White Water	9-I-1979
Ajil Ch'ak	Dawn Red	29-I
Mak	Cover	18-II
Olal Ti'	—	10-III
Jul Ol	Arrival	30-III
Chay K'in	Lost Days	19-IV

20 December 1978 to 1 Batz'ul and from the position of Chay K'in immediately before the first day of the new year, 1 Jok'en Ajaw (Edmonson 1988: 257–58). These dates replicate those of a calendar from Tenejapa reported in 1966 (table 2.34) and thus suggest that San Cristóbal froze the count in 1964–67. As elsewhere, Edmonson substitutes successive numerals for the nonnumerical members of compound names in a series without explanation; for example, Ch'in J'uch and Muk' J'uch denote "Small Opossum" and "Large Opossum," but he writes "1-opossum" and "2-opossum."

SUMMARY: The Tzeltal froze their calendar in 1584, starting their next year with 1 B'atz'ul on 17 January 1585. Between 1588 and 1600 several variants beginning one to four days earlier were apparently immobilized. In 1848–51, for reasons still not clear, many local versions stopped adjusting for leap years and began to recede. In 1916–19 some communities again froze their counts and have kept them so until the present. The others, ignoring bissextiles, continued receding. Decades later, in 1964–67, a few of these finally bound their correlations into lockstep with the Western count. The Late Postclassic invasion by the Pochtecas, the Aztec warrior-merchants, probably prompted the Tzeltals to shift their initial month from Okinahaw to Batz'ul, their equivalent to Izcalli, the first score of days among the Aztec

(Edmonson 1988: 105, 221–22). The five-day span Ch'ay K'in nearly always immediately precedes Okinahaw.

Calendrical Documents in Tzotzil

1972: proto-Tzeltalan in reference to Tzotzil. Tzotzil ultimately derives from
proto-Tzeltalan. Kaufman (1972) reconstructs the proto-Tzeltalan counterparts to
nearly all the Tzotzil names (table 2.28). The modern forms derive from the 1968 list
(table 2.36) with a more practical orthography. Brackets enclose forms not presented
by Kaufman but based on his data. There are no dates. All subsequent calendars are
Tzotzil.

1688. The library of the Casa Blom in San Cristóbal de las Casas houses the earliest
Tzotzil calendar manuscript, written down in 1688 in Guitiupa by the Franciscan
parson Juan de Rodaz and referred to as the calendar of Guitiupa (Edmonson 1988:
133–34). As it allows for no 29 February, the roster cannot refer to the leap year 1688
and must be generic. In 1723 a prelate drew up a copy of this document and clearly

TABLE 2.36

Modern Tzotzil and proto-Tzeltalan month names

Modern Tzotzil forms	Proto-Tzeltalan	Page in Kaufman 1972
Bats'ul	*B'ats'ul	95
Sisak	*. . . Sak . . .	115
Muk'ta Sak	[*Muk' . . . Sak]	110, 115
Mok	*Mäk	109
O'lal Ti'	*Olal/Ohlal-ti'	113
Ulol	*U'lol	103
Ok'en Kahval	J'ok'-i/E-n Äjäw	123
Uch	*Uch	120
Elech	[*Elech?]**	—
Hnichk'in	[*Jnichk'in]	123, 112, 107
Sbavinkil	[*Sb'ah Winkil]	95, 122
Schibal Vinkil	[*Schibal Winkil]	99, 122
Yoxibal Vinkil	[*Yoxibal Winkil]	113, 122
Schanibal Vinkil	[*Schanibal Winkil]	99, 122
Pom	*Pom	114
Yaxk'in	*Yax-k'in	124
Mux	*Mux	111
Ts'un	*Ts'un or *Ts'uhn	99
Ch'ay K'in	[*Ch'ay K'in]**	107

** Kaufman gives no reconstructed form.

signed it "Dionycio Pereyra"; today it reposes in Paris as "Manuscrit Mexicain 411" (see below). Scholars who worked with it do not know of any precedent (Charencey 1885: 398–99; Berlin 1951: 156; Ruz 1989: 15, 87), so the Casa Blom manuscript appears to be either the original or a copy; it commences with a different month than the 1723 work does and is unsigned. Entitled *Arte de la lengua tzotzlem o tzinacanteca con explicación del año solar y un tratado de las quentas de los indios en lengua tzotzlem*, this work might well represent the calendar of Zinacantán. It appears here in the orthography of Evon Vogt (1969: 603, table 16), with tentative translations by an unnamed contemporary Zinacanteco informant (table 2.37). The original spellings probably resemble those of the 1723 copy.

1723. The Dominican friar Dionicio Pereira copied the 1688 manuscript in 1723 at Comitán. On folio 22 of this work he recorded the name, ordinal position, date, and day total for each month. The prelate begins his roll, however, not with the presumed original's Muctacac but with Batzul. Given this difference, the present study refers to this roster as the calendar of Comitán. Table 2.38 summarizes the facsimile and transcription by Mario Humberto Ruz (1989: 143–44), using his orthography. The

TABLE 2.37
Tzotzil month names of Guitiupa

Month names	Initial dates	Translations
Muk'ta Sak	3-III	
Mok	23-III	Fence
Olalti'	12V	
U'lol	2-V	
Hok'in Ahval	22-V	
H'uch	11-VI	Drinker
H'elech	1-VII	
Nichil K'in	21-VII	Flower Ceremony
Hun Vinkil	10-VIII	1st 20-Day Period
Shchibal Vinkil	30-VIII	2nd 20-Day Period
Yoshibal Vinkil	19-IX	3rd 20-Day Period
Shchanibal Vinkil	9-X	4th 20-Day Period
Pom	29-X	Incense
Yash K'in	18-XI	Green Ceremony
Mush (or Muy?)	8-XII	
Ts'un	28-XII	To Plant
Bats'ul	17-I	
Sisak	6-II	
Ch'ay K'in	26-II	Lost Ceremony

TABLE 2.38

Tzotzil month names given by Pereira

Month names	Initial dates
Batzul	17-I
Cicac [Ciçac]	6-II
Chaiquin	26-II
Muctacac	3-III
Moc	23-III
Olalti	12-IV
Ulol; Hoyoh	2-V
Oquinaghual	22-V
Uch	11-VI
Elech	1-VII
Nichilquin	21-VII
Ghunvinquil	10-VIII
Xchibalvinquil	30-VIII
Yoxibalvinquil	19-IX
Xchanibalvinquil	9-X
Pom	29-X
Yaxquin	18-XI
Muy	8-XII
Tzun	28-XII

Bibliothéque nationale lists this work as "Manuscrit Mexicain 411" (Berlin 1951: 156; Ruz 1989: 15).

1845. In his *Descripción geográfica del departamento de Chiapas y Soconusco* Emeterio Pineda (1845: 111–12) records all the names except Chaiquin, offers activities and variant forms for a few, but provides dates for none. Edmonson (1988: 183–84) assigns this count to Istacostoc then offers the equation 1 Mok = 21-III-1845, the basis for the dates in table 2.39, and designates this as the first day of the year. Pineda, however, begins his list with Tzun, probably because this month includes 1 January.

1902. Frederick Starr (1902: 72) preserves a 1901 calendar as reported by Father Sánchez, who mentions that the intercalation of the Five Days takes place during Holy Week. Sánchez does not give the five days a name or place in his sequence but absorbs them with several *winal*s of unorthodox length, whose totals appear in brackets in table 2.40. This roster preserves the sequence of the 1688 and 1723 calendars. Otherwise it is Europeanized. It begins on 1 January; its shortest month, 18 days, ends with February, the shortest Western month; and all its units of 21 days occur in all Western months of 31 days.

TABLE 2.39

Tzotzil month names given by Pineda

Month names	Initial dates	Comments
Tzun	26-XII-1844	
Batzul	15-I-1845	
Sisac	4-II	
Muctasac	24-II	
[Chaiquin]	16-III	
Moc	21-III	fences should be mended
Olalti	10-IV	sowings must be done
Vlol, [Ulol]	30-IV	
Oquinajual	20-V	
Veh [*Uch]	9-VI	the illnesses of the plants suddenly appear, in particular an insect that like the plant louse weakens and destroys them
Elech, Elch	29-VI	healthy winds come; but if they are not favorable, the loss of many plants, like the potato, is sure
Nichqum, Nichquin	19-VII	blooming
Sbanvinquil, Sbavinquil	8-VIII	fertilization
Xchibalvinquil	28-VIII	"pearl" stage of kernel
Yoxibalvinquil	17-IX	"milk" stage of kernel
Xchanibalvinquil	7-X	"mealy" stage of kernel
Poin [*sic:* Pom]	27-X	honeycombs should be extracted, crops gathered
Mux	16-XI	proximity of cold, frost
Yaxquin	6-XII	[All native sources place this in mid-November.]

1915–20. While teaching in a native school from 1915 to 1920, Marcos Becerra collected calendars from several towns then published and discussed them (Becerra 1933: 40–70). With very few exceptions, the names from all the pueblos match. Becerra does not state the referent year, but Edmonson (1988: 259g–i) calculates it at 1931–32. All calendars are fixed to the Western count since 1548 or 1584.

Letters identify the towns, all in the rugged highlands of Chiapas. A: Santa Marta Yolotepec; B: Magdalenas Tanhobel; C: San Miguel Mitontic, San Pablo Chalchihuitán, and San Pedro Chenalhó; and D: San Andrés Istacostoc. No dates appear in the original for the first two under C. Unmarked names listed are in addition to the A forms, while names marked with an asterisk replace them. The dates in A and B, the standard, are identical to those of the 1688 count, but the dates are three days earlier in C and two in D, which also reverses Chaikin and Muktasak. All data derive from

TABLE 2.40

Tzotzil month names given by Starr

Month names	Initial dates	Total days
Tzim [*sic*]	1-I	20
Batzul	21-I	[21]
Sisác	11-II	[18]
Muctasái	1-III	20
Móc	21-III	[21]
Olalti	11-IV	20
Ulol	1-V	20
Oquin-ajual	21-V	[21]
Uch	11-VI	20
Elech	1-VII	20
Nich-quin	21-VII	[21]
Sba-vinquil	11-VIII	[21]
Schibal-vinquil	1-IX	20
Yoshibal-vinquil	21-IX	20
Chanim-vinquil	11-X	[21]
Póm	1-XI	20
Yashquin	21-XI	20
Mush	11-XII	[21]

Becerra (1933: table 7). The following interpretations, most of them by Becerra, come from his study (1933: table 9, 61–70). Brackets and quotation marks enclose my literal translations, based solely on Becerra's derivations and comments.

Tsun 28-XII [Sow]. Tsun: from *tsunun* hummingbird; perhaps also *tsun* to sow (67)

Batsul 17-I ["First Pigweed"?]. *Ba-tsul*: *ba* on top of, first; *tsul* goosefoot, pigweed [*bledo*] (67)

Sisak 6-II ["White Larvae of Ants"?]. *Tsi-sak*: *tsisim* large red ant; *sak* white; refers to the white larvae of this ant, maybe eaten at this time (68)

Chaikin 26-II "Lost Time" or "Lost Days." *Chai-kin*: *chai* to lose; *kinal* time (70)

Muktasak 3-III ["Big White Larva"?]. *Mukta-sak*: *mukta* big; *sak* white; refers to a more developed larva of the *arriera* ant (69)

Mok, Muk 23-III C Muk ["Close the Year"?]. *Mok*: to close, cover; these are the likely meanings and not "fences," because the precolonial Indians had no livestock and so did not have to build fences to protect their fields from such animals. The name refers perhaps to the close of the year, around the vernal equinox, somewhere between 2 March and 6 April. (69)

Olalti 12-IV ["Half of a Small Meal"?]. *Ol-al-ti*: *olol* half; *al* small; *ti* food; refers to fasting (61)

Ulol, Julol 2-V B, C Julol ["Piercing" or "Eating Half"?]. *Ju-lo-ol* or *jul-lo-ol*: *jul* pierce; *loel* eat; *olil* half; refers to fasting and bloodletting (61)

Okinajual 22-V ["Five Days Lord"?]. *O-kin-ajual*: *o* five; *kinal* season, time; *ajual* lord (61–62)

Uch 11-VI B Unenuch, D Bikituch. *Uch* louse; opossum (62)

Elech 1-VII D *Muktauch ["Thieving Louse or Opossum"?]. *El-ech*: *elek* thief; *elkam* steal; *ech* so perhaps a ritual form of *El-uch* (62)

Nichikin 21-VII "Flower Time." *Nichi-kin*: *nichim* flower; *kinal* time (65)

Sbabinkil 10-VIII "First Companion," referring to the Nine Lords of the Night. *Sba-binkil*: *sba* on top of; first; *binkil* companion (66)

Schibalbinkil 30-VIII "Second Companion" of the Night. *Schibal-binkil*: *schibal* second; *binkil* companion (66)

Yoshibalbinkil 19-IX "Third Companion" of the Night. *Yoshibal-binkil*: *yoshibal* third; *binkil* companion (66)

Schanibalbinkil 9-X B, C *Schanebalbinkil "Fourth Companion" of the Night. *Schanibal-binkil*: *schanibal* fourth; *binkil* companion (66)

Pom 29-X. *Pom*: honey, incense

Yashkin 18-XI "Humid Time." *Yash-kin*: *yashal* humidity; *kinal* time (67)

Mush 8-XII ["Nipple-Navel" or fire stick and hearth]. *Mush*: variant of *mosh*, from the day name *Imosh*, itself from *im* nipple Mam, K'iche', Tzeltal, Yukatek, and *mush* navel Mam or *mushush* navel K'iche'. As "nipple-navel" the name alludes to the fire stick nipple and its hearth hole navel [Becerra 1933: 16–22, 67; tables 2, 4, 6].

1941. Ramón Schulz (1942: 9, 12–13) records a calendar in use in 1941–42 at Chamula, San Andrés Larraínzar, María Magdalena Aldama, Santa Marta Manuel Utrilla, and San Pedro Chenalhó (towns in the mountains of Chiapas). Its months all begin one day earlier than those of the 1688/1723 calendar. Schulz's 1941 count from San Miguel Mitontic starts on the same day as the calendar of 1688/1723, with 1 Tsun at 28 December. As this roster plainly records the equation 1 Muktasak = 29 February 1944 (Barbachano 1945a: 16), it must be fixed with a pair of 1 Muktasak every four years.

1944/55, 1946. Calixta Guiteras-Holmes (1946: 188, 1961: 32–35) reported a calendar from San Pedro Chenalhó that begins all its months one day earlier than the calendar of 1688/1723 (table 2.41). Two months list additional names: Me'Okinahual "Old Woman or Mother Okinahual" for Okinahual and Mol Uch "Old Man or

TABLE 2.41

Tzotzil month names given by Guiteras-Holmes

Month names	Initial dates
Tzum	27-XII
Batzul	17-I
Sisak	5-II
The Five Ch'aik'in	25-II
Muktasak	2-III
Mok	22-III
Olaltí	11-IV
Ulol	1-V
(Me') Okinahual	21-V
Uch; Mol Uch	10-VI
Elech	30-VI
Nichikin	20-VII
Sbabinkil	9-VIII
Schibalbinkil	29-VIII
Yoxibalbinkil	18-IX
Schanibalbinkil	8-X
Pom	28-X
Yaxkin	17-XI
Mux	7-XII

Grandfather Uch" for Uch. The months bracket agricultural and ritual activities. The 260-day sacred cycle is fixed within the year, to the span 1 Sisak–20 Pom (5-II–16-XI). This calendar appears to be frozen to the Gregorian calendar (see Edmonson 1988: 258–61). The 1946 list starts with Muctasác, the other with Tzum.

> Tzum-Batzul 27-XII–4-II. While the last harvests come in, the farmers clear the land in preparation for planting the new cornfields.

On 1 Batzul (16 January) the sun begins to move to the north and the days become longer. At dawn the old men face the sun and order it to take its proper place with *Bats kuhan, Kahwal* or "Go [to that end] my lord" (Carlo Antonio Castro G., in Guiteras-Holmes 1961: 36 n. 1). As the solstice falls on 20–23 December, the date seems to be a relic from when the count began with 1 Tzum on 27 December. The closest fit requires that the five days follow rather than precede this month; thus 1 Tzum would start on 22 December.

> Sisak–Muktasak 5-II–21-III. These months are for stubbling and burning. In Muktasak fires appear on every hillside, and the air is full of smoke.

> 1 Sisak–20 Ulol 5-II–20-V. The farmers plant maize, pole beans, and squash. On 3 Ulol or 3 May the people celebrate the Feast of the Holy Cross and ask for rain.

1 Okinahual–20 Sbabinkil 21-V–28-VIII. They plant ground beans. On the greatly respected 1 Okinahual [21-V] the people must all drink the ritual atole or *ul* so that they can eat the harvested maize.

In the *winal* Uch (10-VI–29-VI) they again consume atole, for the same reason. For bountiful harvests they place above the entrance to each home an offering that consists of a tiny ear of maize, a tortilla the size of a small coin, a chili, a few beans, a cigarette, salt, and a miniature ball of posol tied into little bundles of maize husk.

On 1 Elech (30 June), shortly after the summer solstice of June 21 or 22, it is said that the sun moves toward the south and the days grow shorter. At dawn of this day the old men face the sun and command it to take its proper place: *Elech kuhan, Kah-wal* or "Come [to this end], my lord" (Carlo Antonio Castro G., in Guiteras-Holmes 1961: 36 n. 1). *Elech* in this context would be the most straightforward source for the meaning of the month name.

1 Nichikin–20 Sbabinkil 20-VII–28-VIII. In Nichikin itself the maize flowers.

Schanibalbinkil 8-X–27-X. In this *winal* the souls of the dead light up their graves with large and small balls of fire to announce their impending visit in early Pom.

1/3 Pom–20 Mux 28/30-X–26-XII. They sow wheat. On 5 Pom [1-XI] the people gather flowers and prepare food for the dead. That night the souls of the departed visit their kin, retiring before dawn. Many plant on the days after 5–6 Pom [1–2-XI] because the departed bring good luck and abundance. On 10 Pom [6-XI] the saints are taken out in procession, and everyone celebrates the third, last *mukta mixa* to mark the end of the year for the farmers and the authorities.

1948. Heinrich Berlin (1951: 156–59) gathered calendrical data in three Tzotzil villages and published his findings three years later, comparing his calendars to those of Becerra and Schulz from the same locales. His names and dates from Santa Marta match those of Becerra's list for 1915–20 (D) for San Andrés exactly, with 1 Tsun on 26 December and the aberrant reversal of Chaikin and Muktasak. His list for San Andrés is identical, except that the two months are not reversed. Berlin's only other discrepancy surfaces in his dates at Chenalhó: 1 Tsun falls on 20 December, so all the months begin eight days earlier than in the 1688/1723 manuscript. Berlin (1951: 158) can only suspect that the "more frequent communication with Spanish people" at Chenalhó has somewhat blurred the tradition and caused his informants' confessed uncertainty. The first of two alternatives requires that this count remained unfrozen until 1616–19, over thirty years after daykeepers elsewhere froze most of the others to the Gregorian round in 1584. The second posits that it actually was frozen with the rest and remained so until it was thawed out sometime after 29 February 1916 if Berlin's referent year was 1947 or after the bissextile of 1920 if it was 1949.

1963. In this year William Holland (1963: 295–96 n. 14) published a calendar from San Andrés Larraínzar with the same dates as those in the 1941–42 roster from the same village (Schulz 1942: 9). Holland (296 n. 14) points out that for quite some time the native people had frozen their count to the Gregorian round, adding one day to the Ch'ai K'in every leap year.

1968. Gary Gossen (1974b: 230–52) recorded a roster from San Juan Chamula for 1967–68 (table 2.42). Many of his informants preferred to begin it with 1 Tsun as 26 December, while some allowed 27 December. Its months all commence one or two days earlier than those of the 1688/1723 calendar. This count, however, does not reckon leap years (Gossen 1974b: 241–42). The modern orthography respects phonemic distinctions. Gossen offers etymologies as well as a detailed native account of agricultural activities for the hot country and the cold country. Numerals in parentheses indicate pages in this source.

1981. In a personal communication that year with Karl Taube (1988b: 337) Floyd Lounsbury reports *h'ok'en ahwal* as a variant on the Chamula Tzotzil name, presumably Ok'en Kahval.

1988. Edmonson (1988: 259) attributes his list to Emeterio Pineda (1845) but presents several names that do not appear in that original. His apparent sources are the Tzeltal and Tzotzil lists displayed side by side in Thompson (1971: 106, table 8). Asterisks mark Edmonson's "Tzotzil" names that actually appear only in Thompson's Tzeltal roster; the other forms different from Pineda's all occur in Thompson's Tzotzil column. The one exception is Okin Ahau, a form used in the Tzeltal town of Tenejapa, where it is a loan from Tzotzil; *ok'in* is typically Tzotzil, while *ahaw* identifies Tzeltal. Edmonson adds Chai Kin yet places it not fourth, as most Tzotzil lists do, but last.

    Tzun: Plant

    Batzul: 1 Amaranth

    Zi Zac: 1 White

    Mukta Zac: 2 White

    *Mac: Cover

    Olalti: [no entry]

    Ulol: Arrival

    *Okin Ahau: 5 Day Lord

    *Ala Uch: 1 Possum

    Elech: 2 Possum

    *Muc Uch: 3 Possum

    Hum Uinicil: 1 Score

    Xchibal Uinicil: 2 Score

    Yoxchibal Uinicil: 3 Score

Xchanibal Uinicil: 4 Score

Pom: Incense

Yaxkin: Green Time

Mux: Mud

Chai Kin: Lost Days

TABLE 2.42

Tzotzil month names given by Gossen

Month names	Initial dates	Translations (capitalized) or comments
Ts'un	26/27-XII	To Sow
Bats'ul	15/16-I	Monkey Atole; True Atole
Sisak	4/5-II	White Firewood (?); Little (?) White
Ch'ayk'in	24/25-II	lost fiesta, lost period (229)
Muk'ta Sak	1/2-III	Great Whiteness. This might refer to morning frost or to the Milky Way (234)
Mok	21/22-III	Fence, Wall, Barrier. This refers to the fences and walls the farmers build around the fields right after planting maize to keep out sheep, armadillos, rabbits, and other animals
O'lal Ti'	10/11-IV	Half-Bite
Ulol	30-IV/1-V	Zinacanteco (?); Lorenzo's Month
Ok'en Kahval	20/21-V	The Crying of Our [*sic*: My] Lord. This name refers to the spectacular storms, with hail and lightning bolts, that mark the onset of the rainy season. The Earth Lords send them in cooperation with "Our Lord," the Christ/Sun deity. Sometimes this month is called Me'el Uch Old Woman Opossum (237), which recalls the alternate San Pedro Chenalhó name Me'okinahual (see Tzotzil calendar for 1944/55 above in text and table 2.41)
(H) Uch	9/10-VI	(Male) Opossum. This month sometimes is called Mol H'uch Old Man Opossum/Opossum Hunter, an apparent partner to the preceding variant Me'el Uch (237) and virtual cognate to the San Pedro Chenalhó Mol Uch (1944, 1955 above)
(H) Elech	29/30-VI	an unidentified small mammal living in dry logs (237), perhaps a mole
Hnichk'in	19/20-VII	Flower Festival. The informants observed that "this is the month when the maize flowers or tassels and begins to produce small ears; hence the month name" (238)

(*continued*)

TABLE 2.42

(*continued*)

Month names	Initial dates	Translations (capitalized) or comments
Sbavinkil	8/9-VIII	First Man or Winal. The informants suggest that this and the next three month names allude to the four earthbearers (238)
Schibal Vinkil	28/29-VIII	Second Man or Winal
Oxibal Vinkil	17/18-IX	Third Man or Winal
Schanibal Vinkil	7/8-X	Fourth Man or Winal
Pom	27/28-X	Incense
Yaxk'in	17/18-XI	New, Fresh, or Green Festival
Mux	6/7-XII	negation; bad, evil

SUMMARY: The standard Tzotzil calendar was frozen in 1584, starting the next native year with 1 Tz'un on 28 December 1584. A few variants beginning one to three days earlier were apparently frozen four to twelve years later (1588–96). The Chenalhó version appears to have been unfrozen since 1916–19. The Five Days Ch'ay K'in, almost always appear before Muk'ta Sak and less often as the last month of the year. Three months each count one rare alternate, while another allows two. More than with most other calendars, the Tzotzil sources supply seasonal, primarily agricultural information; as the definitions in chapter 3 and the discussions in chapter 4 demonstrate, though, many of the names do not directly relate to their glosses.

Calendrical Documents in Yukatek

All these Yukatek calendars follow the dates and order in Landa, whose referent year is 1553, as the Yukatek Mayas for reasons not yet clear decided to freeze their count to the Julian calendar from 1 March 1548 forward (Edmonson 1988: 76, 78). Landa's equation 16 July 1553 Julian = 12 K'an 1 Pop (Edmonson 1988: 78, 202) places the Yukatek rosters under the Mayapán calendar. The present study preserves the original orthography throughout, using *ch'* for the barred *ch* and *dch*, and *pp* for the barred *p*. *Dz* and Ɔ are equivalent to *tz'*.

1566. Landa penned or dictated his *Relación de las cosas de Yucatán* in Spain in 1566, perhaps as part of his defense against charges that he had overstepped his inquisitorial powers and perhaps also as a reference and introduction for missionaries. Landa returned, acquitted, in October 1573 with the original or a full copy. After his death in 1579 the work found its place in the chapter library at the Convento Mayor de San Francisco in Mérida (table 2.43). The friars may well have sent copies or summaries to Spain. The chronicler royal of the Indies, Antonio de Herrera y Torsedillas (1952), seems to have had one before him while composing his *Historia general* or *Décadas*, a copy perhaps extant among his uncatalogued papers in the Biblioteca del Palacio at

Madrid. Expelled from Mérida in 1820, the friars probably carried off the original or copy that Landa had brought almost 250 years earlier. It has been lost ever since.

In the Biblioteca de la Academia de la Historia of Madrid the Abbé Charles Étienne Brasseur de Bourbourg discovered the only copy now known in 1863 and published most of it the following year. At sixty-six folios, this manuscript runs less than half the length of the original. Three scribes set it down early in the seventeenth century. Dr. George Stuart kindly provided Xerox copies of the manuscript pages on the months.

The Eighteenth Century: The Books of Chilam Balam

Scribes set the original Books of Chilam Balam to paper between 1550 and 1650, basing many of them on hieroglyphic sources. More specifically, they composed the first texts between 1559 and 1579 and much of the standard material between 1593 and 1629 (Bolles 1990: 85). Each of the manuscripts evolved through several copies between 1650 and 1850. The present editions of the Maní, Tizimín, and Chumayel books (the major works) arose between 1824 and 1837 (Edmonson 1976, 1982: xii).

TABLE 2.43
Yukatek month names in Landa's *Relación de las cosas de Yucatán*

Month	Date	Page
Pop	16-VII	39r
Popp	none	38r–40r
Vo	5-VIII	39v, 40r
Zip	25-VIII	40r, 34v
Tzoz	14-IX	40v
Tzoz	none	41v
Tzec	4-X	41rv, 43r
Xul	24-X	42r, 37v
Yaxkin	13-XI	42v
Mol	3-XII	42v, 43r
Chen	23-XII	43v, 34r
Yax	12-I	34r
Zac	1-II	34v
Ceh	21-II	35r
Mac	13-III	35v, 37v
Kankin	2-IV	36r
Muan	22-IV	36v
Pax	12-V	37r
Kaiab	1-VI	37v
Cumhu	21-VI	38r
Uayeb	11-VII	[supplied]

CHUMAYEL

Edmonson (1986: 1–2, 226) designates 1737 as the referent year for the calendar on page 23 of the Chilam Balam of Chumayel. Glyphs accompany the names but function as "little more than vestigial ornaments" (Miller 1986: 195). Ralph Roys (1967: 2, 85) offers drawings of the signs, while Michael Coe and Justin Kerr (1998: 155) and Mary Miller (1986) display photographs of the entire page. The original text below is followed by Edmonson's (1986: 226) translation. When different, Roys's rendition also appears, after double slashes (//). Insignificant marks such as "=," ";," "__," and "—" are omitted. Single slashes (/) separate Mayan and Spanish texts from English translations. The original spacings and capitalization in Mayan and Spanish words remain.

vtzol vinal ychil hunppel hab lae

The count of *uinal*s in one year is this:

poop 16 Julio / Pop 16 July

Voo 5 Agosto / Uo 5 August

Sip 25 Agosto / Zip 25 August

çoↃ 14: sep tiem bre / Zotz' 14 September

çec = 4 octubre / Tzec 4 October

xul 24 octu bre ti lic ya lancal cayi / Xul 24 October, which is when the fish are spawning.

Ↄeyaxkin = 13 Nob^e . . . ti cuua tz'al nali / Little Yaxkin 13 November, which is the bending of the corn ears. // The corn stalks are bent double.

mol 3 Diziem bre / Mol 3 December

ch'een 23 Dizi em bre / Ch'en 23 December

yaax 12 henero u kin hoch. Vtz / Yax 12 January, the time for the harvest is good.

Sac 1 Febrero licil ulolan cal çacob / Sac 1 February, since it is the blooming of the whites. // When the white (flowers) bloom.

Ceeh 21 Febrero / Ceh 21 February

mac 13. marzo licil yalancal aac / Mac 13 March, as it is the mating of turtles when the turtles lay their eggs.

Kan Kin 2 Abril / Kankin 2 April

Muan 22 abril li cumumtal u nak v caanil ki ni / Muan 22 April as it is the rainy period, the time of hiding the sky. // When there is a ring around the sun in the sky (Roys 1967: 85 n. 4: "Probably due to the smoke from the fields which are burned over at this time"). *Se detiene la carrera del sol en la cintura del cielo. // The running of the sun tarries at the waist of the sky* (Mediz Bolio 1985: 63–65; translation mine).

paax 12 Ma yo / Pax 12 May

Kayab 1 Junio / Kayab 1 June

Cum Ku 21 Ju nio u Vaya yab ho ppel kin / Cumku 21 June the month of the Uayebs, five days.

KAUA

The photostatic copy of the Chilam Balam of Kaua in Brigham Young University's Gates Collection presents six *winal* name sets, none with any significant variants (table 2.44). Comments accompany some names on only the third list, on page 72 of the manuscript. Edmonson (1982) translates *oc* and *yoc* as "germinates" and David Bolles (1990: 87) as "is planted." Victoria Bricker and Helga-Marie Miram (2002: 193) render *ocnal kin* as "at sunset," with no reference to maize (*nal*).

TIZIMÍN

The Chilam Balam of Tizimín preserves two series (20v, 22r–27v). The referent year is unknown for the first, while it is 1626 for the second (Edmonson 1982: 115). The first roster inverts Cip and CoↃ, Queh and Mac, and Paax and Kayab. Edmonson (1982: 120, 126–27) offers the following transcriptions and translations.

s[oↃ]: *u sian ku* the birth of god ["spell" might better translate *sian*; "birth" would be *siyan*]

Seec: *u sian chac* the birth of rain (126 n. 3411)

Xul: *u sian chac* the birth of rain (127 n. 3431)

Moan: *y oc uiil* germinate plants (120 n. 3226)

TABLE 2.44

Yukatek month names in the Chilam Balam of Kaua

Month	Gloss	Translation
Pax	*ocnal kin*	corn is planted/germinates days
Kayab	*ocnal kin*	
Cumku	*yoc chicam*	jícama is planted/germinates
Pop	*ocnal kin; yoc pacha[l]nal*	late corn is planted/germinates
Uoo	*ocnal kin*	
	yoc bul	beans are planted /germinate
Sip	*yoc bul*	
SoↃ	*yoc uinal*	seed/famine corn is planted/germinates

Maní, Codex Pérez

Of the three parts of the Codex Pérez the first two preserve significant portions of the now lost Chilam Balam of Maní. Juan Pío Pérez compiled the third part (pp. 138–76) from many sources (Craine and Reindorp 1979: xvi, 6). Together the three parts record nine lists, but only three present noteworthy glosses. Each appears under its roster number. Miram (1988b) supplies the original text, while Craine and Reindorp (1979) and Edmonson (1982) present translations.

List 2: Miram 1988b: 50

Kayab yoc chicam. Miram 1988b M50, MS page 13

Jícama is planted. Craine and Reindorp 1979: 39

List 3: Miram 1988b: 51–64

Sec u sian chac. Miram 1988b: 61, MS page 10

The birth of rain. Edmonson 1982: 126 n. 3411

Rains begin. Craine and Reindorp 1979: 47

Xul u cian chac. Miram 1988b: 62, MS page 1

The birth of rain. Edmonson 1982: 126 n. 3411

Rains begin. Craine and Reindorp 1979: 47

List 8: Miram 1988b: 140–49

SoꝹ u sian ku. Miram 1988b: 147, MS page 6

The birth of god. Edmonson 1982: 126 n. 3391

Gods are born. Craine and Reindorp 1979: 151

1986, 1988. Along with familiar spellings and well-known interpretations, Edmonson (1986: 34, table 3, index) offered several original translations, marked in table 2.45 with an asterisk. Any translation different from those in his table comes from the index and is listed second (with its page number). Two years later Edmonson (1988: 248) presented several variants (marked in *italics* in table 2.45).

Summary: Next to the Ch'olan tradition, the Yukatek most resembles the glyphic system. Eight of its names virtually duplicate their Classic counterparts. The other eleven, however, represent unrelated innovations. The first month always follows the Five Days. Variation touched just two names: Yaxkin became Ɔe Yaxkin, while Sec replaced Tzec. The Yukatek fixed their count to the Gregorian round in 1548 CE, an event perhaps important enough to generate most of the eleven names not corresponding to the glyphs. By 1850 the system had slipped into oblivion.

TABLE 2.45

Yukatek month names as translated by Edmonson

Names	1986	1988
Pop	mat	mat
Uo	frog	frog
Zip	deer	*stag*
Zotz'	bat	bat
Tzec	*skeleton	*skull*
Xul	end	end
Yaxkin	green sun	*green time*
Mol	*track	*gather*
Ch'en	well	well
Yax	first; green, new	green (309)
Zac	white	white
Ceh	deer	deer
Mac	cover	cover
Kankin	yellow sun	*yellow time*
Muan	macaw; *macaw thunder	*owl* (298)
Pax	*break	*drum*
Kayab	songs; *preach the year	*turtle* (296)
Cumku	*dark god	dark god
Uayeb	*specters; *specter steps (307)	specters

Conclusion

Over the past 450 years fourteen native traditions have preserved earlier calendrical knowledge; some are a handful of names, while others are entire lists with glosses and comments. The normal calendar consists of eighteen 20-day months plus a period of five days. The total of 365 days approximates the annual solar round of 365.2422 days. Usually the five-day unit ends the count; but when the month that starts the year is changed, that brief span's berth in the sequence becomes the third, seventh, or ninth place. This happens only in Tzotzil, Tzeltal, or K'iche', respectively. Most traditions assign the five-day set just one name, but a few add more, which they barely use: Tzeltal has four; Chuj, Ixil, K'iche', and maybe Q'eqchi' three apiece; and Tzotzil two. About half the traditions admit infrequent alternate names for the 20-day months. The total number of months that start an annual count varies. Seven traditions use one, while five use two or three, and one has four. Finally, all the calendars include a significant number of month names involving plants, animals, agriculture, and the weather. Though these could serve in frozen calendars as seasonal indicators, only a handful correspond to the time of year right for their meanings.

CHAPTER THREE

The Maya Month Names

Forms and Meanings

This chapter discusses the major facets of the Maya month names. It focuses specifically on definitions from dictionaries and ethnographies, comparisons of cognates and synonyms, themes, counterparts, and stemmas. Many sources include observations on agriculture, astronomy, calendrics and weather; these appear in the next chapter under the discussion of seasonality.

Definitions

This section proposes a spelling and a translation for every month name. It not only presents pronunciation and meaning for month names but also proposes stemmas (that is, their derivation from other names similar in sound or meaning or both). Most of the support comes from numerous vocabulary lists and dictionaries of the colonial, modern, and current periods, along with recent articles and books. Month names appear first in the spelling of their primary source, then in this study's standard orthography. A few primary sources do consistently distinguish plain from glottalized consonants, *ä* from *a*, long vowels from short, or one tone from another in Yukatek. In nearly every instance these resolve ambiguities common in the other works.

When different spellings of a name have distinct definitions, the analysis here presents the various meanings but privileges the more straightforward ones and those with fewer assumptions, better support, and more parallels in other traditions. Cognates inspire the most confidence, followed by synonyms, then homonyms. Stemmas combining these two seem more likely: the more elements they incorporate, the more constraints they satisfy and so the more likely it is that they do not represent coincidence. Similarity in form or meaning probably indicates relatedness rather than chance.

The stemmas are found in appendixes D and E. The derivations indicate that almost three-fourths of all the present month names descend ultimately from the glyphs of the Late Classic period.

Orthography of the Maya Month Names

Each month name, in **boldface**, appears in one of its original spellings and is immediately followed by one or more interpretive renderings in a uniform modern orthography, as well as a gloss of that rendering. Definitions for each term follow. For example, a standardized entry in the Ch'ol section looks like this:

Ccanazi: *K'änäl Si'* Dry Firewood; *K'än Ah Si'* Yellow Woodchopper; *K'än Asih* Yellow Cicada

k'än(äl) yellow

ah agent

si' firewood

k'an asih cicada; literally "yellow cicada"

k'an yellow, mature

asih cicada; also just *sih*

Ccanazi represents an attested version of the name, in its original spelling; it is the reference form. K'änäl Si', K'än Ah Si', and K'an Asih are the interpretations.

Ch'olan Definitions

"Ch'olan" embraces Ch'ol, Chontal, Ch'olti', and Ch'orti'. Ch'olti' for some represents a separate language extinct since around 1800, but for others is the direct ancestor to modern Ch'orti'. Its earlier form, Classic Ch'olti'an, is the language of the glyphs (Houston et al. 2000: 321–22, 331–37).

Ch'ol Definitions

Ah Qui Cou: *Aj Kikow Cacao* Merchant; Beggar; Roasting-Ear/Royal Chocolate (all Poqom)

aj Poqomam: cane; reed (McArthur and McArthur 1995: 2); reed mat (Teletor 1959: 127); Poqomam: prefix denoting persons or things of high status (Reina 1966: 308); Poqomchi': agent; reed; woven reed mat (Stoll 1888: 147); roasting ear (Mayers 1960: 294, item 76)

kikow Poqomam: cacao (Morán 1725: 468, 1720: 86r); Poqomchi'(Stoll 1888: 183). The K'iche' term <*ah cacaoual*>, glossed as "the poor, needy man; it comes from *cacao* because it is what they give as alms to the poor" (Ximénez 1985: 62), prompts the interpretation "beggar."

Cazeu: *Kasew* Male Pacaya Palm (Q'eqchi', Mopan)

kasew Mopan: male *pacaya* palm (Ulrich and Ulrich 1976: 40); Q'eqchi': swamp palm (Haeserijn 1979: 86)

Pacaya is common in the Maya area for the *macayo*, a palm bearing small edible fruit (Fox 1978: 185, item 146). It also denotes several species of palm trees and their edible shoots (Wisdom 1940: 476).

Ccanazi: *K'änäl Si'* Dry Firewood; *Kän Ah Si'* Yellow Woodchopper; *K'an Asih* Yellow Cicada (Q'eqchi')

kän(äl) yellow (Attinasi 1973: 284)

ah agent (Bricker 1986: 44, table 20)

si' firewood (Aulie and Aulie 1978: 105)

k'an asih Q'eqchi' *cigarra*/cicada; calls to the rain (Haeserijn 1979: 195), literally "yellow cicada"

k'an Q'eqchi': yellow, mature (Haeserijn 1979: 195)

asih Q'eqchi': *chicharra* or *cigarra*/cicada; also just *sih* (Haeserijn 1979: 42, 303, 415); *cigarra, chiquirin, grillo*/cicada, cricket; also *sih* (Sedat 1955: 25, 139)

Chac: *Chäk* Red

chäk red, bound form (Attinasi 1973: 250)

Chacᶜat: *Chäk K'ät* Red Wicker

chäk red, bound form (Attinasi 1973: 250)

k'ät in a crisscross way (Aulie and Aulie 1978: 46)

"Wicker" here derives from the plaited object referred to by T552 (**K'AT**) on Yaxchilan's Lintels 6 (B3) and 43 (A2) (Tate 1992: 150,251). The rare *C*ᶜ indicates *k'*.

Chantemac: *Chan Te' Mähk* Granary on Tall Posts

chan-te' mok Tzotzil: fence with four horizontal bars (Laughlin 1975: 110, 332); fence of long, thin poles that encloses the house compound (Vogt 1969: 90)

chante' tree used for support posts (Aulie and Aulie 1978: 47); Ch'orti': wild shrub *Juana islama* (Wisdom 1949: 694, 895)

chan high, tall (Attinasi 1973: 249); sky (Aulie and Aulie 1978: 47)

chän four (Attinasi 1973: 251)

te' tree, pole (Attinasi 1973: 320)

majk a cover (Aulie and Aulie 1978: 77)

mäk to cover something (Aulie and Aulie 1978: 82)

mäk Chontal: a spoon (Keller and Luciano G. 1997:157); spoon; to close, cover (Knowles 1984: 439)

mahk Ch'orti': anything enclosed or stopped up

mak Ch'orti': an enclosing, a covering, anything that surrounds or encloses (Wisdom 1949: 521); a cover, stopple: from phrases under *mak* (Wisdom 1949: 521–22)

mak-a Ch'orti': to close, cover, enclose, fence in (Pérez Martínez et al. 1996: 134–35)

ka'anche' Lakandon: granary (Bruce 1979: 167)

ká'anche' Yukatek: a large *huacal* or crate-like structure made of interlocking logs, mounted atop four posts some two meters tall, used for storing and shucking ears of corn (Barrera Marín et al. 1976: 29, item 81); large, four-cornered table; a model of the world (Sosa 1985: 331, 337, 346, 378); table, literally "elevated wood," the altar in the *ch'áa cháak* rain ceremony; its four legs represent the *yun-tzil-ó'ob'* ["revered lords"] of the corners of the cornfield; the altar's plan duplicates the layout of the milpa (Hanks 1990: 334, 368, 378); "sky wood," two long poles lashed together, supported by two long legs, one at each end, elevated about a meter (Sosa 1985: 331, 337, 346, 378; detailed description and illustration in Pérez Toro 1942: 27, 31, 34–35, 48). Relevant here are these senses for *che'*: "tree; stick; small box [*cepo*], prison" (Bricker et al. 1998: 64; Barrera Vásquez et al. 1980: 86). For definitions relevant to *mac*, see the Yukatek **Mac**.

Chichin: *Ch'ich'in* A Certain Bird (Q'eqchi'), Bird (Q'eqchi'); *Ch'ich'ib'* Hummingbird (Ch'ol)

chichin Q'eqchi': bird (Wirsing n.d.: 134)

ch'ich'in Q'eqchi': a certain bird (Haeserijn 1979: 144)

ch'ich'ib' Ch'ol: hummingbird (Aulie 1948: 6)

tz'ikin Q'anjob'al, Ixil, K'ichean: bird (Campbell 1977: 48, item 23; Brown and Witkowski 1982: 102, table 2)

*tz'ikin proto-Mayan, proto-K'ichean: bird (Campbell 1988: 342, item 25, Fox and Justeson 1984: 42 n. 16); *tz'ichin* Ixil: day name bird (Townsend et al. 1980: 26, line 170)

In some Ch'ol dialects a final *b'* is also realized as *m'*: for instance, *tahm'/tahb'* tumpline (Aulie and Aulie 1978: 109; Hopkins and Josserand 1988: t2); *yihnam'/ihnab'* wife (Sapper 1907: 440). At times the final *m* becomes *n*, as with *la:chäm*: *lahchän* twelve (Attinasi 1973: 287). The -*b'*/-*m'* shift also occurs at times between Tzeltal and Tzotzil as well as dialectally (Laughlin 1975: 246, 284; Guiteras-Holmes 1961: 338, 340) and conditionally within Tzotzil (Kaufman 1972: 23, 24; Fox and Justeson in Bricker 1986: 94). These data suggest that Chichin is *ch'ich'in* hummingbird, from *ch'ich'im'*, from *ch'ich'ib'*. The most direct solution borrows the name from Q'eqchi'.

Another possibility posits a proto-Mayan **tz'ikin* "bird" (Fox and Justeson 1984: 41–42 n. 16). If **tz'ikin* entered proto-Ch'olan early enough, its *k* could have shifted to *ch*, yielding **tz'ichin*. As this violated Ch'olan morpheme structure constraints, the reflex would have normalized to proto-Ch'olan **ch'ich(')in*; there is no modern Ch'olan form (Fox and Justeson 1984: 42 n. 16).

Alternatively, *tz'ikin* entered Ixil from proto-Mayan, Q'anjob'al, or K'ichean to become the attested *tz'ichin*, which then joined proto-Ch'olan or Ch'ol to normalize into **ch'ich(')in*.

Ianguca: *Yam K'u Ka'* Pour Divine Honey; *Yam Kuk Ha'* Pour Jugs of Water (both Q'eqchi')

Justeson (1988: 14, table 3) regards "ianguca" as a long-recognized copyist error for *yax k'in*, perhaps as *yaxquin*, the glyphic, Ch'ol, and Yukatek name for the month. The many changes implied in this transformation, as well as the rather imperfect spelling evident throughout the original manuscript, indicate that it is more straightforward to regard "ianguca" as a slightly modified *iankuka* or *ian4uka*. The *g* here seems to represent a variant of the Parra *4*, the colonial equivalent for the K'ichean *k'*, employed by a scribe apparently familiar with Q'eqchi', a language that would use *4*.

yam Q'eqchi': pour out (Freeze 1975: 57)

k'u <c'u> Q'eqchi': god (Haeserijn 1979: 122)

kuk Q'eqchi': large earthen jar (Freeze 1975: 57)

kab' Q'eqchi': sweet, sweets; *panela* [brown sugar loaf]; domesticated bees (Haeserijn 1979: 79). A variant *ka* or *ka'* would be no surprise, given the frequent shift word-finally of *b'* to *'* or *o*.

ha' Q'eqchi': water (Haeserijn 1979: 159)

Icat: *Ik' K'ät* Black Wicker

ik' black (Attinasi 1973: 274)

The interpretation posits two spelling adjustments: first, the mistaken use of *c* for *k* or in modern terms of *k* for *k'*, a common error; and second, the absence of another *c* before the given *c*, for the *k'* of *ik'*. The first letter might not have needed recording, as it seems likely that here as in other Mayan languages the first of two adjacent identical consonants, even at word boundaries, often reduces to zero.

*K'an Jalib' Sacred Loom.

See Ch'olti' **Canhalib**

Mahi: *Maay* Poison (Ch'olti')

The entire entry for 9 July reads *Holob cutan MAHI yccaba* or "Nauseated Five Days. *MAHI* is their name." Only the word "Mahi" names this period.

<mai> Ch'olti': poison (Morán 1935: 51)

The source for this form sometimes uses *h* to denote vowel length (Fought 1984: 45). This allows a reconstruction of **mahi* for **maay*

i third person singular pronoun: his, her, its (Bricker 1986: 22, table 4; Kaufman and Norman 1984: 91, table 7)

Analogy with Yukatek suggests that this can serve as an abbreviated form of the third person plural *i . . . oʼ/obʼ*, as does the interpretation of the immediately preceding *holob cutan* that translates the *ob* with "five."

kʼabaʼ name (Aulie and Aulie 1978: 40)

holob cutan: *jol oʼobʼ kutan* nauseated/banished five days

jol Qʼeqchiʼ: nauseated, banished (Curley García 1967: 87)

oʼobʼ Qʼeqchiʼ: five (Haeserijn 1979: 240)

kutan Qʼeqchiʼ: day, time (Haeserijn 1979: 105–6)

Mol: *Mol* Harvest, Pile

mol tornamil it is sown after the harvest of the cornfield of the year. There is no burning (Aulie and Aulie 1978: 81); Chʼortiʼ: mound (in compounds) (Wisdom 1949: 529).

Muhan: *Muhan* Hawk

muhan Chʼoltiʼ: hawk, kite (Morán 1935: 32); Chʼortiʼ: *mujan* eagle (Girard 1940: 311), *muan*: *gavilán*, hawk or kite, generic for birds of prey (Girard 1940: 103)

Olh: *Ol* Heart, Bud, Sprout

In colonial Kʼichean scribal practice *-lh* represents the word-final, voiceless *l* (Campbell 1977: 121); the manuscript's "olh" suggests that its author, in Kʼichean territory, used this convention. Edmonson (1988: 154) renders this name "ohl," perhaps to correct a mistakenly reversed *hl*, maybe to indicate the name as cognate to the Poqomchiʼ *ohl*. This name, *ojl* in the original, has the gloss "a tree with lightweight wood" (Gates 1932a: 75–76), maybe from the Chʼol *ohol* "corkwood tree, pond apple or balsa [*corcho, jolocín*]" (Aulie and Aulie 1978: 90). Perhaps the compiler shortened *ohol* to match the Pokomchiʼ form. In brief, the phonemic shape seems to be *ojl*, with *-lh* either a slip-of-the-pen reversal of *hl* or a scribal convention for the voiceless *-l*. "Ball, heart" seem the more likely senses.

w-olj < *guolg* > Chʼol *bola*/ball (Fernandez 1892: 45)

ol Chontal: to desire (Smailus 1975: 160); discarded food (Pérez González and Cruz 1998: 61, 83)

wol Chontal ball-shaped; to make a ball (Knowles 1984: 478)

-wol Chontal numerical classifier: said of balls of ground cooked corn (Keller 1955: 262, item 43). Prefixed to the term for a body part beginning with a vowel, *w* or *y* converts it into an unpossessed noun or adjective of related meaning.

óol Yukatek: formal [spiritual] heart, not the material one; will, desire (Ciudad Real 1984: 349r); heart, spirit; life, breath (McClaran Stefflre 1972: 61); young leaf (Owen 1968: 5); pellet of the blowgun, crossbow, ball of the arquebus; bud or tender young sprout of herbs or grasses [*hierbas*] as well as of trees and other plants (Ciudad Real 1984: 349r)

yol Tzotzil: heart of tree, embryo of corn kernel (Laughlin 1975: 386); Tzeltal: heart of wood (Sarles 1961: 73, item 1889)

*Pojp: *Pojp* Mat. This form is hypothetical for the 1589 roster because Edmonson (1988: 153) advances it as a possibility for an empty slot, apparently via analogy with Q'eqchi' and Yukatek.

pojp mat (Aulie and Aulie 1978: 95)

*Sutz': *Sutz'* Bat. This hypothetical name is supplied from the glyphs.

suts' bat (Aulie and Aulie 1978: 107)

Uniu: *Uniw* Avocado (proto-Mayan)

Oneu really appears only on the Chuj and Q'anjob'al lists. The original form reads *uniu*.

un avocado (McQuown 1976b: 7, item 379); Chontal (Stoll 1938: 71); *vn, hun* Ch'olti': (Morán 1935: 7); *un* Ch'orti' (Sapper 1907: 448); **un* proto-Ch'olan: avocado (Kaufman and Norman 1984: 135, item 598); *un* Tzotzil: avocado; *um* avocado (Stoll 1938: 49); *um* Ch'orti' (Girard 1949: 105); *um* Ch'ol (Sapper 1897: 419; Stoll 1938: 71); *on* avocado (McQuown 1976b: 7, item 379); Tzeltal (Berlin et al. 1974: 217–20); Tzotzil (Laughlin 1975:69)

iw Tzeltal: wild avocado (Berlin et al. 1974: 109, 135)

oven Tzotzil: avocado (Laughlin 1975: 70)

Uniw, spelled several times in the glyphs as **u-ni-w(a)**, seems to present a relic with Ch'olanized vowels (Fox and Justeson 1980: 210–14). The earliest form, the pre-proto-Mixe-Zoque ***onVw(i)* or **onVw, entered proto-Mayan as a loan to become *oonw. This developed by the Early Classic into the Western Mayan *oonew (Fox and Justeson 1980: 210–14, 216 n. 20; Justeson 1988: 14, table 3). In Q'anjob'al this form eventually turned into *onew*, while in Ch'olan, it changed to *uniw* by the middle of the Early Classic (ca. 450 CE), the shape borrowed later by Q'eqchi' and from there lent into Poqomchi' (Fox and Justeson 1980: 211, 213).

A more speculative interpretation casts *uniw* as the proto-Ch'olan cognate either of the proto-Tzeltalan *oniw, a compound (*on* + *iw*) for a type of avocado, or of the Ch'olanized-by-metathesis proto-Tzeltalan *oven, itself derived from pre-proto-Mixe-Zoque **owin (Fox and Justeson 1980: 210). Today *iw* occurs only in Tzeltal and *oven* only in Tzotzil.

The hypothetical *on-iw parallels the attested *on-tzitz*, a kind of avocado. The name consists of *on*, a type of avocado, and *tzitz*, a much smaller variety of *on* (Berlin et al. 1974: 162, 219–20). Some Tzeltal dialects use *tzitz* and *iw* as synonyms for the small wild avocado (Slocum and Gerdel 1965: 12, 144, 194), which strongly suggests first that *on tzitz* has *on iw* as an obsolete synonym and second that its *o* turned to *u* upon entry into Ch'olan.

Yax: *Yäx* Blue-Green

yäx blue-green (Attinasi 1973: 341)

yäx green, blue; first, from *yäx al* first child, first offspring (Aulie and Aulie 1978: 145)

yax Ch'olti': green, blue, unripe (Morán 1935: 2, 21, 30, 67); first, from <*yax util*> *primicia*/first fruits (Morán 1935: 54); Ch'orti': green, clear, fresh, young, first (Wisdom 1949: 766)

Yazquin: *Yäx K'in* Green Days; Dry Season

yaxk'in Ch'olti': *verano*/dry season, summer (Morán 1935: 8, 66); Justeson (1989b: 27 table 3.1) suggests "green time (spring)" for Ch'olan and Yukatekan. If the solar calendar did begin about 550 BCE, this month spanned 15 April–3 May (Bricker 1982a: 103). Thus for a while the name proved apt, as the first sprinkles fell during this period, while at the end the real rains arrived, turning the parched bush country green again.

k'in sun, day (Attinasi 1973: 285)

Zac: *Säk* White

säk white (Attinasi 1973: 313)

Zihora: *Siholá* Wellspring (Yukatek)

siholá Q'eqchi': water that springs up from the earth, running water (Haeserijn 1979: 160, 303)

Siholá seems to be the Q'eqchi' borrowed form of the Yukatek *síih-hóol-ha'* "origin-hole-of-water," a virtual synonym to *ch'e'en* "well, water hole," the counterpart on the Yukatek roster. The presence of sensible etyma in Yukatek but not Ch'ol and Q'eqchi' bolsters the case for borrowing. The Yukatek original likely shortened to *síih-hóol-a'* then *síih-hóor-a'* (the rare *r* does replace *l* intervocalically in some dialects). Q'eqchi' then shaped this into *sihorá*. Another explanation for the name might stem from one reading for the main sign glyph: **SIHOOM** "flower" (Houston et al. 1998: 282); the clear *zihora* of the manuscript would then represent a misreading of a slovenly written *zihom* in a previous version.

síih Yukatek: to be born (Po'ot Yah and Bricker 1981: 27); to surge or well up (Arzapalo Marín 1987: 276, text II, line 249)

hóol Yukatek: small hole, crack (Fisher 1973: 237, item 281; Ciudad Real 1984: 208r)

<*hom*> Yukatek: sinkhole (Barrera Vásquez et al. 1980: 228)

ha' Yukatek: water (Bricker et al. 1998: 91)

-a Yukatek: compound form of *haʼ* (Barrera Vasquez et al. 1980: 1)

ho(l)-haʼ Lakandon: *manantial*/spring (Bruce 1974: 199, line 13)

Chʼoltiʼ Definitions

Canhalib: *Kʼän Jäl-Ibʼ* Yellow Loom

The main sign and phonetic complements in the glyphic compound indicate this name or a variant as the most likely reading. Its presence in the inscriptions, in Chʼoltiʼ, and in the Poqom calendar (as Kanhalam) makes some form of Kʼän Jäl-Ibʼ a likely candidate for the Chʼol rosters of 1589 and 1971.

The only two instances of this name ending with both **-wa** and **-bʼu** together appear at Palenque, on the Palace Tablet (N8) and on a jamb from Temple 18 (B9). The **-bʼu** perhaps cues the *kʼanjalibʼ* reading of the standard *kʼan jalaw*, while its simultaneous presence with **-wa** suggests that it marks the local ethnic or royal dialect.

kän yellow (Morán 1935: 2)

jäl weave from *halbil* woven (Morán 1935: 65)

-ibʼ instrument (Morán 1935: 3)

Chacat: *Chäk Kʼät* Red Wicker

chäk red, from *chak chak* vermilion (Morán 1935: 12)

kät plaited, from *kätäl* criss-crossed (Morán 1935: 2)

Chʼortiʼ Definitions

Tsikin: *Tsʼikin* Bird (Chʼolan, Kʼichean)

tzʼikin day of the saints [1 November] (Pérez Martínez et al. 1996: 230)

*tzʼikin proto-Mayan: bird (Fox and Justeson 1984: 42 n. 16)

Wisdom (1949) and Girard (1949: vols. 1, 2) record variants or reinterpretations for *tzikin* as follows:

sikin <*siquin*>, <*sikín*> festival of the dead (Girard 1949: 1:215, 217; 2:703–4)

si kʼin <*si qʼin*> festival of the dead (Girard 1949: 1:99); Days of the Saints (November), Days of the Dead (31 October, 1 November) (Wisdom 1949: 504, 632)

si series, line of objects, row (Wisdom 1949: 632)

kʼin day, sun, period (Wisdom 1949: 504)

Girard relates *sikin* to the Kʼicheʼ *tzʼi-kin* bird, apparently because it resembles that term and because the Chʼortiʼs equate the bird to the soul of the departed in the "Ascent of the Bird" ceremony at the end of Siquin.

SUMMARY: The one calendar and miscellaneous sources for the four languages of Chʼolan manage to preserve most of the hieroglyphic month names (table 3.1). Conspicuously absent are only **SUUTZʼ** "bat"; the base of the four color months

TABLE 3.1

Summary of Ch'olan month names

#	Names in source	Interpretations	Translations
1	Canhalib (Ch'olti')	K'än Jäl-Ib'	Yellow Loom
2	Icat	Ik' K'ät	Black Wicker
3	Chacc^cat	Chäk K'ät	Red Wicker
3	Chacat	Chäk K'ät	Red Wicker
4	*Sutz'	Sutz'	Bat
5	Cazeu	Kasew	Male Pacaya Palm
6	Chichin	Ch'ich'in	A Certain Bird
		Chichin	Bird
6	Tsikin (Ch'orti')	Ts'ikin	Bird
7	Yazquin	Yäx K'in	Green Days; Dry Season
7	Ianguca	Yam K'u Ka'	Pour Divine Honey
		Yam Kuk Ha'	Pour Jugs of Water
8	Mol	Mol	Harvest, Pile
9	Zihora	Siholá	Wellspring
10	Yax	Yäx	Blue-Green
11	Zac	Säk	White
12	Chac	Chäk	Red
13	Chantemac	Chan Te' Mähk	Granary on Tall Posts
14	Uniu	Uniw	Avocado
15	Muhan	Muhan	Hawk
16	Ah Qui Cou	Aj Kikow	Cacao Merchant
			Beggar
			Roasting-Ear Chocolate
			Royal Chocolate
17	Ccanazi	K'änäl Si'	Dry Firewood
		K'än Ah Si'	Yellow Woodchopper
		K'an Asih	Yellow Cicada
18	Olh	Ol	Heart; Bud, Sprout
19	Mahi	Maay	Poison

Note: All unspecified names are Ch'ol.

(probably **HA'** "water, rain" or **HAB'** "season"), **PAX** "upright drum" and **WAYAB'** "bed." These all show up slightly or not at all modified in other traditions.

Chuj Definitions

In the general consensus Chuj with Tojolab'al forms Chujean (Kaufman 1974b: 959). A few allot Tojolab'al to Tzeltal and Tzotzil instead, with the trio defining Tzeltalan (Robertson 1977).

Bac: *B'ak* Bones, Corncob Worms

b'ak bone (Diego 1998: 33); *p'ak* bone; worm in corncob (Hopkins 1968: 73); white with black head and black spots, the *gusano barrenador* (Hopkins 1980: 37)

Bex: *Wex* Loincloth

wex pants (Hopkins 1968: 109)

H Oye' K'u, Hoyeb' Ku: *Hoyeb' K'uh* Five Days. Both spellings appear in Edmonson (1988: 95, 157).

hoye' five (Hopkins 1968: 30)

o'e' five (Williams n.d.: entry 1237)

-e'/-eb' plural marker (Judith Maxwell, personal communication, 2002)

-y- intervocalic euphonic

k'uh sun, day (Hopkins 1968: 44); *k'u* day, sun (Diego 1998: 123)

Huatziquin: *Wa' Tz'ikin* Bird That Eats [Crops]; *Watz' Tz'ikin* Squeak Bird or Hummingbird; *Wak Tz'ikin* [Month/Day] Six: Bird

wa' to eat (Hopkins 1967a: 175); root of *wa'in* to eat (intransitive verb) (Hopkins 2012: 372)

watz' watz', adj. squeaking as if to break (Hopkins 2012: 378)

tz'ikin bird (Recinos 1954: chart opposite 224); day name; also proto-Mayan **tz'ikin* bird (Hopkins 1980: 20, 2012: 363–64)

Kanal: *K'anal* Star, Dance

k'anal Chuj: star; Chuj of San Juan Ixcoy: dance (Termer 1930: 393 n. 125)

Meak: *Mej Ak* Fallen-Over or "Sleepy" Grass; Long Thatch-Grass?

ak grass (Hopkins 1968: 1), thatch, grass (Diego 1998: 4)

mej limp, fallen over asleep, under *<mexan mexan>* (Hopkins 1968: 54)

Mo: *Mo'* Hawk; Hawk-like Owl (Ixil)

mo' Ixil: Swainson's hawk/*azacuán*. A black bird like a vulture that screams like a baby, and lives on the coasts (Chel and Ramírez 1999:164); Ixil: owl of the hot country, almost identical to the hawk (Kaufman 1974: 426)

Mol: *Mol* Group

mol group, from *molo'* gather (Hopkins 1968: 55); group, pile (Diego 1998: 145)

Nabich: *Nab'ich* Rainy Season

nab'ich Q'anjob'al: winter, rainy season (Diego Antonio et al. 1996: 205)

Oja Kwal: *O-y-e' K'u-al* Five Day Period. See **H Oye' K'u**. In Stadelman (1940: 255) *j* is the present *y*.

-al abstractive (Judith Maxwell, personal communication, 11 July 2000)

Oneu: *Onew* Avocado

oŋ avocado (Hopkins 1968: 7); **oonw* or **ooneew* proto-Mayan avocado See Ch'ol **Uniu**.

Oyebin: *Hoyeb' k'inh* Five Days/Winals

k'iŋ period of 20 days; fiesta (Hopkins 1968: 42); *k'iŋ* fiesta, holiday; *k'inh* month of twenty days (Diego 1998: 119). This reduces to *iŋ* due to the juxtaposed *b'*, perhaps because both are glottalized.

In its earlier sense as "sun, day" *k'iŋ* is synonymous with *k'uh*, the more recent term for "sun, day," but now it also designates the period of twenty days. Oyebin parallels the Q'anjob'al Oyeb K'u in both its meaning, "Five Days," and its position, before Bex/Wex, so it remains a plausible candidate for the short terminal unit of the year.

Sacmay: *Sak May* White Tobacco

sak white (Hopkins 1968: 81)

may Chuj of San Mateo Ixtatan: tobacco (Termer 1930: 392)

Savul: *Sak P'u'ul* White Maize Kernels, Hominy; *Sak B'ul* White Piles

sak white (Hopkins 1968: 81)

b'ul <p'ul> pile (Hopkins 1973: 171)

*p'u'ul, assumed original form of *p'u'uj <p'u'ux>*: maize grains boiled without lime (Hopkins 1973: 179)

"Boiled without lime" points to "hominy" and to synonymy with Tzeltal-Tzotzil Mush. The assumptions are, first, that *v* represents *b'*, as it frequently does in naive Spanish spelling; second, that *b'* in some orthographies or dialects corresponds to *p'*; and third, that word-final *l* weakens to *j*, as it does in many Mayan languages, even in citation forms. The required reduction of *sak* to *saj* and finally *sa* also qualifies as an occasional but attested progression.

Sivil, Siwil: *Siiwiil* Woodchopper (K'iche')

siiwiil K'iche': woodchopper (Ajpacaja Tum et al. 1996: 362); Kaqchikel: *<çivil>* woodchopper (Coto 1983: 310)

Tam: *Tam* Splitting Blows (Hopkins 1968: 92)

Tap: *Tap* Crab (K'iche')

tap K'iche': crab (Campbell 1977: 49, item 37); a K'iche' loan into Ixil (Kaufman 1974: 679). The form seems to be only K'ichean (Dienhart 1989: 2:160–61).

Xujim: *Xuj-Im* Threader [Bloodletter]

xuj-u' to thread, string together (Diego 1998: 268)

-im possibly a variant of *-Vm*, agentive (*-Vm* agentive: Judith Maxwell, personal communication, June 2000)

Yaxaquil: *Ya'ax Ak-il* Green Grass Times

ya'ax green, blue (Hopkins 1968: 120); green (Diego 1998: 269)

ak grass (Hopkins 1968: 1)

-il place of, time of

Yaxul: *Yaxul* Blue Jay (Q'anjob'al); *Ya'ax Xul* Green/Blue Bird

yaxul blue jay (Diego Antonio et al. 1996: 375)

xul bird (Hopkins 1968: 90)

SUMMARY: Nearly all the Chuj names appear to come as loans from the nearby, related Q'anjob'al rosters; they might also derive from the same sources that contributed to Q'anjob'al, but this arrangement would be a bit less parsimonious, as Q'anjob'al would lie between them and Chuj. Meak, Savul, Tam, Oja Kwal, and Oyebin are unique to Chuj (table 3.2).

IXIL DEFINITIONS

The frequent Ixil word *q'ih*, represented in table 3.3 with Ki and K'i, means sun, day (Kaufman 1974: 551).

Akmór: *Aq' Mo'l* Work Group, Reed Bundle, Root Cluster; *Ak' Mol* New Companion; *Ak Mol* Honorable Companion

aq' root for work, give (Kaufman 1974: 30–35); root; tongue, liana, or reed (*bejuco*) (Chel and Ramírez 1999: 5)

ak' new (Kaufman 1974: 19)

ak honorable; term of respect used with people advanced in age (Chel and Ramírez 1999: 2)

mor: *mo'l* group of about 100 birds or animals; *mol* companion, countryman (Kaufman 1974: 426; 429)

Table 3.2

Summary of Chuj month names

	Names in source	Interpretations	Translations
1	Bex	Wex	Loincloth
2	Sacmay	Sak May	White Tobacco
3	Nabich	Nab'ich	Rainy Season
4	Mo'	Mo'	Hawk; Hawk-like Owl
5	Bac	B'ak	Bones; Corncob Worms
6	Tam	Tam	Splitting Blow
7	Huatziquin	Wa' Tz'ikin	Bird That Eats
		Watz' Tz'ikin	"Squeak Bird": Bat?
			Hummingbird
8	Kanal	K'anal	Star; Dance
9	Yaxaquil	Ya'ax Ak-il	Green Grass Times
10	Yaxual	Yaxul	Blue Jay
		Ya'ax Xul	Green/Blue Bird
11	Savul	Sak P'u'ul	White Maize Kernels
			Hominy
		Sak B'ul	White Piles
12	Xujim	Xuj-im	Threader
13	Oyebin	Hoyeb' K'iŋ	Five Days/Winals
14	Mol	Mol	Group
15	Meak	Mej Ak	Fallen-Over Grass
			Long Thatch-Grass?
16	Oneu	Onew	Avocado
17	Sivil, Siwil	Siiwiil	Woodchopper
18	Tap	Tap	Crab
19	H Oye' K'u, Hoyeb' Ku	Hoyeb' K'uh	Five Days
19	Oja Kwal	O-y-e' K'u-al	Five-Day Period

A Ki: *A' Q'ih* Water Days; *Aaq'ii* Diviner, Astrologer

a' water (Kaufman 1974: 2)

aaq'ii diviner, astrologer (Chel and Ramírez 1999: 12)

Awax Ki: *Awax Q'ih* Corn-Sowing Days

aw sow corn (Kaufman 1974: 40–41)

-ax verbal nominalizer; compare *molax* gathering (Nachtigall 1978: 314)

Cajab'ki: *Kahab' Q'ih* Honey(comb) Days, Bee Days, Bee of Venomous Honey. The (*-ha-*) here and in Kaháb below appears to be a Nebaj feature.

kab' bee, honeycomb, sweets (Kaufman 1974: 303–4)

kap bee, honey (Nachtigall 1978: 119)

kab' q'ih a bee that makes venomous honey (Kaufman 1974: 305)

Chantemák: *Chan Te' Mähk* Granary on Tall Posts. See Ch'ol **Chantemac**.

Hui Ki: *Ju'il Q'ih* Mount; *Ju'il* Days; *Wi' Q'ih* Tip Days

ju' nose, beak, snout; *ju'il* a hill north of Chajul with remains of temples where the ancestors used to carry out offerings, rites, and dances (Chel and Ramírez 1999: 108); *wi'* head, tip (Kaufman 1974: 875). Perhaps this refers to emerging plants.

Kaháb: *Kahab'* Bee, Honey(comb), Sweets

Kahabtsé: *Kahab' Tzé'* Log Beehive

tsé: *tzé'* tree, log; prison (Kaufman 1974: 824)

Koj Ki: *Koj Q'ih* Clear-the-Land Days, *Koj Q'ih* Puma Days

koj clearing, from *koj-om'* to clear land (fell trees, cut brush) (Kaufman 1974: 337); jaguar, lion (Rodríguez Sánchez et al. 1995: 81)

Kucham: *Kucham* Tree Trunks

kucham tree trunk (Rodríguez Sánchez et al. 1995: 88)

K'olk'óy: *K'ol K'oy* Oak Monkey; *K'o'l K'oy* Lump Monkey; *Q'ool K'oy* Headscarf or Performer Monkey

k'ol bellota, chicharra (palo) oak tree (Chel and Ramírez 1999: 131)

k'o'l lump of mud, sand or dough (Chel and Ramírez 1999: 134)

q'ool: *tocador* performer; head scarf (Kaufman 1974: 557)

k'oy: *mono, mico* (Kaufman 1974: 292)

"Headscarf Monkey" recalls the monkeys and simian twins sporting headbands on the Classic ceramics and in the head- and full-figure variants of the glyphs for "sun, day" (see Thompson 1971: figs. 27, 29).

   "Performer Monkey" brings to mind the simian artisan half-brothers of the Hero Twins as well as the general association linking monkeys with the arts and crafts.

Lajab Ki: *Lajab' Q'ih* Deadfall-Trap Days

lajab' old-style trap for killing animals (Rodríguez Sánchez et al. 1995: 93); trap for hunting, made of wood (Chel and Ramírez 1999: 139)

laj, of *laj-om'* press down on, crush; prop or push up (Asicona Ramírez et al. 1998: 79)

-ab' instrumental

Lem: *Lem* Maize Worm or Bug, Bean Beetle, Water Mosquito

lem: *gallina ciega*, a worm that eats the maize field when it is sprouting (Kaufman 1974: 373); a bean bug (Asicona Ramírez et al. 1998:82); small black worm attacking the maize plant (Stadelman 1940: 259); a type of mosquito that "dances" over the water (Nachtigall 1978: 203)

Some of these definitions, linked with those given next from Chuj and Tojolab'al, link this name to the Kaqchikel month Tumusus, a term that at least in recent times has come to designate a blue bean beetle and possibly, in the sense of "maize pest," synonymous with the Chuj and Q'anjob'al *bac*. The meanings for *lem* are as follows: Chuj: bug or worm that attaches itself to the leaves of beans; *lorito* (Diego 1998: 130); Tojolab'al: a species of very small beetle; blue, black, or red; it destroys bean fields; beetle that attacks bean fields (Lenkersdorf 1979: 224), known as *lorito* (Ruz 1982: 307).

Maháb: *Majab'*

majab' dam (Chel and Ramírez 1999: 153)

Mam A Ki: *Mam Ah Q'ih* Grandfather Prayermaker; *Mama' Q'ih* Prayermaker Days; *Mama' Ah Q'ih* Prayermakers

mam grandfather, great-grandfather; man's grandchild; sir; male; large (Kaufman 1974: 406–8)

mama' prayermakers, those who do the rituals and know the days; big (Kaufman 1974: 408)

ah agent (Kaufman 1974: 14–18)

ah q'ih sorcerer, sage, diviner, seer; he knows the days, does not do evil (Kaufman 1974: 16)

Mamak'i Kejepk'í: *Mama' Q'ih, Kehep Q'ih* Prayermaker Days, Open Days

mamak'i see immediately above.

qep wide open, from *qepkin* (Kaufman 1974: 568)

"Prayermaker Days" suggest that "wide open" refers not to clear skies but to days especially propitious for communication with the gods or ancestors.

Masnimchó: *Max Nim Choj/Tx'oj* Many Large Payments/Rats

Variant: Moxnimchó

*mas a probable variant of *max*

max a variant of *mox* all, many (Chel and Ramírez 1999: 164)

See **Nimchó** below.

Mek'aj: *Meq' Ah* Bent-Over Stalks

meq' folded, lined (Kaufman 1974: 421)

aj stalk of ditch reed (Kaufman 1974: 13)

Metch'ki: *Metch' Q'ih* Point Days [tips of maize sprouts]

metch' pointy, protruding (Kaufman 1974: 422–23)

Edmonson's (1988: 186) Mech Ki represents an erroneous reading or reduction of the original name, from Chajul, where *tch* and *tch'* are phonemic, contrasting with *ch* and *ch'* (Academia de las Lenguas Mayas de Guatemala 1988: 16).

Mo: *Mo'* Hawk

Mo name of a month in the Ixil calendar (Chel and Ramírez 1999: 162)

mo' Swainson's hawk/*azacuán* (Chel and Ramírez 1999: 164)

Mochnimchó: *Moch Nim Choj/Tx'oj* Last Large Payment/Big Rat

mox all, many (Chel and Ramírez 1999: 164)

motx last (Chel and Ramírez 1999: 168); perhaps a variant is *mox*.

See **Nimchó** below.

Mol: *Mo'l* Group

mo'l group (ca. 100) of birds or animals (Kaufman 1974: 426)

Molch'ú: *Mo'l Ch'uh* Flock of Grackles

mo'l ch'uh flock of grackles (Kaufman 1974: 426)

ch'uh: zanate (Stoll 1887: 114). The Great-Tailed Grackle is the *clarinero* (Land 1970: 309, Smithe 1966: 262). Louise C. Schoenhals (1988: 496) identifies *zanate* as the general term for "blackbird" and "grackle," and as the particular term for "great-tailed grackle."

Mol Tche: *Mo'l Tcheh* Herd of Deer

See **Mol**. *tchej* deer (Kaufman 1974: 688); *chee* deer, in Nebaj (Rodríguez Sánchez et al. 1995: 98)

Mol Masat: *Mo'l Masat* Herd of Deer

masat Nahuatl: from *masatl* deer (Molina 1977: 50)

Moochó: *Moocho* Large Brown Hawk

moochó large brown hawk, synonymous with *pekxítx* (Nachtigall 1978: 307)

Mu: *Muj* Oak or Palo Negro, shade

muj shade (Kaufman 1974: 435); *muu* shade (Rodríguez Sánchez et al. 1995: 102)

Muen: *Muen* Seedbed, Seedlings (Kaqchikel)

muen Kaqchikel: vegetables or garden produce or a plant for transplanting (Anonymous 1675: 74r); nursery or seedbed that they make for transplanting (Coto 1983: 421)

Muen Chin: *Muen Chin* Seedling Bag (Kaqchikel). See **Muen** above.

chin Kaqchikel: net bag (Vico 1555: 62v); *tyin* perhaps a local Ixil form of *chin*

chiim Kiché: small net bag (Ajpacaja Tum et al. 1996 :50)

Ne'chó: *Ne' Choj/Tx'oj* Baby or Small Payment/Rat

ne' baby of up to five years (Kaufman 1974: 455). See **Nimchó**.

Nimchó: *Nim Choj/Tx'oj* Big Payment / Rat

nim big (Stoll 1887: 120; Rodríguez Sánchez et al. 1995: 105)

choj carrying cloth, in Cotzal (*cho* in Chajul, *chok'* in Nebaj) (Nachtigall 1978: 76–78)

choj lake, lagoon, in Cotzal (Kaufman 1974: 191); *choo*, in Cotzal (Rodríguez Sánchez et al. 1995: 44)

choj payment, in Chajul and Nebaj (Kaufman 1974: 191); *choo* payment, candle, in Nebaj (Nachtigall 1978: 64)

ch'oh small, short (Kaufman 1974: 155); *ch'oo* small (Rodríguez Sánchez et al. 1995: 51)

tch'oj small, short (Asicona Ramírez et al. 1998: 159)

tx'oj rat (Kaufman 1974: 741); *tx'oo* mouse (Rodríguez Sánchez et al. 1995: 150, 168)

All the Ixil lists represent just the dialects of Chajul and Nebaj. All the names ending with *-cho*, *-choo*, and *-choh* come from Nebaj for certain, with Chajul origins uncertain. *Choh* and *choo* in Nebaj both mean "payment," so this translation serves for all the listed names. The *-cho* of the names seems to result from the presence of a fronted, perhaps short, preceding vowel. Cho as Tx'oj "Rat" deserves serious consideration because Nachtigall (1978: 17) indicates a difficulty in distinguishing *ch* from *tx* (if not *ch'* from *tx'* as well). Dialectal phonology lies beyond the scope of this work.

Nol: *Nool* Earthworms; *Nor* Slippery-Wet

nool large white earthworm, the size of a snake (Kaufman 1974: 464)

no'r earthworm (Chel and Ramírez 1999: 178)

nor slick, slippery, watery (Chel and Ramírez 1999: 178)

Nol Ki: *Nool Q'ih* Earthworm Days; *Nor Q'ih* Slippery-Wet Days

Och Ki: *Otch' Q'ih* Unripe Corn Days

och': *elote* (Asicona Ramírez et al. 1998: 95)

otch' unripe or green corn or *elote* (Kaufman 1974: 82)

O Ki: *O' Q'ih* Five Days

o'q'ii last month of the Maya calendar, composed of five days (Rodríguez Sánchez et al. 1995: 113)

ooq'ii five days at the end of the year in the Ixil calendar (Chel and Ramírez 1999: 184)

o' five (Townsend 1986: 32)

Okok'í: *Oq O' Q'ih* Foot Five Days; *O' Ko' Q'ih* Five Owner Days; *O' K'oo Q'ih* Five Mask/Horsefly/Patch Days

ok: *oq*: foot (Kaufman 1974: 80)

ko' owner of any domestic animal (Kaufman 1974: 333), perhaps a reference to the Earth Lords

k'oo mask; horsefly (Rodríguez Sánchez et al. 1995: 88); *k'oj* mask; green fly; patch (Asicona Ramírez et al. 1998: 75)

Onchil: *Oonchil* Arrive; *Oon Chi'l* Chills or Colds

oonchil arrive (Chel and Ramírez 1999: 183). [Perhaps this denotes the emergence of little animals out of the earth in mid-November, as recorded in the gloss for this name in the 1942 calendar from Nebaj.]

oon cold, cough (Kaufman 1974: 79); cold, sneezes and cough; flu (Nachtigall 1978: 340, 341)

chi'l body, skin (Kaufman 1974: 183)

Onchív: *Oon Chi'l* Chills, Colds

Oval K'í: *O'-Va'l Q'ih* Five Days

oval five (Nachtigall 1978: 300); *owal* (Lincoln 1942: 107); *o'va'l* (Townsend 1986: 32). See **O Ki** above.

-va(')l dialect variant of *-wal*, which Thompson (Lincoln 1942: 126 n. 27) designates as a numeral classifier

Pac Tzi: *Pa'k' Tzi'* Spoon Beak; *Paq Tzi'* Curve Beak; *Paq' Tzi'* Hole Beak (woodpecker?)

The native informant of Nebaj defines Pactzi as a kind of bird (Lincoln n.d.: 111 n. 1).

pa'k' gourd bowl (Rodríguez Sánchez et al. 1995: 117); gourd spoon, bowl (Asicona Ramírez et al. 1998: 99)

paq- root of verb: to bend, fold, roll up (Asicona Ramírez et al. 1998: 100)

paq' opening, hole (Asicona Ramírez et al. 1998: 101)

tzi' mouth (Rodríguez Sánchez et al. 1995: 171)

Paksí: *Paq Si'* Bend Over [Break Off?] Firewood

paq fold, bend over (Asicona Ramírez et al. 1998: 100)

si' firewood (Asicona Ramírez et al. 1998: 128)

Pekxítx: *Peq Xitch'* Cacao Wing [large brown hawk]

pekxítx bird name, synonym for *moochó* (Nachtigall 1978: 307). [The *tx* is Nachtigall's version of *tch'*.]

peq cacao, *pataxte* (Kaufman 1974: 498); *patashte* inferior-quality cacao (*Theobroma bicolor*) (Schoenhals 1988: 87)

xitch' wing; large hawk (Kaufman 1974: 495)

Petzetz Ki: *Petzetz Q'ih* Very Twisted Days; Flailing Days

petzetz very twisted; from *petz* twisted (Kaufman 1974: 359, 499)

*petz-etz beat-beat, flail, thresh? from *petzu'm* beat with a small stick (Chel and Ramírez 1999: 198)

Soojnóy: *Sooj No'y* Esophagus Worms; Disappearance of Vermin

sooj root for vanish, disappear (Kaufman 1974: 634–35)

sooj trachea, larynx, esophagus, pharynx, throat (Asicona Ramírez et al. 1998: 100–130)

nóy, no'y intestinal worm; any bug or little animal; the *choconoy* worm—it doesn't bite (Kaufman 1974: 463)

The base *sooj* "disappear" suggests equivalency with Zoj Ki.

Soonchój: *So'n Choj/Tx'oj* Smallest Payment/Rat

soon: *so'n* little tiny one; the last fruit (Kaufman 1974: 634); last *criatura*/baby (Asicona Ramírez et al. 1998: 141)

See **Nimchó** above.

Talchoo: *Tal Choj/Tx'oj* Small Payment/Rat

tal small (Kaufman 1974: 674–77)

Tchoch'ol: *Tchotch'ol* Plant

tchotch'ol a type of plant that produces bulbs and is eaten (Asicona Ramírez et al. 1998: 151)

Tchoochcho: *Tchooch Choj/Tx'oj* Useless Payment/Rat

See **Nimchó** above.

   This name is the last of eight -*cho* months. The 1942 Chajul and Nebaj calendars record two, Tal Choh/Tx'oj and Nim Choh/Tx'oj, "Small Payment/Rat" and "Big Payment/Rat." In its 1973 roster Nebaj features the third, So'n Choh/Tx'oj "Tiny or Last Payment/Rat," adds a fourth, Ne' Cho/Tx'oj "Baby Payment/Rat," and modifies Nim Choh/Tx'oj with Mas, Moch, and Mox. Maybe these names, if meaning "Payment," preserve relics of a tribute schedule.

*mas as a variant of *max*, itself a variant of *mox* all, many (Chel and Ramírez 1999: 164)

motx last (Chel and Ramírez 1999: 168). [Perhaps a variant is *mox*. All three would then denote "last."]

Tsanakváy: *Tzanak'b'ay* Yellow Worm; *Tz'an Ak' Way* Flat, Wide Unripe-Corn Tortillas

tzanak'b'ay a type of yellow worm (Chel and Ramírez 1999: 297) *tz'an* from *tz'ankin* flat and wide (Kaufman 1974: 790)

ak' moist; fresh; new (Kaufman 1974: 19)

ak' way tortilla or tamale of fresh green or unripe corn (*elote*) (Kaufman 1974: 18); *ak'vay* tortilla or tamal of *elote* or fresh green corn; eaten on at least the Day of the Dead [31 October–2 November?] (Rodríguez Sánchez et al. 1995: 17)

Tzijep: *Tzihe'p* Omens of Destruction

tzi message, warning, command (Kaufman 1974: 838)

e'p destruction, from *e'pu-m'* take apart, destroy (Kaufman 1974: 44)

Tzil Ki: *Tz'il Q'ih* Filthy Days; Split Firewood Days

tz'il filthy; split piece of wood, from *tz'ilu-m'* split narrow pieces of firewood (Kaufman 1974: 799)

Tzu Ki: *Tzu' Q'ih* Loom Bar Days; Stench Days; *Tzuh Q'ih* Gourd Days

tzu' thick, lower loom bar (Nachtigall 1978: 83); dirty, a spot (Asicona Ramírez et al. 1998: 173); bad odor (Chel and Ramírez 1999: 304)

tzuj calabash; gourd bowl for tortillas (Kaufman 1974: 849)

Tzunun Ki: *Tz'unun Q'ih* Hummingbird Days

tz'unun hummingbird (Kaufman 1974: 809)

Tz'ikin Ki: *Tz'ikin Q'ih* Bird Days

tz'ikin bird (Kaufman 1974: 798); <*tsik'ín*> bird (Nachtigall 1978: 85); <*tsikín*> quail (German *Wachtel*) (Nachtigall 1978: 304, day 15)

Xet Ki: *Xeet/ Xe't Q'ih* Beginning Days

xeet or *xe't* beginning, from *xeet/xe'-itchil* begin (Kaufman 1974: 916)

Xukul (Ki): *Xu'k'ul (Q'ih)* Piece of Firewood (Days)

xu'k'ul trimmed piece of wood, piece of firewood (Rodríguez Sánchez et al. 1995: 211)

Yax Ki: *Yax Q'ih* Crab Days; *Yax Q'ih* Green Days or Summer

yax crab (Kaufman 1974: 972). See Ch'ol **Yazquin**.

Yowal: *Yowal* Obligation; *Yawal* Head-Ring

yowal obligation, from *yowalil* necessary, obligatory (Diego Antonio et al. 1996: 385)

yawal <*yagual*> (padded ring for carrying things on the head) (Kaufman 1974: 635), under *sok*

Zil'ki: See **Tzil Ki**.

Zoj Ki: *Sooh Q'ih* Cough Days; *Sok Q'ih* Nest or Head-Ring Days

sooj root for vanish, disappear (Kaufman 1974: 634–35)

sooh cough (Nachtigall 1978: 342)

sok nest of bird or animal; *yagual* (head-ring) (Kaufman 1974: 635); *sok* with *k* reduced to *h* by neighboring *q'*, analogous to the *q-h* shift in *saq/sah* "white, clear" (Rodríguez Sánchez et al. 1995: 135)? The base *sooj* "disappear" suggests equivalency with Soojnóy.

Summary: Ixil draws attention for the number and diversity of its names and internal subtraditions. Of the six rosters from 1942, two are complete and a third nearly so. Two more full lists come down from just one site, Nebaj, in 1973, along with a third, virtually entire. These total nine, with four complete and two virtually intact. The rosters together report fifty-six different names, the most for any calendar group. Of these, fifteen or 27 percent are purely Ixil, with no apparent derivation from any other source, while the rest stem from various rosters, principally the glyphs and K'iche'. The armies of the K'iche' king Q'uqumatz subjugated the Ixil between 1400 and 1425, providing the first recorded opportunity for the influx of K'iche' terms. In calendrical affairs at least, the Ixil might not have been as isolated as Benjamin

TABLE 3.3
Summary of Ixil month names

Names in source	Interpretations	Translations
Akmór	Aq' Mo'l	Work Group
		Reed Bundle
		Root Cluster
	Ak' Mol	New Companion
	Ak Mol	Honorable Companion
A Ki	A' Q'ih	Water Days
	Aaq'ii	Diviner, Astrologer
Awax Ki	Awax Q'ih	Corn-Sowing Days
Cajab'ki	Kahab' Q'ih	Honey(comb) Days
		Bee Days
Chantemák	Chan Te' Mähk	Tall-Posts Granary
Hui Ki	Ju'il Q'ih	Mount Ju'il Days
	Wi' Q'ih	Tip Days
Kaháb	Kahab'	Bee; Honey(comb); Sweets
Kahabtsé	Kahab' Tzé'	Log Beehive
Koj Ki	Koj Q'ih	Clear-the-Land Days
	Koj Q'ih	Puma Days
Kucham	Kucham	Tree Trunks
K'olk'óy	K'ol K'oy	Oak Monkey
	K'o'l K'oy	Lump Monkey
	Q'ool K'oy	Headscarf or Performer Monkey
Lajab Ki	Lajab' Q'ih	Deadfall-Trap Days
Lem	Lem	Maize Worm or Bug
		Bean Beetle
		Water Mosquito
Maháb	Majab'	Dam
Mam A Ki	Mam Ah Q'ih	Grandfather Prayermaker
	Mama' Q'ih	Prayermaker Days
	Mama'ah Q'ih	Prayermakers
Mamak'i	Mama' Q'ih	Prayermaker Days
Kejepk'í	Kehep Q'ih	Open Days
Masnimchó	Max Nim Choj/Tx'oj	Many Large Payments/Rats
Mek'aj	Meq' Ah	Bent-Over Stalks
Metch'ki	Metch' Q'ih	Sprout Tip Days
Mo	Mo'	Hawk
Mochnimchó	Moch Nim Choj/Tx'oj	Last Large Payment/Big Rat
Mol	Mo'l	Group
Molch'ú	Mo'l Ch'uh	Flock of Grackles

(continues)

TABLE 3.3   Summary of Ixil month names (*continued*)

Names in source	Interpretations	Translations
Mol Tche	Mo'l Tcheh	Herd of Deer
Mol Masat	Mo'l Masat	Herd of Deer
Moochó	Moocho	Large Brown Hawk
Mu	Muj	Oak or Palo Negro Shade
Muen	Muen	Seedbed, Seedlings
Muen Chin	Muen Chin	Seedling Bag
Ne'chó	Ne' Choj/Tx'oj	Baby Rat Small Payment
Nimchó	Nim Choj/Tx'oj	Big Payment/Rat
Nol	Nool	Earthworms
	Nor	Slippery-Wet
Nol Ki	Nool Q'ih	Earthworm Days
	Nor Q'ih	Slippery-Wet Days
Och Ki	Otch' Q'ih	Unripe Corn Days
O Ki	O' Q'ih	Five Days
Okok'í	Oq O' Q'ih	Foot Five Days
	O' Ko' Q'ih	Five Owner Days
	O' K'oo Q'ih	Five Mask Days Horsefly Days Patch Days
Onchil	Oonchil	Arrive
	Oon Chi'l	Chills or Colds
Onchív	Oon Chi'l	Chills, Colds
Oval K'í	O'-va'l Q'ih	Five Days
Pac Tzi	Pa'k' Tzi'	Spoon Beak
	Paq Tzi'	Curve Beak
	Paq' Tzi'	Hole Beak (Woodpecker?)
Paksí	Paq Si'	Bend Firewood
Pekxítx	Peq Xitch'	Cacao Wing [Hawk]
Petzetz Ki	Petzetz Q'ih	Very Twisted Days Flailing Days
Soojnóy	Sooj No'y	Esophagus Worms Disappearance of Vermin
Soonchój	So'n Choj/Tx'oj	Smallest Payment/Rat
Talchoo	Tal Choj/Tx'oj	Small Payment/Rat
Tchoch'ol	Tchotch'ol	A Plant
Tchoochcho	Tchooch Choj/Tx'oj	Useless Payment/Rat
Tsanakváy	Tzanak'b'ay	Yellow Worm
	Tz'an Ak' Way	Flat, Wide Unripe-Corn Tortillas
Tzijep	Tzihe'p	Omens of Destruction

Names in source	Interpretations	Translations
Tzil Ki/Zil'ki	Tz'il Q'ih	Filthy Days
		Split Firewood Days
Tzu Ki	Tzu' Q'ih	Loom Bar Days
		Stench Days
	Tzuh Q'ih	Gourd Days
Tzunun Ki	Tz'unun Q'ih	Hummingbird Days
Tz'ikin Ki	Tz'ikin Q'ih	Bird Days
Xet Ki	Xeet/Xe't Q'ih	Beginning Days
Xukul (Ki)	Xu'k'ul (Q'ih)	Piece of Firewood (Days)
Yax Ki	Yax Q'ih	Crab Days
	Yax Q'ih	Green Days or Summer
Yowal	Yowal	Obligation
	Yawal	Head-Ring
Zoj Ki	Sooh Q'ih	Cough Days
	Sok Q'ih	Nest or Head-Ring Days

Colby (1976) thought; Ilom, for that matter, lies on regular Highland-Lowland trade routes.

Kaqchikel Definitions

Kaqchikel is a K'ichean language, along with K'iche', Poqomam, Poqomchi', Q'eqchi', and several others (Campbell 1984b: 2, fig. 1; Kaufman 1974b: 959). In the list below parentheses enclose source languages for words absent in Kaqchikel. The more frequently repeated words or parts of words are defined here. Smailus presents his study of verbs in a grammar (1989b), chapter 4.7, especially pages 137–42 and 149–57.

am spider (Campbell 1977: 7, chart 2); bubo (Smailus 1989b: 2:37)

-am past participle on Transitive 2 verbs, Voices 1 and 2 (Smailus 1989b: 1:172)

-an variant of *-Vn* to be, become; intransitive verbalizer with adjectives (Smailus 1989b: 1:125–26)

-an variant of *-am* (by analogy to examples in Campbell 1977: 7–8, chart 2; and Sper 1970: 37, 38)

*-eb' plural marker, fused in some languages with the root (Fox 1978: 251, item 257; 257, item 268)

ik chile (Vico 1555: 97r)

-i-k verbal nominalizer (Smailus 1989b: 1:64)

ik' moon, month, menstrual period (Smailus 1989b: 2:292)

-in converts many adjectives into intransitive verbs meaning "to be or become *x*" (Smailus 1989b: 1:125)

-in antipassive for Transitive 2b verbs; these take objects (Smailus 1989b: 1:150, 154)

-in root intransitive marker, *rección*/paradigm 4b (Smailus 1989b: 1:156)

-in verbal nominalizer for Transitive 2 verbs, *rección*/paradigm 1 or 2, with or without patient; without ergative pronoun (Smailus 1989b: 1:181). The verbal noun, next to a noun, can function as a qualifier.

*ka'/*ka'-eb' two; predicative alternates in proto-Mayan: this fossilized form seems to represent a fusion of the root with the old plural suffix **-Vb'* or **-eb'* (Fox 1978: 251, item 257; 257, item 268).

kab' two, attributive or dependent, from <cablahuh> twelve ["2 + 10"] (Vico 1555: 28r; Coto 1983: xcviii); <caba> two years (Coto 1983: 172); and <cab-ichal> (Vico 1555: 28r; Coto 1983: xcviii, 29). The second glottal stop of the independent predicative form *ka'i* (Ruyán Canú and Coyote Tum 1991: 371; Smailus 1989b: 2:310–11) and the K'iche' parallel *ka'ib'* (Ximénez 1985: 146; Basseta 2005: 169v) indicate a Kaqchikel **ka'ib'. Kab'* consists of *ka'* + *-ib'* (Judith Maxwell, personal communication, 2000).

kam' <cam> two, variant of *kab'*. Clearly attested here, the change of *b'* to *-m'* takes place not only in Kaqchikel but in other languages and dialects too, such as Ch'ol, Tzeltal, and Tzotzil (Fox and Justeson 1984: 52 n. 30, in Bricker 1986: 94; Kaufman 1972: 23, 24). *Kam'/kab'* means not only "two" but sometimes "second" (Judith Maxwell, personal communication, 2002).

ka'n <can> variant of *kam'*, itself a form of *kab'* two. In many dialects word-final *m* is realized as *n* (Campbell 1977: 7–8, chart 2, 11–12).

nab'ey first, preeminent (Smailus 1989b: 3:567; Coto 1983: 440)

-om completive or past participle marker on verbs, Transitive 1, Voices 1 and 2 (Smailus 1989b: 1:170)

q'ij sun, day, time (Vico 1555: 233v)

ru preconsonantal E3s (ergative, third person, singular); his, her, its. Prefixed to cardinal numbers for ordinals (Smailus 1989b: 1:69–70)

ru kam' <ru cam> second (Smailus 1989b: 2:310)

ru ka'n second (Blair et al. 1981: 175)

Kaqchikel Definitions

Bota: *B'ot A'* Roll of the Year [Calendar on Paper/Hide]

b'ot to hill or bank, to cover something with earth; to roll up a blanket, mat, paper (Smailus 1989b: 2:81)

a' year (Vico 1555: 1r)

Botam: *B'ot-Am* Rolled Up; Doubled Over [Maize Ears]; *B'ot Am* Cotton-Spider [Spindle Whorl] (K'iche') *b'otom* a thing rolled up like this (blanket); folded or doubled over thing (Smailus 1989b: 2:8)

b'otam K'iche': rolled (mats) (Adrián Recinos, in Edmonson 1965: 16)

b'ot combed cotton (Ximénez 1985: 120)

The putative *b'ot-am* "combed-cotton spider" derives from an imagined similarity of a spider's abdomen with filament to a spindle whorl with thread, recalling the parallel between the Tzotzil *tzek* scorpion and *tzek* spindle whorl (Laughlin 1975: 90).

Cakam: *Käq Am* Red Spider, Red Bubo; *Kaq-am* Stabbed; *K'aq-am* Pierced with Arrows; *Kaj Q'a'm* Sky Pole

käq from original *kaq* red (Vico 1555: 33r); vermilion; angry; fine-grained, minute (Smailus 1989b: 3:316); to stab a bull with a lance; to pierce (Sáenz de Santa María 1940: 180)

k'aq to stone or hit with stones or the bow and arrow; for the sun to damage seriously; to wound; appoint or declare a person something (Vico 1555: 246v; Smailus 1989b: 2:372); to stab a bull with a lance; to pierce (Sáenz de Santa María 1940: 180)

kaj sky (Ruyán Canú and Coyote Tum 1991: 13)

q'a'm ladder made of a straight log with notches; a bridge of one or more logs (Coto 1983: 204, 447)

Cakan: *Käq-Q'än* Red-Yellow; *Käq-an* To Become Red; *Käq Ajn* Red or Finely Grained Ear of Tender Corn; *Kaj K'an* Sky Cord; *Kaj Q'a'an* Sky Pole

This spelling presents the most frequent form and that of Vico, who meticulously applies *c*, *4*, *k*, and *Σ* to distinguish *k*, *k'*, *q*, and *q'*, respectively (Carmack 1973: 115).

käq-är <ca*Σ*ar> to turn angry; for the sky in the evening to have red tinted clouds; to end up red (Anonymous 1813: 8r; Ángel b 18th century: 31v)

ajn ear of tender corn (Smailus 1989b: 2:25)

k'am string, cord, tumpline, net (Teletor 1959: 187); cord, string, rope, leather strap (Vico 1555: 242v)

k'an: *mimbre*; *pita* [century plant; cord] (Rodríguez Guaján et al. 1990: 65); *pita* cord, a measure of land (Ruyán Canú and Coyote Tum 1991: 21)

q'a'n ladder, bridge (Ruyán Canú and Coyote Tum 1991: 54)

q'an yellow, mature, ripe (Santo Domingo 1693: xx–r)

Çibix: *Sib'-ix* There Is Smoke

sib'ix to become smoky (Smailus 1989b: 3:693)

sib' smoke (Vico 1555: 42v)

-ix nominal verbalizer: derives intransitive verbs from root nouns (Smailus 1989b: 1:157, 159)

Çibixic: *Sib'-ix-ik* Smoking [Being Smoky]; *Sib'-ix Ik'* Smoky Month

See **Çibix** above.

-i-k verbal nominalizer (Smailus 1989b: 1:64).

sib'ixik under *ahumar* to make something smoky: "And because, after clouds of smoke and burnings, they sow their fields, at the time of sowing they say: *mi-x cuke çibixic avexabal Σih*, as soon as the smoke settles, it will be the time of sowing'" (Coto 1983: 20).

Hun Mam: *Jun Mam* First Grandchild; First Grandfather

See **Mam** below. *hun* first (Flores 1753: 332)

Ibota: *Il B'ot A'* Large Roll of the Year; To See the Roll of the Year [Divination]

See **Bota** above for definitions. It remains unclear whether *ibota* assimilated to *obota* or *obota* dissimilated to *ibota*. Variants of other words display the identical contrast and are sometimes listed side by side, like *oboyel/iboyel* messenger (Vico 1555: 142v).

il fault, misfortune; large, much (Vico 1555: 99r); to meet, learn, know (see Sáenz de Santa María 1940: 131)

Itzcal Σih: *Itzkal Q'ij* Side Days; *Izkal Q'ij* Sprout Days

itzcal Nahuatl: *itzcal-li* side (right or left) (Andrews 1975: 446)

izcal Nahuatl: *izkal-li* Sprout; *Is Qal* Cotton Tribute Blankets (K'iche')

izcal-li Nahuatl: it is a sprout; sprout, bud (Andrews 1975: 372, 403–4, 448)

is hair, fur (Ruyán Canú and Coyote Tum 1991: 40); K'iche': hair, wool, blanket, cotton (Edmonson 1965: 49)

qal tribute blanket four "legs" wide [*pierna*/leg denotes a variable measure of width] (Vico 1555: 111r); cotton cloth (Sáenz de Santa María 1940: 185)

K'atic: *Kat-ik* Abstinence; *K'at-ik* Burning the Underbrush

kat-ej to abstain from any activity whatever (Coto 1983: 7r)

kat-i to abstain from something (Smailus 1989b: 2:326)

k'at to burn or fire something; to have a great celebration, for weddings, house-raisings or harvests; for the cornfields, grasses, or other crops to be burned by the sun, wind, or ice; to set fire to brush or cornfield (Coto 1983: 241, 456); to burn, braise, scorch; to wither, wilt, dry up, parch (Sáenz de Santa María 1940: 196) *k'atik* burn off in savannas and brush, a verbal noun (Smailus 1989b: 2:390 n. 19; 482 n. 2b); a fire, a burn (Blair et al. 1981: 217)

LiΣinΣá: *Lik'-in Kaaj* Stretched Out Sky [Creation]; *Lik'-in Q'a'aj* Scattering Hand [Sowing, Incensing], Spreading Out Hand [Divination; Drying Maize, Cacao]

lik' to unfold something, like clothes, paper and such; to stretch something out; to mix cacao, maize, cotton, etc., spreading it out so that it dry in the sun (Coto 1983: 5, 147, 218, 467, 544)

liq' root of *liq'e*, written *liΣe*, to lie down, to stretch out; to go all out; to droop, get beaten down (corn, etc.) (Vico 1555: 118r)

kaaj sky (Campbell 1977: 58, item 180)

q'a' hand, arm, possessed and compound forms (Vico 1555: 225v) [Vico points out that Kaqchikel drops the final *b'* while K'iche' retains it]; hand, arm, compound and possessed (Smailus 1989b: 3:458)

Mam: *Mam* Grandchild of a Man; Grandfather

mam grandfather's term for grandson or granddaughter (Vico 1555: 123v)

mam lord, elder (Edmonson 1988: 145, 217, 237)

mama Nahuatl: to carry another on the back; to rule and govern others; to carry a load on the shoulders (Molina 1977: 51v); perhaps derived from *ma* "to take someone captive," as warriors bound prisoners and bore them off on their shoulders (Kartunnen 1992: 134)

Moh: *Mo'* Crow, Hawk. The grapheme *h* often just marks voicelessness.

mo crow (Ruyán Canú and Coyote Tum 1991: 63); big black hawk (Rodríguez Guaján et al. 1990: 74)

Nabey Mam: *Nab'ey Mam* First Grandchild, First Grandfather

Nabey Pach: *Nab'ey Pach* First Incubation/Mistletoe. See **Pach**.

Nabey ToΣ: *Nab'ey Toq'* First Jab, First Harvest Cacao; *Nab'ey Tok'* First Flint. See **ToΣic** below.

Nabey ToΣic: *Nab'ey Toq'ik* First Short One; *Nab'ey Toq'-ik* First Lancing, First Bloodletting, First Cacao Harvest. See **Nabey ToΣ**.

toq'-i-k <toΣic> little or short of stature (Guzmán 1984: 101), under *<to4om>*

Nabey Tumuzuz: *Nab'ey Tumusus* First Bean or Maize Bug, First Winged Termite or Ant (K'iche'). See **Tumuzuz**.

Obota: *O('ob') B'ot A'* Five Rolls of the Year [Calendars]. See **Bota** and **Botam** above.

o'o' five (Vico 1555: 142v), under *obota*, 145v

o'ob' <*oob*> five (Schultze-Jena 1972: 261)

Pach: *Pach* Incubation (Kaqchikel); Moss/Mistletoe (Pipil Nahuatl or Gulf Coast Nahua)

pach to incline, bow, lean; to cover; for hens to sit on eggs (Smailus 1989b: 3:607 nn. 1–3; Coto 1983: 174–75, 244, 293; Vico 1555: 151v, 152r)

pach Pipil Nahuatl: moss (Edmonson 1988: 145, 217, 222, 237); Gulf Coast Nahua: moss (Campbell 1977: 106:37)

pach-oaa tla- Nahuatl: to press down on something, to control something; to brood (said of a hen), to sit on eggs (Andrews 1975: 461)

pach-oaa tee Nahuatl: to govern someone (Andrews 1975: 461)

pach-oaa mo Nahuatl: to stoop over, to bend over, to bow down (Andrews 1975: 461; Kartunnen 1992: 182; Molina 1977: 78–79)

pach-tli Nahuatl: parasitic plant that grows on trees and with which one decorates above all the temples during ceremonies (Campbell 1977: 106, item 37; Forrest and Brewer 1962: 172); mistletoe, hay, plant refuse (Karttunen 1992: 183); *malhojo*: moss or a certain shrub [*yerua*] that grows and hangs in the trees (Molina 1977:79)

Pariche: *Par-i Che'* Covered Enclosure or Granary; *Pa R-ij Che'* At the Back of the Trees, In the Granary

par kind of cover or shawl against rain water, often of palm leaves (Coto 1983: 96, 463)

-i form of *-V*, attributive on adjectives and nouns, mostly *-a* and *-i* (Smailus 1989b: 1:110, 119)

pa in, to, near (Sáenz de Santa María 1940: 300); within, in, on, at (Blair et al. 1981: 267; Rodríguez Guaján et al. 1990: 80)

r his, her, its; prevocalic 3Es (Vico 1555: 176v; Blair et al. 1981: 274; Smailus 1989b: 1:69)

-ij back; shell, bark, peel; fish scale (Coto 1983: 209, 204); rind, shell, back, pod (Blair et al. 1981: 240)

che' stick, wood, tree, forest or bush country (Vico 1555: 60v); log, beam, pole (Smailus 1989b: 2:151); jail (Blair et al. 1981: 229)

Pay: *Pay* Trickster, Hoja de Santa María [an edible herb]

pay: *la hoja de Santa María* (Guzmán 1984: 11)

pay vulgar comic, person who tells dirty jokes; a wit (Vico 1555: 149v); jester, maker of funny faces, trickster, buffoon (Smailus 1989b: 3:621); jokester, who laughs a lot and makes much about what he says (Coto 1983: 134)

Payriché Pay Ri' Che' Joker the Log. See **Pariche** and **Pay**.

The sense "log" and this month's position immediately next to the last five days of the year recall the Yukatek deity Mam "Grandfather," lord of the last five days, worshipped as a dressed up old log (Cogolludo 1971: 1:255). In any particular year of the K'iche' calendar, one of the four Mams oversees the first day of the year, functioning thus as the yearbearer, as well as the first day of each month and the first day of the last five days. If this name represents not a slip of the pen but a native variant, then it might indicate the Mam and his log.

Qammam: *Kam' Mam* Two/Second Grandfather, Two/Second Grandchild. Variant: Can Mam.

The source (Sáenz de Santa María 1940: 262) lists this name right after <*nabeimam*> "first grandfather" and consistently uses *q* for the colonial *c* or modern *k*. This renders <*qam*> a virtually certain *kam'* "two."

Qam Pach: *Kam' Pach* Two/Second Incubation/Mistletoe. Variant: Qan Pach. See **Pach** above for definitions.

Ru Cab Mam: *Ru Kab' Mam* Second Grandchild; Second Grandfather. See **Mam** above.

Ru Cab Pach: *Ru Kab' Pach* Second Incubation/Mistletoe. For definitions see the section introduction and **Pach**.

Ru Cab Tokic: *Ru Kab' Toq'ik* The Second Short One; *Ru Kab' Toq'-ik* Second Lancing, Second Bloodletting, Second Cacao Harvest. See **ToΣic** below.

Rucab ToΣ: *Ru Kab' Toq'* Second Jab, Second Bleed with a Lancet, Second Cacao Harvest. For definitions see the section introduction and **ToΣic**.

Rucab Tumuzuz: *Ru Kab' Tumusus* Second Winged Termite or Winged Ant (Kiché'), Second Bluish Bean Beetle or Maize Bug. See **Tumuzuz**.

RucactoΣiΣ: *Ru Kaq' Toq-ik* His Guava Harvest. See **ToΣic**.

kak <*cac*> This variant for *kab'* is attested only in the 1685 calendar, in both instances of the month name on pages 5 and 18, though three other names include <*cab*> or its variant <*can*>. The seventeenth-century version of a late sixteenth-century dictionary represents *b'* with the graphemes *b*, *bp*, and *pp* (Smailus 1989b: 1:9–10, 13; 2:2). This suggests that some dialects might vary a syllable-final or word-final *b'* for *p*; *p'* does not exist (Smailus 1989b: 2:2). The *p* might then parallel the word-final variation of *p* with *k* and *'* attested among several unspecified Kaqchikel dialects (Sper 1970: 37–38). In short, the development would read *kab'* > **kap* > **kak*.

kaq': *guayabas* (Vico 1555: 33r); guava (Schoenhals 1988: 55)

Tacaxepegual: *Tlaca-xipehua-liz-tl*i Flaying of Men; *Tlacoh-Xipehua-liz-tli* Flaying of Slaves

The Nahuatl dialects of Guatemala ultimately derived from Gulf Coast Nahua, so *tl* became *t*.

Tlaca-xipehua-liz-tli Nahuatl: "It Is the Man-Flaying Action" (Andrews 1975: 403)

tlaca-tl Nahuatl: person (Karttunen 1992: 253); man, person, lord (Molina 1977: 115v)

xipehu(a) Nahuatl: to flay, skin, peel something (Karttunen 1992: 325; Molina 1977: 159r)

-liz-tli Nahuatl: absolutive verbal nominalizer; with *-liz* the verbal nominalizer and *-tli* the absolute marker after consonants other than *l* (Sullivan 1988: 15, 87; Andrews 1975: 144, 228–30)

*To'q: Loincloth

to'q loincloth (Cojti Macario et al. 1998: 349)

ToΣic: *Toq'ik* The Short One; *Toq'-ik* Lancing [Scaffold Sacrifice]; Bloodletting; Cacao Harvest

toq' to cut, wound, stab (Vico 1555: 192r); flint <*toΣ*> (Smailus 1989b: 3:778); to lower fruit striking, it with the point of a pole; to prick, piercing; to slit or jab; to bleed with a lancet or to open a swelling or boil; to lance; to collect cacao pods from the tree (Coto 1983: 143, 213, 271, 306, 327)

Tumuzuz: *Tumusus* Winged Termite or Ant (K'iche'); Bluish Bean Beetle, Maize Bug

tumusus worms or bugs [*gusanos*] that appear at the beginning of winter (Sáenz de Santa María 1940: 373; Rodríguez Guaján et al. 1990: 101)

tumusus, ch'eken K'iche': the flying insect [*palometi(ll)a*] that breaks the earth at the start of winter (Anonymous 1787: 205v) (under *volatiles*); very common winged house termite (Stoll 1889: 60)

A K'iche' informant relates that the *tumusus* are *azulitos* (light blue), smaller than a housefly, winged, and fond of maize plant leaves (Cipriano Alvaredo in Gates n.d. [1922]: 35). Stoll (1889: 60) deduces that *tumusus* refers at least to the common winged house termite. Schoenhals (1988: 215) confirms both winged and wingless termites.

timusús [*sic*]: *oruga de los frijolares o tortuguitas de color azul tornasol*/bean patch caterpillars or sunflower-blue bean beetles; weevil or snout beetle [*gorgojo*] on beans when dry or on maize (Herbruger and Díaz Barrios 1956: 234, 237).

My translation assumes *tortuguita* equates with *tortuguilla*, the "Mexican bean beetle" (Schoenhals 1988: 219, 229).

4,Api Σih: *Tz'api Q'ij* Shut Door Days; *Tz'ap-i Q'ij* Disaster Days, Beehive Days, Enclosure Days; *Tz'apij Q'ij* Lock Up Days

tz'api door for closing (Vico 1555: 268v); door made from canes or something else that are tied and made into a cane frame or wattle with which they close the doors of their houses or corrals (Coto 1983: 447)

tz'ap disaster (Smailus 1989b: 2:123); domestic beehive (Coto 1983: 98); closing or an instrument for closing or a thing that is closed (Coto 1983: 127)

tz'apij to fasten, close, lock up, seal, block, cover (Blair et al. 1981: 289)

4,iquin Σih: *Tz'ikin Q'ij* Hawk, Bird or Hummingbird Days; Penis Bloodletting Days

tz'ikin eagle, hawk (Vico 1555: 270v); bird in general; member [*piquita*] of a boy or little dog (Coto 1983: 57, 348, 388, 401); *pájaro*; *gorrión* (Cadena 1788 in Fernández 1892: 19, 22); bird; sparrow, or in Chiapas and Guatemala, hummingbird (Schoenhals 1988: 417)

As the most common K'iche' metaphor for penis, *tz'ikin* "bird" can refer to genital bloodletting (Tedlock 1993: 19–20, 238–39).

Uchum: Uchum Opossum (Q'anjob'al)

uchum Q'anjob'al: *tlacuache (tecuasín)* (Andrade 1946: 750, in Dienhart 1989: 476; Fox 1978: 180, item 139)

utxum Q'anjob'al: opossum (Diego Antonio et al. 1996: 330)

uchum Chuj: opossum (Diego 1998: 248)

SUMMARY: Of the thirty-one distinct Kaqchikel names, nineteen have K'iche' cognate counterparts; of the remaining twelve, four come from K'iche' indirectly, as synonyms; one derives from Aztec Nahuatl, another from Q'eqchi', and two from Q'anjob'al. Only four names are strictly Kaqchikel in origin. Like most others, this tradition proves heavily derivative (table 3.4).

TABLE 3.4

Summary of Kaqchikel month names

	Names in source	Interpretations	Translations
1	Tacaxepegual	Tlaca-Xipehua-Liz-Tli	Flaying of Men
2	Nabey Tumuzuz	Nab'ey Tumusus	First Bean or Maize Bug
			First Winged Termite/Ant
3	Rucab Tumuzuz	Ru Kab' Tumusus	Second Winged Termite/Ant
			Second Bluish Bean Beetle
			Second Bluish Maize Bug
	Tumuzuz	Tumusus	Winged Termite/Ant
			Bluish Bean Beetle
4	Çibix	Sib'-ix	There Is Smoke
4	Çibixic	Sib'-ix-ik	Smoking
		Sib'-ix Ik'	Smoky Month
5	Uchum	Uchum	Opossum
6	Nabey Mam	Nab'ey Mam	First Grandfather
			First Grandchild
6	Hun Mam	Jun Mam	First Grandchild
			First Grandfather
7	Ru Cab Mam	Ru Kab' Mam	Second Grandchild
			Second Grandfather
7	Qammam	Kam' Mam	2/Second Grandfather
			2/Second Grandchild
	Mam	Mam	Grandchild of a Man
			Grandfather
8	LiΣinΣá	Lik'-in Kaaj	Stretched Out Sky
		Lik'-in Q'a'aj	Scattering Hand
			Spreading Out Hand
9	Nabey ToΣ	Nab'ey Toq'	First Jab
			First Harvest Cacao
		Nab'ey Tok'	First Flint
9	Nabey ToΣic	Nab'ey Toq'ik	First Short One
		Nab'ey Toq'-ik	First Lancing
			First Bloodletting
			First Cacao Harvest
10	Rucab ToΣ	Ru Kab' Toq'	Second Jab
			Second Lancet-Bleed
			Second Cacao Harvest
10	Ru Cab Tokic	Ru Kab' Toq'ik	Second Short One
		Ru Kab' Toq'-ik	Second Lancing
			Second Bloodletting
			Second Cacao Harvest
10	RucactoΣiΣ	Ru Kaq' Toq-ik	His Guava

	Names in source	Interpretations	Translations
	ToΣic	Toq'ik	Short One
		Toq'-ik	Lancing
			Bloodletting
			Cacao Harvest
	*To'q	To'q	Loincloth
11	Nabey Pach	Nab'ey Pach	First Incubation
			First Mistletoe
12	Ru Cab Pach	Ru Kab' Pach	Second Incubation
			Second Mistletoe
			Bluish Maize Bug
12	Qam Pach	Kam' Pach	2/Second Incubation
			2/Second Mistletoe
	Pach	Pach	Incubation
			Moss/Mistletoe
13	4,iquin Σih	Tz'ikin Q'ij	Hawk or Bird Days
			Hummingbird Days
			Penis Bloodletting Days
13	Moh	Mo'	Crow, Hawk
14	Cakam	Käq Am	Red Spider, Red Bubo
		Kaq-am	Stabbed
		K'aq-am	Pierced with Arrows
		Kaj Q'a'm	Sky Pole
14	Cakan	Käq-Q'än	Red-Yellow
		Käq-an	To Become Red
		Käq Ajn	Red or Finely Grained Ear of Tender Corn
		Kaj K'an	Sky Cord
		Kaj Q'a'an	Sky Pole
15	Bota	B'ot A'	Roll of the Year
15	Botam	B'ot-am	Rolled Up
			Doubled Over [maize]
		B'ot Am	Cotton-Spider
15	Ibota	Il B'ot A'	Large Roll of Year
			Seeing the Roll of the Year
16	K'atic	Kat-ik	Abstinence
		K'at-ik	Burning Underbrush
17	Itzcal Σih	Itzkal Q'ij	Sprout Days
18	Pariche	Par-i Che'	Covered Enclosure Granary
		Pa R–ij Che'	At the Back of the Trees
			In the Granary
18	Payriché	Pay Ri' Che'	Joker the Log

(continues)

TABLE 3.4

Summary of Kaqchikel month names (*continued*)

	Names in source	Interpretations	Translations
18	Pay	Pay	Trickster
			Hoja de Santa María
			(unidentified plant)
19	4,api Σih	Tz'api Q'ij	Shut Door Days
		Tz'ap-i Q'ij	Disaster Days
			Enclosure Days
			Beehive Days
		Tz'apij Q'ij	Lock Up Days
			Beehive Days

K'iche' Definitions

As many authors employ an inconsistent or incomplete orthography to represent *k*, *k'*, *q* and *q'*, the spellings here derive from first modern K'iche' then modern Kaqchikel sources. Where necessary they depend on the 1555 Kaqchikel dictionary by Vico, who "meticulously applies the Parra phonemic symbols" (Carmack 1973: 115); and on two other early (1550–1600) Kaqchikel dictionaries. Modern editors have improved the original complete, fairly correct, consistent orthographies (Coto 1983: l–li; Smailus 1989b: 1:29–30). The more frequently repeated parts of names are defined here.

-a- desubstantive transitivizer (Edmonson 1967: 255)

-a-j form of *-V-j*, transitive infinitive (Edmonson 1967: 260)

am spider (Ximénez 1985: 70)

-aa-m nominalizer with simple or stative intransitive (Mondloch 1978: 145)

-am transitive perfective (Mondloch 1978: 124)

-am past participle marker for verbs of the subgroup Transitive 2, with Voices 1 and 2 (Smailus 1989b: 2:172); form of *-om*: perfective, with transitive verbs of Classes 1 and 2, in voices 1 and 2 (Mondloch 1978: 124)

-an indefinitive with deverbative noun stems [like past participle] (Edmonson 1967: 261); absolutive for root transitives

-an Kaqchikel: variant of *-Vn* be, become [inchoative]; intransitive verbalizer with adjectives (Smailus 1989b: 1:125–26)

-b' purposive (Edmonson 1967: 258)

-e desubstantive transitivizer (Edmonson 1967: 255)

-i form of *-V*, attributive on adjectives, nouns: by analogy to Kaqchikel (Smailus 1989b: 1:110, 119)

-e-j form of *-V-j*, transitive infinitive (Edmonson 1967: 260)

-i-j desubstantive [= denominal] transitive *-V-* with utterance-final transitive marker *-j* (Edmonson 1967: 254–55, Mondloch 1978: 46, 54)

-il generalizer (Edmonson 1967: 262); form of *-Vl*, active, with deverbative noun stems (Edmonson 1967: 261); "neuter" or absolutive

-il abstractive nominalizer for adjectives (Mondloch 1978: 160)

-i-x passive marker on transitive verb roots or stems and n transitivized nouns, deverbative as a terminal (Edmonson 1967: 258). See Kaqchikel *-ïx*: either a nominal verbalizer, deriving intransitive verbs from root nouns (Smailus 1989b: 1:157, 159, 184); or a passivizer (Blair et al. 1981: 436)

-j transitive marker (Edmonson 1967: 255); termination marker for voice 1 on class 2 transitive verbs (Mondloch 1978: 26, 54)

k- their (Mondloch 1978: 176)

kab' two (Silva Leal 1875: 8r); two, in compounds (Mondloch 1978: 176)

-l- stative intransitive root, from CVC-*i'* active root (Mondloch 1978: 138)

naab'e first (Mondloch 1978: 184)

nab'ey first (Wick and Cochojil-González 1966: 2:472)

ox three (Schultze Jena 1972: 261)

q'iij sun, day (Campbell 1977: 58:179)

r- preconsonantal E3s; his, her, its (Mondloch 1978: 24); ordinal prefix (Charencey 1883: 15)

r-ox third (Anonymous 1787: 182v; Schultze Jena 1972: 274)

u E3s, preconsonantal; his, her, its (Mondloch 1978: 22); ordinal prefix (Charencey 1883: 152)

u kab' second (Mondloch 1978: 189)

-x same as *(-V)-x*, the *-i-x* above

K'iche' Definitions

Ajau Chap: *Ajaaw Ch'ab'* Lord of the Arrow, Archer

ajaaw owner (Mondloch 1978: 174); lord, master (Campbell 1977: 58, item 192)

ch'ab' ray, shaft, arrow, blade, lance (Edmonson 1965: 19); bow, arrow (Schultze Jena 1972: 233)

Batam: *B'at-aa-m* Hardening [Kernels]; *B'at-am* Rolled Up

b'at become stone hard; roll up something like a mat (Ximénez 1985: 92)

Botam: *B'ot-am* or *B'ot-om* Rolled Up; Hilled [Mounds at Base of Maize Plants]; *B'ot Am* Combed-Cotton-Spider [Spindle Whorl]. See Kaqchikel **Botam** above.

b'otam rolled (mats) (Edmonson 1965: 16)

b'ot-om wrapped by rolling up (Anonymous 1787: 81r)

b'ot combed cotton; grab like rolling up a mat; to hill or bank, to cover something with earth (Ximénez 1985: 120)

Botan: *B'ot-an* Rolled Up; Roll Up. See Kaqchikel **Botam** above.

Caam Kij: *K'a'aam Q'iij* Cord Days, Plot of Land Days; *Q'am' Q'iij* Ladder or Bridge Days, Pole-Ladder Sun

k'a'aam vine, string, one *cuerda* [cord], a measure of land (Mondloch 1978: 177); *k'a'm* vine, string (Campbell 1977: 50, item 65); *k'a'am* rope, cord (Saquic Calel 1989: 306)

q'a'm bridge (Campbell 1977: 57:175); Kaqchikel: ladder made of a straight log with notches; a bridge of one or more logs (Coto 1983: 204, 447)

"Cord Days" suggests a time for making hemp ropes or measuring off plots of land with them. "Pole-Ladder of the Sun" evokes an upright notched pole that serves as a gnomon to fix the two days of zenithal passage and then as a platform for ritual performances on those days.

Cakam: *Kaq-Am* Red Spider or Bubo; *K'aq-am* Wounded, Installed; *K'aq Q'a'm* Pierce Ladder [Scaffold Sacrifice], Install Ladder [Accession Ladder]. For definitions of Kam see **Caam Kij**. Both Cakam and Cakan (below) contrast their first and third letters and fail to mark glottalization. This allows interpretations with respective counterparts of *k/k'* and *q/q'*.

kaaj sky (Campbell 1977: 58, item 180)

kaq red (Campbell 1977: 56, item 153; Mondloch 1978: 176)

k'aq shoot with arrows, throw at with stones; for the sun to damage seriously; pierce; throw stones at (Anonymous 1787: 21v); appoint, designate (Edmonson 1965: 50); Kaqchikel: declare a person something; appoint, install (Vico 1555: 246v)

Cakan: *Kaq-Q'an* Red-Yellow; *Kaq-an* Reddened or Reddening; *Kaaj Q'an* The Sky Is Yellow; *K'aq-an* Wounding. For definitions see also **Cakam** above.

<can> tree of red beans (Basseta 2005: 170r)

q'an yellow (Campbell 1977: 56:154); yellow, mature (Ximénez 1985: 156, 158)

Cak Che: *Kaq Chee'* "Red Tree" or Red Prison; *Kaq' Chee'* Guava Tree, Dance Log; *K'aq Chee'* Piercing Tree [Sacrifice], Installation Tree [Accession Scaffold]. For definitions of Cak see **Cakam**, and for Che see **Chee** below.

kaq chee' Kaqchikel: *el árbol de carreta* / "the cart tree" (Guzmán 1984: 5). Literally the name translates as "red tree."

Only León (1945) lists Cak Che. León (1954: 91) also applies it to the Pleiades but does not explain why or define the term.

Chee: *Chee'* Tree; Log or Wooden Statue; Granary; Prison. For suggestions of a link between this name as "Log" and the four Mam presiding over the last days of the year, see **Nabey Mam**.

chee' tree, wood, pole, cudgel, footstool, wooden object (Edmonson 1965: 25); jail (Mondloch 1978:1 78); bin, granary (Campbell 1977 :50:58)

Chi Il Lacan: *Chi Il Laqan* In-The-Zenith Banners; *Ch'il Laqan* Flaying Banners [Ritual Flags]

chi inside, by, at, to, with, for, in (Edmonson 1965: 26)

il see, watch; big (Ximénez 1985: 314); zenith, place of viewing (León 1954: 34)

laqam/laqan flag, pennant *<lacam>* (Ximénez 1985: 338)

Ch'ab: *Ch'ab'* Arrow [Arrow Sacrifice]

ch'ab'. See **Ajau Chap** above.

In Yucatán dancing archers aimed for the painted white area over the heart of the victim, whom the priests had tied to a stake (Tozzer 1941: 117–18; Pagden 1975: 83); a native song about this ritual survives (Barrera Vásquez 1980: 122–25), as do a few possible references in the Chilam Balam literature (Roys 1967: 67, 76–77; Edmonson 1986: 1ines 709–14, 1295–1304). In their month Uchum the Kaqchikel reenacted the dance and arrow-sacrifice of the demigod Tolgom [Tolk'om] (Recinos and Goetz 1953: 73–75). The K'iche's might have celebrated similar rites in Ch'ab', including perhaps the dance-sacrifice called *tum teleche* "twisted-trumpet captive" (Edmonson 1971: 136, line 4395). At Late Classic Chinkultic, Monuments 1, 7, 17, 18, and 20 each depict a prisoner standing against a post (Taube 1988b: 348; Navarrete Cáceres 1984: figs. 9, 24–29, 50–62, 64).

Cox Xe Puatl: *Kos-xiip-eehua-tl* Yellow Flayed Skin (Nahuatl). The interpretations based on the Nahuatl words below remain speculative. Doubled vowels are long.

cooztíc something yellow, golden. The *coz* of this appears as an element in some compounds (Kartunnen 1992: 43).

xiip-ee-hua flay, peel, skin, bark, hull (Andrews 1975: 484)

xiip- element in compounds and derivations referring to peeling, flaying (Kartunnen 1992: 325)

eehua-tl skin, pelt, hide (Andrews 1975: 436)

Huno Bix Gih: *Juun (J)O('Oob) B'iix Q'iij* Five Days from Now (past); Five Songs Days

jun-o'-b'ix Kaqchikel: five days, from *<chupam hun obix>* within five days (Ángel 18th century b: 105v)

The *o'-b'ix* "five days from now" gloss recalls Günther Zimmerman's (1954: 61) translation of this term as "period of five days." Analysis suggests *jun-(j)o'(ob')-b'ix* "one-five-days past," with "one-five" indicating a group of five.

juun one; a, an (Mondloch 1978: 182); root of *hun-in*: for birds to warble or trill (*gorjear*)

jo'oob'/job' five (Mondloch 1978: 182)

o, oob' five (Edmonson 1965: 81)

ob'ix five days from today (Ximénez 1985: 426); it has been five days (Basseta 2005: 52)

b'iix song (Campbell 1977: 59, 207)

Ibota: *Il B'ot-Ab'* Large Roll of the Year [Calendar On Hide] (see Roll of the Year); *Il B'ot Aaj* Big Bundle of Reeds. For definitions see **Botam** above and Kaqchikel **Ibota**.

il big, much (Edmonson 1965: 46); see, watch; much, big (Ximénez 1985: 314)

ab' [short form] year (Edmonson 1965: 3)

aj fresh corn (Mondloch 1978: 174)

aaj reed (Mondloch 1978: 174)

Iquim Kij: *Ikim Q'iij* Low Sun

ikim down below (García Hernández and Yac Sam 1989: 94)

Jun Bix Kij: *Juun B'iix Q'iij* One Song Day; Trill Song Day. See **Huno Bix Gih**.

Nabé Mamkuk: *Nab'ey Maam Q'uuq'* First Grandfather Quetzal; First Dance-Feathers. For definitions and comments see **Nabey Mam**.

mamq'uq quetzal (Anonymous 1787: 200r; Basseta 2005: 249r)

q'uuq' quetzal (Campbell 1977: 48, item 28); the long, outer, gold-green wing and tail feathers of the quetzal, used in celebratory dances (Schultze Jena 1972: 230); green feathers with which they dance (Ximénez 1985: 234)

Nabey Lik'in Ka: *Nab'ey Lik'-in Kaaj* First Stretched Out Sky [Creation]; *Nab'ey Lik'-in Kaj* First Scattered Powder

lik collect with a gourd or the hand (Ximénez 1985: 350)

lik' open the hand (Ximénez 1985: 350); sprinkle, scatter; to spread out in the sun (Silva Leal 1875: 27r)

kaaj sky (Campbell 1977: 58, item 180)

k'aj flour, powder, lime, bits, crumbs (Edmonson 1965: 95)

Nabey Mam: *Nab'ey Maam* First Grandfather; First Grandchild *maam* grandchild, grandfather (Ximénez 1985: 374); grandfather, ancestor, elder, old man; grandchild (Edmonson 1965: 70); old man or grandfather; grandson or granddaughter, a term used by the grandfather (Silva Leal 1875: 118v)

"Grandfather" and "grandchild" could indicate the practice of the *ke'x* naming custom, recorded by ethnographers (Mondloch 1980: 9, 21). Parents name a child after one of its grandparents, living or dead, using the first or last name. The child and the grandparent address each other with the reciprocal term *ke'x* "substitute, replacement." The child also uses the same kin terms of address and reference with many members of its family that the grandparent does. Many believe the *ke'x* child is the actual replacement and alter ego of the grandparent and that they are essentially one and the same person. This custom provides a social mechanism for replacing the ancestors and obtaining personal immortality (Mondloch 1980: 9–11, 21).

The four calendrical Maam or "Old Men, Grandfathers" go by the day names Keej "Deer," Eej/Eey "Tooth," No'j "Thought/Resin," and Iiq' "Wind," ranked from first and greatest to last and least in the same sequence. In order of rank, a different one each year "carries in" or inaugurates the new solar cycle, serving as "yearbearer." The K'iche' annual count ignores the leap year correction, always totaling only 365 days, so in the Gregorian calendar it starts one day earlier every four years, presently in late February. The yearbearers' counterparts in the Codex Dresden New Year almanac, pages 25–28, go by the name glyphs **ma** + **ma** or *mam* "maternal grandfather" and "grandchild, child of a man's daughter" in Yukatek; they sport full-body opossum costumes. In the *Popol Vuh* the epithet *mama* "old man" links the opossum making streaks in the dawn sky to the yearbearers. The sobriquet *hunahpu uuch* [*uu4h*], defined as "One Blowgunner Opossum" and glossed with "a name they gave the Creator in their heathendom" (Ximénez 1985: 290), also ties the opossum to the yearbearers because it designates the one hero twin Hunahpu when both perform as vagabond minstrels, precisely the role that the yearbearers play among the Lowland Maya (Tedlock 1982: 89–90, 100; D. Tedlock 1985: 44, 287–88, 290, 294, 1993: 34, 221–23). For the Mam-Yearbearers the opossum guise may mask a Venus aspect already intimated by their relation to the Hero Twins, as the early Kaqchikel dictionary by Vico (1555: 211v–212r) translates *uu4h* not only as "opossum" but also as "morning star."

Since perhaps 1773 the Mam presiding over the dreaded final Five Days has accepted the adoration of the multitudes in his form as the masked wooden statue called "Maximon" or "Mam Simon" (Tedlock 1993: 221–22). This recalls the Yukatek Mam or "Grandfather," lord of the last five days, revered by the crowds as a dressed up old log (Cogolludo 1971: 1:25). The log and the wooden statue suggest that the eighteenth K'iche' name Chee' translates as "log, statue of wood" in reference to preparation of the Mam effigy for use in the next period.

Nabey Pach: *Nab'ey Pach* First Incubation; First Moss (Gulf Coast Nahua). For Nahuatl definitions see Kaqchikel **Pach**.

pach bow, lean, cover, for hens to sit on eggs (Ximénez 1985: 438)

Nabey Zih: *Nab'ey Si'j* First White Flowers; *Nab'ey Sii'-ij* First Firewood Cutting; *Nab'ey Tziij* First Word; First Kindling or Setting-on-Fire

si'j flower or blossom of any plant that changes to fruit and is not used for adornment (Ajpacaja Tum et al. 1996: 365); certain white flowers (Ximénez 1985: 644)

sii'-ij <*ziih*> to cut firewood (Ximénez 1985: 644); *sii'* firewood (Mondloch 1978: 187)

tziij set afire, light; fire; word (Edmonson 1965: 134); word, truth (Mondloch 1978: 188); word; speak (eloquently); light like a candle, illuminate (Ximénez 1985: 566)

Obota: *O('Ob') B'ot-Ab'* Five Rolls of the Year [Calendars]; *O('ob')-B'ot-aj* Five Rolling Ups; *O('ob') B'ot-Am* Five Cotton-Spiders [Spindles]

See Kaqchikel **Bota**, **Botam**, and **Obota** and Kiché' **Botam**.

Pubaix: *Pub'ah-ix* Shoot with a Blowgun

pub blowgun, Kaqchikel (Ximénez 1985: 467); blowgun or arquebus (*escopeta*) (Vico 1555: 161v)

puba to strike as with a rag (Ximénez 1985: 467)

pubagh to shoot with it (blowgun, arquebus) (Vico 1555: 161v)

pubah to shoot with a blowgun or arquebus (Basseta 2005: 472; Coto 1983: 43, *arcabuz*); to shoot with the blowgun; to spew out mouthfuls like water (Ximénez 1985: 467)

pubaix sudden death; because they imagined that [for] Hun Ahpu, with just a puff on the blowgun the birds dropped dead (Ximénez 1985: 467; Basseta 2005: 473 n. 57); unfortunate day (Coto 1983: 59, *aziago*; Vico 1555: 161v)

R Ox Zih: *R-Ox Si'j* Third White Flowers; *R-Ox Sii'-i-j* Third Firewood Cutting; *R-Ox Tziij* Third Word; Third Kindling or Setting-on-Fire. See **Nabey Zih**.

The attested variants Urox Çih and U Rox Tzij each prefix both variants of the third person singular ergative pronoun, *u-* and *r-*, to form the ordinal, instead of just the *r-*. One possible explanation analyzes *r* not as a pronoun but as an epenthetic, euphonic consonant attached to *u-* before vowels that in time replaced the morpheme *u-*. Another account suggests that the ordinal is formed by adding *u* to *r-ox* just because *u* joins *kab'* "two" for *u kab'* "second"; linguists call this assimilation between adjacent items on a list "contamination."

Sac Imuj: *Saq-il Mu(')j* Canopy or Throne of Splendor

muj defense; shadow, shade; throne (Edmonson 1965: 74)

mu(')j shade or canopy (Álvarez n.d.: 121r); shade, throne (Anonymous 1787: 179v, 187r); Kaqchikel: a specter and what appears in the mirror, and in another time they considered it the soul; the canopy that they carry when processions are underway

(Smailus 1989b: 2:552); principal seat where only the lords, the bishop, the governor, etc., sit; backed chair or throne, royal seat; arbor (Coto 1983: 50; 210, 462, 529). For Sac see **Sak**.

Sak: *Saq* White; Lightning; *White Flowers (from Yukatek, via Poqom)

saq white, bright, light, dawn, splendor, lightning, crystal-clear water (Edmonson 1965: 158)

Tecoxepual: *Tlaca-xipehua-liz-tli* Flaying of Men. See the Kaqchikel **Tacaxepegual**.

Tz'api Q'ih: *Tz'api Q'iij* Closed Door Days; *Tz'ap-i Q'iij* Misfortune Days; Beehive Days; *Tz'ap-ij Q'iij* Shelter Days

tz'ap-ij to enclose or imprison; to cover or stopple (Álvarez n.d.: 251v; Silva Leal 1875: 251v)

tz'api closed door (Ximénez 1985: 558)

tz'ap beehive; fault, error, disgrace or misfortune (Ximénez 1985: 558)

Tz'iba Pop: *Tz'ib'aa Poop* Mat for Writing or Painting; *Tz'ib'-am Poop* Painted Mat; *Tz'ib'-aa-m Poop* Painting Mat

tz'ib'-am poop <4,ibam pop> Kaqchikel: the painted reed mat (Coto 1983: 219); *estera labrada* worked or painted mat (Smailus 1989b: 2:129)

tz'ib'-an poop <4,ibanpop> *estera labrada*/worked or painted mat (Anonymous 1787: 100v)

tz'ib'-am painted, written (Edmonson 1965: 130)

poop mat (Campbell 1977: 60:219); straw mat (Mondloch 1978: 186); Kaqchikel: men who have held offices and are respected for them (Coto 1983: 248); noble, of good lineage (Smailus 1989b: 2:643)

Kaqchikel and K'iche' dictionaries record both *tz'ib'-am* and *tz'ib'-an*, but the calendars offer only *tz'ib'-a*. The *-a* functions as a ligature forming adjective-noun compounds (Judith Maxwell, personal communication, 2002). Only two calendars (1698 and 1935: see chapter 2) present *tz'ib'-e* or *tz'ib'-é* instead of *tz'ib'-a*. Some evidence could ascribe this *a-e* shift to dialectal variation (Richards 1985).

Tz'iquin Q'ih: *Tz'ikin Q'iij* Bird Days; Penis Bloodletting Days. For definitions and comments see Kaqchikel **4,iquin Σih** above.

tz'ikin q'iij <tzikin kih>: *el águila* eagle or hawk (Silva Leal 1875: 74v)

Tz'izil Lacam: *Tz'is-il Laqam* Sprout Flag; Sewing or Bloodletter Banner

The most direct interpretation "Sprout Flag" finds support first in the Tzotzil practice of sticking little red tissue paper flags in the fields of green maize ears in

August to frighten birds (Guiteras-Holmes 1961: 45) and second in the Ch'orti' parallel of erecting in the ground all over the milpas sticks with thin zigzag strips of plantain bark to keep birds from lighting during the sprouting stage in mid-May (Wisdom 1940: 47). This suggests that the very next month, Tz'ikin Q'iij, alludes to these dreaded pests and not to seasonal migrants such as Swainson's hawks.

The gloss "Bloodletter Banner or Cloth" evokes the blood-flecked or soaked strips of cloth worn by principals on polychrome pots and monuments. Sewing and the breechcloth serve as a metaphor about bloodletting and the penis (Tedlock 1993: 238–39).

tz'isil <tzizl> the little tips that begin to appear (Ximénez 1985: 570); sprouts, shoots just coming up (Edmonson 1965: 135) Kaqchikel: *<4,izl>* tip of the stem that comes out of the earth; *4,is* is to sew with a needle, and as the point of the needle pokes out [like the tip of the stem], from there they take the metaphor (Coto 1983: 449)

tz'is <tziz> thread, pierce; to string beads (Ximénez 1985: 569); root of *tz'is-oj <tzizo>* to sew seams, to thread or string; and anciently to sacrifice to the idols, to sacrifice one's blood (Basseta 2005: 226v); Kaqchikel: sew with a needle (Coto 1983: 116); to pierce, thread, or string beads, flowers (Coto 1983: 196)

tz'itz <tzitz> to nail (Edmonson 1965: 135); utensil for bloodletting (León 1954: 61; Alvarado López 1975: 70). Glosses under *tz'is* show that this variant strengthens the final *s* to *tz*.

laqam/ laqan flag, pennant *<lacam>* (Ximénez 1985: 338)

U Cab Mamkuk: *U Kab' Maam Q'uuq'* Second Grandfather Quetzal; Second Dance Feathers. See **Nabé Mamkuk**.

U Kab Likin Ka: *U Ka' Lik'-in Kaaj* Second Stretched Out Sky; *U Kab' Lik'-in Kaj* Second Scattered Powder. For definitions see **Nabey Lik'in Ka**.

U Kab Mam: *U Kab' Maam* Second Grandfather; Second Grandchild. See **Nabey Mam**.

U Kab Pach: *U Kab' Pach* Second Incubation; Second Moss (Gulf Coast Nahua). See **Nabey Pach** above.

U Kab Zih: *U Kab' Si'j* Second White Flowers; *U Kab' Sii'-i-j* Second Firewood Cutting; *U Kab' Tziij* Second Word; Second Kindling or Setting-on-Fire. See **Nabey Zih** above.

SUMMARY: The K'iche' set draws attention for its diversity (table 3.5). Three names are direct loans from outside the Maya sphere: Tequexepual and the Pach pair. Two different months start the year: Nabe Çih, in the 1698 calendar only, and Nabe Mam,

TABLE 3.5
Summary of K'iche' month names

	Names in source	Interpretations	Translations
1	Tecoxepual	Tlaca-xipehua-liz-tli	Flaying of Men
1	Cox Xe Puatl	Kos-xiip-eehua-tl	Yellow Flayed Skin
2	Tz'iba Pop	Tz'ib'aa Poop	Mat for Writing or Painting
		Tz'ib'-am Poop	Painted Mat
		Tz'ib'-aa-m Poop	Painting Mat
3	Sak	Saq	White; Lightning
3	Sac Imuj	Saq-il Mu(')j	Throne of Splendor
4	Ch'ab	Ch'ab'	Arrow
4	Ajau Chap	Ajaaw Ch'ab'	Arrow Lord, Archer
5	Huno Bix Gih	Juun (J)O('Oob) B'iix Q'iij	Five Days from Now
			Five Songs Days
5	Nabey Mam	Nab'ey Maam	First Grandfather
			First Grandchild
5	Nabé Mamkuk	Nab'ey Maam Q'uuq'	First Grandfather Quetzal
			First Dance-Feathers
6	U Kab Mam	U Kab' Maam	Second Grandfather
			Second Grandchild
6	U Cab Mamkuk	U Kab' Maam Q'uuq'	Second Grandfather Quetzal
			Second Dance-Feathers
7	Nabey Lik'in Ka	Nab'ey Lik'-in Kaaj	First Stretched-Out Sky
		Nab'ey Lik'-in Kaj	First Scattered Powder
8	U Kab Likin Ka	U Ka' Lik'-in Kaaj	Second Stretched-Out Sky
		U Kab' Lik'-in Kaj	Second Scattered Powder
9	Nabey Pach	Nab'ey Pach	First Incubation
			First Moss
10	U Kab Pach	U Kab' Pach	Second Incubation
			Second Moss
10	Iquim Kij	Ikim Q'iij	Low Sun
11	Tz'izil Lacam	Tz'is-il Laqam	Sprout Flags
			Sewing or Bloodletter Banner
11	Chi Il Lacan	Chi il Laqan	In-the-Zenith Banners
		Ch'il Laqan	Flaying Banners
12	Tz'iquin Q'ih	Tz'ikin Q'iij	Bird Days
			Penis Bloodletting Days
12	Caam Kij	K'a'aam Q'iij	Cord Days
			Plot of Land Days
		Q'am' Q'iij	Ladder or Bridge Days
			Pole-Ladder Sun
13	Cakam	Kaq-am	Red Spider or Bubo
		K'aq-am	Wounded; Installed
		K'aq Q'a'm	Pierce Ladder
			Install Ladder

(continues)

TABLE 3.5
Summary of K'iche' month names (*continued*)

	Names in source	Interpretations	Translations
13	Cakan	Kaq-Q'an	Red-Yellow
		Kaq-an	Reddened, Reddening
		Kaaj Q'an	Sky Is Yellow
		K'aq-an	Wounding
14	Ibota	Il B'ot-ab'	Large Roll of the Year
		Il B'ot Aaj	Big Bundle of Reeds
14	Obota	O('Ob') B'ot-ab'	Five Rolls of the Year
		O('Ob')-B'ot-aj	Five Rolling Ups
		O('Ob') B'ot-am	Five Cotton-Spiders
14	Botam	B'ot-am/B'ot-om	Rolled Up
			Hilled
		B'ot Am	Combed-Cotton-Spider
14	Batam	B'at-aa-m	Hardening
		B'at-am	Rolled Up
15	Nabey Zih	Nab'ey Si'j	First White Flowers
		Nab'ey Sii'-ij	First Firewood Cutting
		Nab'ey Tziij	First Word
			First Kindling
			First Setting-on-Fire
16	U Kab Zih	U Kab' Si'j	Second White Flowers
		U Kab' Sii'-i-j	Second Firewood Cutting
		U Kab' Tziij	Second Word
			Second Setting-on-Fire
17	R Ox Zih	R-Ox Si'j	Third White Flowers
		R-Ox Sii'-i-j	Third Firewood Cutting
		R-Ox Tziij	Third Word
			Third Kindling
			Third Setting-on-Fire
18	Cak Che	Kaq Chee'	Red Tree
			Red Prison
		Kaq' Chee'	Guava Tree
			Dance Log
		K'aq Chee'	Piercing Tree
			Installation Tree
18	Chee	Chee	Tree, Log
			Wooden Statue
			Granary
19	Tz'api Q'ih	Tz'api Q'iij	Closed Door Days
		Tz'ap-i Q'iij	Misfortune Days
			Beehive Days
		Tz'ap-ij Q'iij	Shelter Days
19	Jun Bix Kij	Juun B'iix Q'iij	One Song Day; Trill Song Day

in the rest. The Five Days goes by two distinct terms, Hunobixquih in the 1698 list, and 4,api Σih in the others. The K'iche' group includes more series than any other, four altogether. Three (Liquin Ca, Mam, and Pach) order their members as "First-Second" and the fourth (Tzih or Sih) arranges the names as "First-Second-Third."

Poqom Definitions

Poqom consists of Poqomam and Poqomchi'. Scholars disagree on whether these represent separate languages (Kaufman 1978: 103, 111) or just dialects (Campbell 1977: 32–33).

Frequent word: *q'iij* sun, day (Campbell 1977: 58:179)

Poqomam Definitions

Mol: *Mohl* Family, Group
mohl group, family (Zinn and Zinn n.d.: 42)

Muan: *Muan* Butcher Bird (Ch'ol). See Ch'ol **Muhan**.

*Utzumckigh: *Uhtz'um' Q'iij* Flower Days, perhaps an obsolete month name
uhtz'um' flower (Fernández 1892: 27); Poqomchí: *uhtz'ub'* flower (Mayers 1956: 58)

Poqomchi' Definitions

Canazi: *Q'an Aj Siij* Yellow Gift-Offerer; *Q'an Aj Si'* Yellow Woodchopper
q'anal attributive variant of *q'an* yellow, mature (Stoll 1888: 172)
aj agent (Fernández 1937: 2, various entries)
si' firewood (Stoll 1888: 187)
siij gift (Mayers 1956: 51, 98); <ah zi> denotes a priest, who presents gifts to God (Miles 1957: 750)

Chab: *Ch'ab'* Bow and Arrow
ch'ab' bow, arrow (Sapper 1897: 412)

Cham: *Chaam* Weeds; *Ch'am'* Arrow
chaam Poqomam: weeds (Zinn and Zinn n.d.: 14)
ch'am' Poqomam: bow and arrow (Zinn and Zinn n.d.: 18)

Chantemak: *Chante' Mak* Granary on Tall Posts (Ch'ol). See Ch'ol **Chantemac**.

Kanajal: *Q'an-a Jal* Yellow, Ripe Corn Ear
q'an jal large variety of corn (Fernández 1937: 64, Termer 1930: 394)

q'an(al) yellow, ripe (Stoll 1888: 172); *q'an-a* attributive form

jal corn ear (Stoll 1888 :167)

Kanjalam: *Q'an Jalam'* Yellow Loom (Ch'olan). See Ch'olti' **Canhalib**.

Kasew: *Kasew* Palm Tree (Mopan, Q'eqchi'). See Ch'ol **Cazeu**.

Kaxik-Laj-Kij: *K'axik-laj Q'iij* Very Painful Days; *Kajxik-laj Q'iij* All-Finished Days

k'axik-laj-q'iij Holy Week (difficult days) (Fernández 1937: 65)

k'axik difficult, hard (Mayers 1958: 139)

kajxik past, finished (Fernández 1937: 64)

-laj intensifier (Judith Maxwell, personal communication, 2002)

Kcham: *Ch'am'* Arrows. See **Cham**.

Kchip: *Ch'ip* Small or Last Child (a thunder-lightning god)

k- The *k* indicates glottalization in this source's orthography.

ch'ip small, young; last child (Stoll 1888: 163); *ch'ib'* little one (Mayers 1958: 139)

In Poqomchi' tales the twelve large thunder and lightning gods failed to split open the cliff that hid corn, but the smallest one (*ch'ip*) succeeded (Mayers 1958: 9–10, 13–14).

Makux: *Ma Q'uux* Lord Moss; *Ma Qux* Lord Crow; *Ma K'uux* Lord Palo de Amate; *Ma Kux* Lord End

ma lord (Mayers 1956: 37, 99)

q'uux Poqomchi': moss, mold (Mayers 1956: 36, 89)

qux crow (Stoll 1888: 171)

k'uux Spanish: *palo de amate* [no English equivalent found] Poqomchi' (Mayers 1956: 20)

kux end (Fernández 1937: 57)

Mox-Kij: *Mox Q'iij* Dung Beetle Days

mox dung beetle (Fernández 1937: 69)

Muan: *Muhan* Hawk (Ch'olan). See Ch'ol **Muhan**.

Oj: *Ooj* Avocado

ooj avocado (Morán 1725: 440r); likely slip for **Ojl**.

Ojl: *Ojl* Heart (Ch'olan); Bud (Yukatekan). See Ch'ol **Olh** above.

Pet Cat: *Pet Q'at* First Chop; *Pet K'aht* First Burn

pet first (Mayers 1956: 44, 96)

q'at chop (Mayers 1958: 142)

k'ahtel a burn (Mayers 1956: 18, 97)

Sac: *Saq* White; *White Flowers (Yukatek)

saq white (Stoll 1888: 186)

Sac-Gojk: *Saq Q'ojq'* White Chilacayote Gourd

saq white (Stoll 1888: 186)

q'ojq' chilacayote/white squash (Campbell 1977: 51, item 73) green, smooth gourd-like squash; malabar gourd (Schoenhals 1988: 31, 41, 141)

sakil k'ojk' pepita, pepitera (Mayers 1956: 51)

Tik-Che-ik: *Tik Chee-ik* Planting with a Stick

tik-oj to plant; a planting

chee' tree, [stick] (Mayers 1958: 139)

-ik verbal nouns (Owen 1968: 39, 227)

Tsi: *Tz'i'* Dog
tz'i' dog (Stoll 1888: 194)

If <tokgüik> or *toq'-ik* is correctly glossed "piercing object, piercing" as well as "dog" (Teletor 1959: 167; Fernández 1937: 193), it might derive from the Kaqchikel ToΣic months, as Kaqchikel *toq'-ik* means "bloodletting." *Toq'-ik* led to *tz'i'* "dog."

Tzikin Kij: *Tz'iqin Q'iij* Bird Days
tz'iqin bird (Morán 1720: 514r). See Ch'ol **Chichin**.

Uniw: *Uniw* Avocado (Ch'ol). See Ch'ol **Uniu**.

Yax: *Yax* Green (Ch'ol). K'ichean "green" is *rax*.

SUMMARY: Of the twenty-one Poqomam and Poqomchi' names (table 3.6), eleven or twelve are apparently borrowed from Ch'olan. So much adoption indicates a considerable Ch'olan prestige or presence. Present data are insufficient to identify the mechanism as trade, warfare, politics, or religion. Of the very few Kaqchikel loans to other traditions, two or three enter Poqomam and Poqomchi'. Chronicles record that as allies and kin of the expansionist K'iche's the Kaqchikels first established

TABLE 3.6

Summary of Poqomchi' month names

	Names in source	Interpretations	Translations
1	Kanjalam	Q'an Jalam'	Yellow Loom
2	Makux	Ma Q'uux	Lord Moss
		Ma Qux	Lord Crow
		Ma K'uux	Lord *Palo de Amate*
		Ma Kux	Lord End
2	Pet Cat	Pet Q'at	First Chop
		Pet K'aht	First Burn
3	Kasew	Kasew	Palm Tree
4	Canazi	Q'an Aj Siij	Yellow Gift-Offerer
		Q'an Aj Si'	Yellow Woodchopper
5	Kanajal	Q'an-a Jal	Yellow, Ripe Corn Ear
6	Tzikin Kij	Tz'iqin Q'iij	Bird Days
7	Mox-kij	Mox Q'iij	Dung Beetle Days
8	Tik-che-ik	Tik Chee-ik	Planting with a Stick
9	Kaxik-laj-kij	K'axik-laj Q'iij	Very Painful Days
		Kajxik-laj Q'iij	All-Finished Days
10	Yax	Yax	Green
11	Sac	Saq	White; White Flowers
12	Tsi	Tz'i'	Dog
13	Kchip	Ch'ip	Last Child
14	Chantemak	Chante' Mak	Granary on Tall Posts
15	Uniw	Uniw	Avocado
16	Muan	Muhan	Hawk
17	Chab	Ch'ab'	Bow and Arrow
17	Cham	Chaam	Weeds
		Ch'am'	Arrow
17	Kcham	Ch'am'	Arrow
18	Sac-gojk	Saq Q'ojq'	White Chilacayote Gourd
19	Ojl	Oojl	Ball

contact with Poqom upon expelling the Poqomams from the Rabinal Valley in about 1350 (Fox 1978: 255). The next major clash took place in 1470 or 1480, when the Western Kaqchikels conquered five major Poqomam towns, following with several more a few years later (Fox 1978: 229).

Q'ANJOB'AL DEFINITIONS

In the prevalent classification, Q'anjob'al joins Chuj, Tojolab'al, and several other related languages to form Greater Q'anjob'alan (Kaufman 1974b: 959). Most relevant here is that Q'anjob'al and Chuj are not only related but close geographically. Most Chuj names come from Q'anjob'al. Frequently used words are listed below.

kaq red; a flower, in Solomá (Diego Antonio et al. 1996: 138–39)

q'an yellow, ripe (Diego Antonio et al. 1996: 242)

q'eq' black (Diego Antonio et al. 1996: 247)

saq white; clean (Diego Antonio et al. 1996: 255)

sax saj, preconsonantal *saq*

ul inside, interior (Diego Antonio et al. 1996: 327)

yax blue, green (McQuown 1976b: 58)

c one source's grapheme for the modern *x*

Q'anjob'al Definitions

Bak: *P'aq* Bones; Corncob Worm

p'aq b'aq bone; corncob worm (Diego Antonio et al. 1996: 24, 26)

Bak': *P'aq'* Seeds

p'aq' seed (McQuown 1976b: 41)

Baktan: *P'aq Tan* Bone Ash. This perhaps represents a fusion of two consecutive names.

tan lime, ash (McQuown 1976b: 33, 108)

Cubihl: *Xub'al* Whistle

xub'al whistle (Diego Antonio et al. 1996: 360)

Kanal: *Kanal* Dance; *Q'an Nal* Yellow Ear of Corn

kanal dance (Diego Antonio et al. 1996: 135)

q'annal [*sic*] yellow ear of corn (Diego Antonio et al. 1996: 243)

nal unshelled corncob (McQuown 1976b: 34)

Kanul: *Q'an Ul* Yellow Interior

K'an Sihom: *Q'an Sihom* Yellow Soapberry See **Sihom** below.

K'aq Cuhem: *Kaq Xuj-Im* Red Threader [Bloodletter]. See **Xuhem** below.

K'aq Sihom: *Kaq Sihom* Red Soapberry. See **Sihom** below.

K'eq Sihom: *Q'eq' Sihom* Black Soapberry. See **Sihom** below.

Mak: *Mak* Cover, Enclosure

mak cover, from *maq-an* enclosed, stoppled (Diego Antonio et al. 1996: 190)

Mo: *Mo'* Hawk-like Owl, Sparrow Hawk (Ixil; K'iche'). See Chuj **Mo** above.

Mol: *Mol* Group

Nabitc: *Nab'ich* Winter, Rainy Season

nab'ich winter [rainy season] (Diego Antonio et al. 1996: 205)

nab' rain (Diego Antonio et al. 1996: 205)

Oneu: *Onew* Avocado (Western Mayan). See Ch'ol **Uniu**.

Oyeb K'u: *Oyeb' K'u* Five Days

oyeb' five (Diego Antonio et al. 1996: 221)

o five, from *ob'ixi* five days ago (Diego Antonio et al. 1996: 216)

-y- intervocalic euphonic

-eb' generic numeral classifier for inanimates (Diego Antonio et al. 1996: 35)

k'u day, sun (McQuown 1976b: 35, 37)

Saqmay: *Saq May* White Danger

may danger (Diego Antonio et al. 1996: 196)

Sihom: *Sihom* Soapberry (Yukatek)

sihom Yukatek: certain tree which bears a little fruit that serves as soap (Ciudad Real 1984: 103r); soapberry (Roys 1931: 309); *sijom* Mopan: *jaboncillo, amole, jabón de monte*/jaboncillo, wild soap, wild soap tree (Hofling 2011: 385)

Sivil: *Siwil* Woodchopper (K'iche'). See Chuj **Sivil** above.

Tap: *Tap* Crab (K'iche'). See Chuj **Tap** above.

tap a month of the agricultural calendar or corresponding to February, it contains twenty days (Diego Antonio et al. 1996: 265–66)

Etap: *E Tap* It Is Crab (K'iché)

e 3sA he, she, it [third person singular; absolutive] (Judith Maxwell, personal communication, 2000)

Wa Tsikin: *Watz' Tz'ikin* Squeak Birds: Bats or Hummingbirds; *Watx' Tz'ikin* Good Bird. See Chuj **Huatziquin** above.

Thompson's (1971: 106, table 8) Huachsicin is an error, as the original (LaFarge 1947: 168) has <*watsikin*>. Gwatsikin might have come from Q'eqchi, as most words that are *w-* initial words in other forms of Mayan "strengthen" phonetically there to *kw-* (Haeserijn 1979: 94, 478).

watz' squeak, screech? from *watz'xi* screech, squeak (Diego Antonio et al. 1996: 335). By the rule of geminate deletion (Kaufman 1971: 21; Slocum 1948: 78 n. 8), the first *tz'* would reduce to 0, for *wa-tz'ikin*.

watx' good (Diego Antonio 1996: 334). *Tx'* is the retroflex form of *ch'* (Academia de las Lenguas Mayas de Guatemala 1988: 4, 9).

tz'ikin bird (Diego Antonio et al. 1996: 319)

Wex: *Wex* Loincloth

wex loincloth, pants (Diego Antonio et al. 1996: 339)

Xuhem: *Xuj-em* Threaded, Threader

xuj-u' to pierce, thread, sew; *xuj-an* pierced, riddled (Diego Antonio et al. 1996: 361)

-Vm resultative, agentive (Judith Maxwell, personal communication, 2000)

Yacul: *Yaxul* Blue Jay

yaxul blue jay (Diego Antonio et al. 1996: 375); *xul* bird; St. John's mushroom, a small, yellow variety (Diego Antonio et al. 1996: 36). The term might consist of *yax + xul* (blue + bird).

Yaxakil: *Yax Ak'-il* Green Grass Times

ak' grass, from *ak'al* plain, grassland (McQuown 1976b: 40); *ak'al* plain, meadow, valley (Diego Antonio et al. 1996: 6). *Ak'il* would be a dissimilated form of *ak'al* or the compound of *ak'* plus *-il*, which indicates "place of, time of."

Yax Sihom: *Yax Sihom* Green Soapberry. See **Sihom** above.

Zaq Sihom: *Saq Sihom* White Soapberry

Summary: Numerous names in this tradition are cognates to glyphic counterparts; only here do the color or Sihoom months survive. Nearly all of the remaining terms present synonyms for names in the glyphic calendar. Kanul might well be the sole member of purely Q'anjob'al origin. The extremely derivative nature of this tradition argues for intense, long contact with the bearers of Classic culture. As the beginnings and the order of the months vary so much, the table 3.7 lists the names alphabetically and assigns no dates.

Table 3.7

Summary of Q'anjob'al month names

Names in source	Interpretations	Translations
Bak	P'aq	Bones; Corncob Worm
Bak'	P'aq'	Seeds
Baktan	P'aq Tan	Bone Ash
Cubihl	Xub'al	Whistle
Kanal	Kanal	Dance
	Q'an Nal	Yellow Ear of Corn
Kanul	Q'an Ul	Yellow Interior
K'an Sihom	Q'an Sihom	Yellow Soapberry
K'aq Cuhem	Kaq Xuj-im	Red Threader
K'aq Sihom	Kaq Sihom	Red Soapberry
K'eq Sihom	Q'eq' Sihom	Black Soapberry
Mak	Mak	Cover, Enclosure
Mo	Mo'	Hawk-Owl, Sparrow Hawk
Mol	Mol	Group
Nabitc	Nab'ich	Winter, Rainy Season
Oneu	Onew	Avocado
Oyeb K'u	Oyeb' K'u	Five Days
Saqmay	Saq May	White Danger
Sihom	Sihom	Soapberry
Sivil	Siwil	Woodchopper
Tap	Tap	Crab
Etap	E Tap	It Is Crab
Wa Tsikin	Watz' Tz'ikin	Squeak Birds
	Watx' Tz'ikin	Good Bird
Wex	Wex	Loincloth
Xuhem	Xuj-em	Threaded, Threader
Yacul	Yaxul	Blue Jay
Yaxakil	Yax Ak'-il	Green Grass Times
Yax Sihom	Yax Sihom	Green Soapberry
Zaq Sihom	Saq Sihom	White Soapberry

Q'eqchi' Definitions

Q'eqchi' is one of the many K'ichean languages, more closely related to Poqomam and Poqomchi' than to K'iche' or Kaqchikel.

Auuck (Aauc): *Aawk* To Sow or Transplant

auc/aauc to sow, transplant; *auic* act or time of sowing (Haeserijn 1979: 47)

Chantemac: *Chan Te' Mahk* Granary on Tall Posts (Ch'ol). See Ch'ol **Chantemac** above.

Cheen: *Ch'en* Mosquito
ch'en mosquito (Freeze 1975: 36)

Gkalec (C'alec): *K'al-K* To Clear or Weed Fields
c'alec to clear land (Haeserijn 1979: 102); *k'alek* clear fields, do weeding (Pinkerton 1976: 123)

Gkan (K'an): *Q'an* Yellow, Ripe
k'an yellow, ripe (Haeserijn 1979: 195)

Gkatoc (C'atoc): *K'at-ok* To Set on Fire
c'atoc to set on fire (Haeserijn 1979: 116)

Gkoloc (K'oloc): *Q'ol-ok* To Harvest Corn
k'oloc to harvest corn (Haeserijn 1979: 203)

Mayejick/ Mayejic': *Mayej Ik'* Offerings Time
mayej offering, sacrifice (Haeserijn 1979: 222)
ik' moon (Haeserijn 1979: 167) [I assume derived senses of "month, time."]

Moloc: *Mol-ok* To Collect
moloc to collect (Haeserijn 1979: 226)

Muan: *Muan* Hawk (Ch'ol). See Ch'ol **Muhan** above.

Oxlajuel: *Oxlaju-al* Group of Thirteen
oxlaju thirteen (Haeserijn 1979: 247); *-al* group of two to ten, of indefinite composition (Stewart 1980: 114)

Pop: *Pojp* Reed Mat
poop mat (Haeserijn 1979: 263); *pohp* (Campbell 1977: 60, item 219)

Rail Cutan: *Rail Kutan* Painful Days
rail painful (Pinkerton 1976: 129)
kutan, kutank clear; day; time (Haeserijn 1979: 106–7)

Rakol: *Raqol* Merchant; *Raqol* Marking Off (of the Cornfield)
rakol merchant, dealer, trader (Stoll 1896: 169)

rakoc to determine the limit (of a surface, time or action), to terminate, to trace out (Haeserijn 1979: 279)

-ol suffix on verbal or nominal roots for words that realize the thing or quality expressed by the root (Haeserijn 1979: 245)

Raxgkim (Raxk'im): *Rax K'im* Green Thatch-Grass

rax green, wet, fresh, unripe, clear (Haeserijn 1979: 282)

gkim/q'uim straw, leaves for the roof (Haeserijn 1979: 275)

Tap: *Tap* Crab

tap crab; name of the last month (Haeserijn 1979: 317). Haeserijn gives no source for this name, found otherwise only in the Chuj and Q'anjob'alan lists. In the former it is the last winal before the five days called Oja Kwal, Q'eqchi for "Five Sick, Malevolent Ones" (Haeserijn 1979: 317). In the latter it is the last winal before Oyeb K'u Five Days.

Tzaack (Tzaak): *Tzak* Game from the Hunt

tzac game (from hunt) (Haeserijn 1979: 342)

*Xul: *Xul* Animal; Bird?; hypothetical as a month name

xul unbaptized [wild] animal, booty from the hunt; penis (Haeserijn 1979: 374); animal, wild game; prone to bite, wild (Wirsing n.d.: 388r)

*Xulul: *Xulul* Partridge, Quail; Animal; Ridge Pole

xul-ul animal; certain bird (Haeserijn 1979: 374); *perdiz*/partridge of the area; piece of wood that goes in the ridge [of the house] (Curley García 1967: 187)

Yax: *Yax* Crab Pincers

yax pincers of the crab (Curley García 1967: 193); pincers (Haeserijn 1979: 380)

This etymon would make the <yax> of the Lanquín calendar a Q'eqchi' term rather than the Ch'olan word for "green, blue" and so explain why the compiler did not use the Q'eqchi' *rax* "green, blue" instead. As such the name would preserve another rare instance of semantic transfer through homophony. In other words the Q'eqchi' heard the Ch'olan *yäx* or *yax* but equated it to their own *yax* (homophony), so changing the meaning from "green" to "pincers." This Yax recalls the Q'anjob'al and Chuj units called Tap "Crab."

Yulic: *Yul-ik* Anointment Time

yul-uc to anoint, to smear with salve (Haeserijn 1979: 386); *-ik* moment of action (Haeserijn 1979: 167)

TABLE 3.8

Summary of Q'eqchi' month names

	Names in source	Interpretations	Translations
1	Pop	Pojp	Reed Mat
2	Rakol	Raqol	Merchant
			Cornfield Marking-Off
3	Gkalec (C'alec)	K'al-ek	Clear or Weed Fields
4	Gkatoc (C'atoc)	K'at-ok	Set on Fire
5	Auuck (Aauc)	Aawk	Sow, Transplant
	Chantemac	Chan Te' Mahk	Granary on Tall Posts
6	Cheen	Ch'en	Mosquito
7	Mayejick/Mayejic'	Mayej Ik'	Offerings Time
7	Cutan	Rail Kutan	Painful Days
8	Tzaack (Tzaak)	Tzak	Game from Hunt
9	Raxgkim (Raxk'im)	Rax K'im	Green Thatch-Grass
10	Gkan (K'an)	Q'an	Yellow, Ripe
11	Gkoloc (K'oloc)	Q'ol-ok	Harvest Corn
12	Moloc	Mol-ok	To Collect
13	Yulic	Yul-ik	Anointment Time
14	Oxlajuel	Oxlaju-al	Group of Thirteen
?	Muan	Muan	Hawk
?	Tap	Tap	Crab
?	*Xul	Xul	Animal; Bird
			Animal
			Ridge Pole
?	*Xulul	Xulul	Partridge, Quail
?	Yax	Yax	Crab Pincers

SUMMARY: Most of these names constitute the 1979 calendar, a result of major revisions effected sometime after 1788 (table 3.8). These modifications dropped, altered, or replaced nearly all the names cognate to Ch'olan or glyphic counterparts, reduced the tally of months from eighteen to thirteen, increased their length to twenty-eight days, and exchanged the five last days for one day. A handful of Q'eqchi' names survive in the Ch'olan calendar from Lanquín, specifically Chichin, Ianguca, and Mahi if not Yax and Ah Qui Cou as well. Almost every name on the 1979 roster relates to maize agriculture.

TZELTAL DEFINITIONS

Tzeltal and Tzotzil together form Tzeltalan (Kaufman 1974b: 959). Tojolab'al might belong to this complex instead of with Chuj (Robertson 1977). The more frequently repeated parts of names are defined here.

-ab instrument; noun stems from transitive, positional, and derived transitive roots (Kaufman 1971: 75, item 71)

-al nominalizer: noun stems from noun roots; absolutive or unpossessed on noun stems (Kaufman 1971: 77, item 77, 111, item 13)

alal weight; grief (Ara 1986: 244); child, baby (Ara 1986: 498; Robles Urribe 1966: 18)

-eb general noun classifier (Kaufman 1971: 79–80, item 85)

-el verbal nouns from most verbs (Kaufman 1971: 71, item 60)

-em variant of *-eb*, a version of *-Vb*, instrument

ho' compound form of *ho'e'b* five (Kaufman 1971: 91–92)

hul <*ghul*> to drill, pierce, puncture; a puncture (Ara 1986: 297); to leech blood (Sarles 1961: 72, item 1870)

-il attributive, adjectivizer; place of abundance; class marker in absolute noun stems (Kaufman 1971: 81, item 89, 85–86, item 101, 111, item 13)

-in transitive stems from radical or derived noun stems (Kaufman 1971: 50, item 11); intransitive roots from noun roots and stems, from transitive and positional roots (Kaufman 1971: 56, item 27)

k'in fiesta; divination, fortune (Ara 1986: 373)

ok' cry, howl; sing, warble (Laughlin 1988: 152); call, weep, sing; beat a drum (Laughlin 1975: 67)

-ol noun stems from transitive verb roots (Kaufman 1971: 75, item 73)

ti' mouth, bite; entrance. Caves and springs represent portals to the netherworld.

uch opossum (Hunn 1977: 204); *h'uch* opossum (Berlin and Kaufman 1962: 99); *chahuistle*/maize smut, rust, or bunt (Slocum and Gerdel 1965: 38, 197)

winik person, people; twenty (Ara 1986: 264, 409; Kaufman 1971: 93). In the common dialectal variant *binik* the *b* replaces *w* (Kaufman 1972: 35–36)

Tzeltal Definitions

Agelchac: *Ahil Ch'ak* Marsh Flea (Tzotzil); Prickly Flea (Tzotzil) Variant: *Ch'ak*

ahil <*aghil*> Tzotzil reed; prickly; "fledged with very small or young feathers" (under *emplumecido*) (Laughlin 1988: 130)

ch'ak flea (Ara 1986: 262; Hunn 1977: 306)

Alajuch: *Alal Uch* Small Opossum

h- class marker on noun stems, when absolute (Kaufman 1971: 110, item 12). This name derives from *alal uch* via *alah uch*, with the final *l* weakening to *h* (Kaufman 1971: 21).

The Tzeltal recognize two kinds of opossum: the large dark one called *muk'ul uch* "large opossum" and the little pale one called *tsail uch* "small opossum" (Hunn 1977: 202–3), apparently the referent for Alal Uch.

Alalti: *Alal Ti'* Small Mouth

ti' mouth, tip, edge (Slocum 1953: 58)

Batsul: *B'a Ts'ul* Tips of Amaranth

b'a end, tip, beginning, first (Ara 1986: 248)

ts'ul Spanish: *bledo* (Ara 1986: 401) denotes amaranth, green amaranth, pigweed (Schoenhals 1988: 22)

Bikituch: *B'ik'it Uch* Small Opossum

b'ik'it small (Slocum 1953: 7)

Chanbinkil: *Chan Winikil* Four Score

chan four, compound form of *chaneb* (Ara 1986: 264–65)

Chaikín: *Ch'ay K'in* Lost (Feast) Days; Completion (Feast) Days

ch'ay lose, be lost, lapse (Ara 1986: 263)

Chayquim Hohel Cacal: *Ch'ay K'in Ho-Hel K'ak'-Al* Lost Days, Five Successive Days

hel numeral classifier: enumeration of times certain items are changed (Berlin 1968: 198)

k'ahk'al sun, day (Slocum 1953: 29)

Chinuch: *Ch'in Uch* Small Opossum

ch'in small, little (Slocum 1953: 14)

Guac-Binquil: *(K)Wak Winikil* Six Score

The *b* is an orthographic or dialect variant of *v*, itself a variant of *w*. The optional *g-/k-* indicates strengthening of the *w*, rare in Tzeltal but common in Q'eqchi'.

wak compound form of *wakeb'* six (Slocum 1953: 89)

Guincil: *Winikil* Score, Month

Ho Binkil: *Ho' Winikil* Five Score

ho' compound form of *ho'eb* five (Ara 1986: 301)

Ho'eb Sore Ahtal: *Ho'eb' Sore Ahtal* Five over the Count

ho'eb five (Ara 1986: 301); *sore* from Spanish *sobre* over, above

ahtal count (Ara 1986: 246)

Hokin Ahaw: *Hok'in Ahaw* Waterhole Lord; *Hok' K'in Ahaw* Water Hole Day Lord

Jok'en Ajaw: *Hok'-En Ahaw* Water Hole(–Device) Lord; *H'ok'e'n Ahaw* Drummer Lord

hok' well, water hole (Slocum 1953: 23). The senses of the examples for the usative *-in* (Kaufman 1971: 50, item 11.7) suggest *hok'-in* could mean "make, protect, inhabit the waterhole."

k'in fiesta; divination, fortune (Ara 1986: 373). Hok' K'in reduces to Hok'in by the rule of geminate consonant deletion (Kaufman 1971: 21).

ahaw king (Kaufman 1971: 115)

-en variant of *-in?* or variant of *-em'/-eb'*, instrumental?

Hoquén-Hajab: *Ok-Em' Ahaw* Staff King; *Ok'-Em' Ahaw* Decoy Whistle Lord

The orthography of this source (Pineda 1887) uses *h* for ' and often *b* for *w*. The Tzotzil apply their cognate to a certain caterpillar and to a starving demon whose wails and tears for food explain the thunder and downpours of the rainy season. See the Tzotzil H'ok'en Ahwal below.

okem <oquem> handle, stick, staff (Ara 1986: 353)

Okem Ahaw Staff Lord might also gloss "Office Holder," as the staff symbolizes office. This recalls the "Staff Kings" at Tikal and other sites. These rods might depict ceremonial fire drills (Stuart 1998).

ok'em <oquem> whistle in shape of a decoy; call made with such a whistle; complaint, claim (Ara 1986: 353)

If *-em* represents a variant of the instrumental *-eb'*, the literal sense of *ok'-em'* is "call-device." "Drum Lord" becomes possible because *ok'* also denotes the beat on a drum (Laughlin 1975: 67).

Huk Binkil: *Huk Winikil* Seven Score

huk compound form of *hukeb* seven (Ara 1986: 303)

Hulol: *Hul-ol* Piercing or Bloodletting. See the Tzotzil **Julol** below.

Many Tzeltal dialects formerly contrasted *h*, written *h* or *j* in various sources, to *j*, recorded with *gh*, *j*, or *x*. Most have since merged *j* into *h* in all or most positions (Kaufman 1972: 22–23, passim). The phoneme *h* does not seem to merge into *j*, so most modern dialects use just *h*. For clear and possible examples of *j* this study uses *j* and presents the original spelling in angle brackets.

Joeb K'aal Ch'ay K'iin: *Ho'-eb' K'aal Ch'ay K'in* Lost Days Ch'ay K'in
k'aal variant of *k'ahk'-al* day (Slocum 1953: 29)

Jul El: *Hulel* Piercing or Bloodletting; *Hulel* Arrival
hulel inject, injection (Díaz Olivares 1980: 302)
<*julel*> arrival (Slocum 1953: 24); *ulel* <*ulel*> arrival (Ara 1986: 410)

Kwak Binkil: *Wak Winkil* Six Score. See **Guac-Binquil** above.
k- The Tenejapa *kwak* represents a strengthening of the rare Bachajon *hwak* (Blom and LaFarge 1927:2:478), a variant of the normal *wak* "six."

Mac: *Mak* Cover, Wall
mak to cover, close (Ara 1986: 327–28)

Muc-uch: *Muk' Uch* Big Opossum
muk' big, important (Ara 1986: 33). "Big" suggests the large dark opossum.

Mucul-uch: *Muk'ul Uch* Big Opossum
muk'ul big (Ara 1986: 338)
muk' big (Slocum 1953: 41)
-ul on monosyllabic radical adjective (Kaufman 1971: 106)

Mush: *Mux* Hominy
mux corn, cooked without salt (Slocum 1953: 41)

Olalti: *Olal Ti'* Heart Mouth [Bowl for Hearts]. See Tzotzil **Olalti** below.
olil half; middle, center (Ara 1986: 352)
olal, *olil* Tzotzil heart (Laughlin 1988: 154). The *olal* in this name might record an early Tzeltal cognate. The entire name, alternatively, could have entered from Tzotzil.

Oshbinkil: *Ox Winikil* Three Score
ox compound form of *oxeb'* three (Ara 1986: 354)

Pom: *Pom* Incense (Ara 1986: 367)

Sakilab: *Sak-il Ajb'* White Reed; *Sak Il-ab'* Clear/White Seer
sakil white (Ara 1986: 422)
ajb' reed (Slocum and Gerdel 1965: 115)

sak white, clear, clean (Ara 1986: 420)

il-ab' thing with which to see (Ara 1986: 310)

The literal sense of *ilab'* is "see-instrument." With *sak* as "clean, clear" the name might gloss as "clear seer," evoking the shaman's looking-quartz, a large clear stone that brings the understanding needed to cure illnesses (Sosa 1985: 308–9). *Ilab'* also recalls the K'iche' *ilb'al* "instrument for seeing," used for the codex ancestral to the *Popol Vuh* (D. Tedlock 1985: 242–43 n. 7, 319 n. 219d). With *sak* alluding to the pages, *sak il-ab'* could denote "white seer, codex." Sakilab never appears as Sakil Hab, suggesting that *hab'* constitutes no part of the name, that Sakilab means only Sak-il Ajb' or Sak Il-ab'.

Sakil Ha': *Sak-il Ha'* White Water; *Sak-il Ha* White Fly; White Bowl

ha' water (Slocum and Gerdel 1965: 139)

ha fly (Ara 1986: 299)

ha gourd bowl (Slocum 1953: 18); *ha, hay* (Berlin and Kaufman 1962: 102: for *tol*)

Tsun: *Tz'un* Sowing; *Tz'um* Drying Out

The *m* represents an assimilation of the *n* to the labiality or roundedness of the *u*.

ts'un to sow (Robles Urribe 1966: 74); numerical classifier, "actions of planting (with stick)," from *-ts'un* "to plant it" (Berlin 1968: 195, 224, item 40)

ts'um to dry slightly; a month (Sarles 1961: 58, item 1425)

Uec-Cachaquin: *B'ek K'a' Ch'ay K'in* Bundled Up Rotten Useless Days

b'ek <uec> to swaddle, as with a child (Ara 1986: 406)

k'a' old, rotten, used (Ara 1986: 254).

See **Chaikín.**

Yahal-Uch: *Yal Uch* Small Opossum. See **Alajuch** above.

yahal sources (Pineda 1887); variant for *yal*

yal her child (Slocum 1953: 3, 44); small, by extension. In the "big/ small" contrast of the "opossum" month names, *yal* best makes sense as "small," as it does in several other Maya languages.

Yajuch: *Yal (H')Uch* Small Opossum

Yashkín: *Yax K'in* Green Days; Crab Days

yax green, blue (Slocum 1953: 74); *yax* crab (Ara 1986: 307)

SUMMARY: Tzeltal receives many names from the glyphs, most rather altered but a few still cognate (table 3.9). Nearly all proceed into Tzotzil, while some reach one

TABLE 3.9

Summary of Tzeltal month names

Names in source	Interpretations	Translations
Agelchac	Ahil Ch'ak	Marsh or Prickly Flea
Alajuch	Alal Uch	Small Opossum
Alalti	Alal Ti'	Small Mouth
Batsul	B'a Ts'ul	Tips of Amaranth
Bikituch	B'ik'it Uch	Small Opossum
Chanbinkil	Chan Winikil	Four Score
Chinuch	Ch'in Uch	Small Opossum
Guac-binquil	(K)Wak Winikil	Six Score
Ho Binkil	Ho' Winikil	Five Score
Hokin Ahaw	Hok'in Ahaw	Water Hole Lord
	Hok' K'in Ahaw	Water Hole Day Lord
Jok'en Ajaw	Hok'-en Ahaw	Water Hole-Device Lord
	H'ok'e'n Ahaw	Drummer Lord
Hoquén-hajab	Ok-em' Ahaw	Staff King
	Ok'-em' Ahaw	Decoy Whistle Lord
	H'ok'en Ahwal	Wailer Lord (Tzotzil)
Jul El	Hulel	Piercing, Bloodletting
	Hulel	Arrival
Kwak Binkil	Wak Winkil	Six Score
Huk Binkil	Huk Winikil	Seven Score
Hulol	Hul-ol	Piercing, Bloodletting
Mac	Mak	Cover, Wall
Muc-uch	Muk' Uch	Big Opossum
Mucul-uch	Muk'ul Uch	Big Opossum
Mush	Mux	Hominy
Olalti	Olal Ti'	Heart Mouth
Oshbinkil	Ox Winikil	Three Score
Pom	Pom	Incense
Sakilab	Sak-il Ajb'	White Reed
	Sak Il-ab'	Clear/White Seer
Sakil Ha'	Sak-il Ha'	White Water
	Sak-il Ha	White Fly
	Sak-il Ha	White Bowl
Tsun	Tz'un	Sowing
	Tz'um	Drying Out
Yahal-uch	Yal Uch	Small Opossum
Yajuch	Yal (H')Uch	Small Opossum
Yashkín	Yax K'in	Green Days
		Crab Days
Chaikín	Ch'ay K'in	Lost Days
		Completion Days
Chayquim Hohel Cacal	Ch'ay K'in Ho-Hel K'ak'-al	Lost Days, Five Successive Days
Ho'eb Sore Ahtal	Ho'eb' Sore Ahtal	Five over the Count
Joeb K'aal Ch'ay K'iin	Ho'-eb' K'aal Ch'ay K'in	Lost Days Ch'ay K'in
Uec-cachaquin	B'ek K'a' Ch'ay K'in	Bundled Up Rotten Useless Days

or two other traditions. As a whole the Tzeltal names show little variation in form or meaning. Three, however, serve as the first month of the year, each in different villages: specifically Ajelch'ak, Batz'ul, and Jokenajau, The Five Days go by one form, but two epithets occasionally find themselves on a roster. Only Agelchac and the series 7–3 Binkil seem wholly Tzeltal in sense and sound, neither cognates to nor derived from outside names.

Tzotzil Definitions

Tzotzil and Tzeltal, perhaps along with Tojolab'al, constitute Tzeltalan. The more frequently repeated parts of names are defined here.

-al relational or attributive, in frame with *y-* and *-ib'-* to form ordinals: *y*-NUMBER-*ib'-al*

-em stative aspect on intransitive (Haviland 1981: 101–5)

h a grapheme: glottal voiced spirant, a colonial phoneme present in some modern dialects, absent in others (Laughlin 1988: 84–85)

h- agent (Delgaty and Ruiz Sánchez 1978: 386)

-ib' generalized classifier (Haviland 1981: 165, 1988: 106); attributive or relational (Laughlin 1975: 25–28); "indefinitely possessed" or absolute (Haviland 1988: 99, item ii)

-il attributive, generalizer (Laughlin 1975: 25); nominalizer for adjectives; indefinite, absolute, abstractive (Haviland 1981: 66–68, 1988: 99)

k'in fiesta, fate, bewitchment, omen, knowledge or prophecy of diviner (Laughlin 1988: 232). By analogy to its reflexes in most other languages, *k'in* once meant "sun, day, time."

muk' big, fat, greater (Laughlin 1988: 263) [*mayor*: older?]

muk'ta large, great (Laughlin 1975: 243)

-ol nominalizer on just CVC + *im* intransitive verbs (Cowan 1968: 107, item 25); no restrictions on root-stem shape (Colby 1966: 389)

s- 3sA, preconsonantal; ordinal with numerals (Haviland 1981: 47, 168–70); "With temporal expressions, ordinal numerals imply past time" (Haviland 1981, 169; 1988, 106)

sak white, clean, light, bright (Laughlin 1975: 302)

uch opossum; black sapote; haze, drizzle; yellowed; cornfield searing (Laughlin 1975: 72; Breedlove and Laughlin 1993: 232)

vinik man, woman, person; twenty, score (Laughlin 1988, 328, 1242). "Twenty" apparently derives from the total of a person's digits (Haviland 1981: 166).

vinkil man (Laughlin 1975: 371); *vinikil* shortened

x- variant of *s-* as 3sA assimilated to adjacent *ch/ch'/x* (Haviland 1981: 52; Hopkins 1967a: 13)

y- 3sA possession (Haviland 1981: 187–96); ordinals in frame "y-NUMBER-*ib'-al*" (Haviland 1981: 165–68)

Tzotzil Definitions

Ala Uch. This is a Tzeltal and not Tzotzil name. See Tzeltal **Alajuch** above.

Bats'ul: *Ba Ts'ul* Top of Amaranth

ba best, first; head, top (Laughlin 1975: 75–76)

ts'ul Tzeltal: *bledo* Spanish (Ara 1986: 401) amaranth, green amaranth, pigweed (Schoenhals 1988: 22)

Bikit Uch: *Bik'it Uch* Little Opossum; Little Drizzle; Little Searing (of the Cornfield)

bik'it small, little (Laughlin 1988: 166)

Ch'ay K'in: *Ch'ay K'in* Fiesta of the End; Days Lost; Days of Destruction; Days of Smoke

ch'ay be absent from work; be lacking; get lost; be depopulated; be destroyed; end, die (Laughlin 1988: 196); smoke (Laughlin 1975: 132)

Elech: *Elech* Small, Log-Dwelling Mammal; *El-'ech* "Greasy Back." Variant: *Eluch* (Seler 1902: 1:707)

Perhaps the hypothetical name *el-'ech* "Greasy Back" designates the log-dwelling mammal.

elech small unidentified mammal living in dry logs (Gossen 1974b: 237)

el greasy, sweaty (Laughlin 1975: 51)

ech back (Laughlin 1975: 50)

Ghunviniquil: *Hun Vinikil* One Score. See **Sbabinkil** below.

hun one (Laughlin 1975: 160); *hum, hun,* with *m* for an *n* assimilated to the labiality of the *u* or the *w* or both. The *v* represents /*v*/ as a phonetic variant of [*w*].

Hoyoh: *Hoyoh* Area or Object Being Enclosed

hoyoh area or object being enclosed (Laughlin 1988: 215)

H'uch: *H'uch* Male Opossum; Opossum Hunter; Drizzle Maker

Uch seems to name Venus as Morning Star. A native describes the *uch* as a fiery red, predawn light distinct from sunlight (Guiteras-Holmes 1961: 195–97, 215,

292): "Uch is greatly respected because it has fire, because at dawn it lights up the hills. It is not the sunlight, for the sun rises later" (195). "[I]t seemed as if the hills were on fire. It was the Uch. . . . Then it slowly disappeared and gave place to the white light of dawn. . . . I woke up when it was still night . . . and I saw that . . . the sky toward the northeast was red" (206). "Uch is as God, because it has light, a red light, that later disappears to give place to the God, the sun. Neither Okinahual [a caterpillar and month name] nor Uch are [*sic*] evil, but they must be respected as a God" (196).

The Tzotzil describe Venus, too, as *b'atz'i tzoj* "bright or truly red" (Laughlin 1977: 253). The earliest Kaqchikel dictionary (Vico 1555: 211v–212r) seems to confirm the suggested Tzotzil identification of Venus with the opossum by glossing *wuch'*, its cognate to *uch*, with "opossum" and "morning star." In their *Popol Vuh* the K'iche' also link the opossum to the red sky at morning and maybe to Venus as well (D. Tedlock 1985: 145, 149, 287–88, 290, 342).

Julol: *Hul Ol* Pierce the Heart; Pierce the Child/Corn [Infant Sacrifice]; *Jul Ol* Arrival of Corn or Offspring; Arrival at Midpoint [Zenith Passage]

hul come (Laughlin 1988: 342)

hul bleed, chisel, engrave, pierce, prick, sew, stab; a puncture, stitch (Laughlin 1988: 215–16)

ol female's child, female animal's offspring, in compounds (Laughlin 1975: 68), heart, in compounds (Laughlin 1988: 154); noon, midnight [literally "middle"], compound form of *o'lol* center, middle, half (Laughlin 1975: 64). *Ol* "child" might designate maize, because in San Pedro Chenalhó Tzotzil *yaxkin baol*, literally "the dry season's first child," denotes early maize (Guiteras-Holmes 1961: 34).

Hul Ol as "Pierce the Heart" suggests a term for heart sacrifice. See the K'iche' **Ch'ab** above for ethnographic and archaeological details. It might refer to cardiac excision too. Stelae 11 and 14 at Late Classic Piedras Negras display the victim draped supine over an altar (Tozzer 1957, 12, fig. 389a, b; Maler 1901, pl. 20, nos. 1, 2; Clancy 2009: fig. 6.1, 138). The practice continued well into the colonial period (Tozzer 1941; Helfrich 1973; Garza et al. 1983).

If *ol* refers also to the stomach or entrails, the name Hulol could denote the disembowlment sacrifice, often carried out on a victim tied to a scaffold (Coe 1973: 76: #33, 1975: #16; Schele and Miller 1986: 111–12, pl. 92, 94; Taube 1988b: fig. 12.22c, d: Tikal Altars 8, 10; Trik and Kampen 1983: fig. 38a).

Hulol as "Pierce the Child" suggests the well-attested practice of infant sacrifice (Helfrich 1973; Tozzer 1941; Coe 1982: 17: #2; Robicsek and Hales 1982: 40: #12). As "Arrival of Corn or Offspring" it might refer to the first appearance of the corn ears or to a time of year noted for the arrival of newborn animals. "Arrival at Midpoint" suggests the zenith passage, when the sun attains the exact center of the sky at midday and casts no shadow; at these latitudes this occurs twice a year.

Me'el Uch: *Me'el Uch* Old Woman Opossum/Drizzle/Cornfield Yellowing

me'el old woman, widow (Laughlin 1975: 232)

Me'okinahual: *Me' Ho' K'in Ahval* Mother Rain Days Lord. See **Ok'in Ahwal** below.

me' mother, female; large (Laughlin 1975: 232)

Moc: *Mok* Wall

moc wall; corral (Hurley-Delgaty and Ruiz Sánchez 1978: 84)

Mol Uch: *Mol Uch* Old Opossum; Old Man Drizzle, Old Man Yellowing of the Cornfield

Mol H'uch: *Mol H'uch* Old Opossum Hunter; Old Man Drizzle Maker; Yellower of the Cornfield

mol man, husband, elder; old; large; eldest (of male kin) (Laughlin 1975: 239)

Muctacac: *Muk'ta Sak* Great White Corn

muk'ta sak apparent synonym for *muk'ta sakil ixim*, the name for several strains of Indian corn, planted in April or May, harvested in December or January (Laughlin 1975: 63, 244). Muk'ta Sak "Big White" refers to three specific varieties of highland corn (Breedlove and Laughlin 1993: 234–35).

Muctasái: *Muk'ta Sakil* Great White Corn

sakil whiteness; unattested lowland squash (Laughlin 1975: 304)

Muk: *Muk* Bury, Plant; *Muk'* Big

muk bury, sink, plant (Laughlin 1975: 243)

Muktauch: *Muk'ta Uch* Big or Old Opossum; Big Drizzle; Big Yellowing of the Cornfield

Mush: *Mux* Hominy (K'iche')

The word *mux* as "corn kernels" is present in Ch'ol, Chontal, Ch'orti', Lakandon, Mopán, Yukatek, Kaqchikel, K'iche', Poqom, and Tojolab'al, as well as Tzeltal, but Tzotzil has no cognate. *Mux*, then, appears to be a loan from K'iche' into Tzeltal and from there into Tzotzil, first because it does not gloss in the extensive Tzotzil sources and second because it turns up only rarely in the dictionaries for languages other than K'iche'.

mux Tzeltal: corn cooked or boiled without lime (Slocum 1953: 41, 52); when the silks darken and fall away, the corn enters the roasting ear stage. These ears, roasted, are then boiled to produce *mux* (Berlin et al. 1974: 113).

Muy: *Muy* Chico Sapote Tree; Finely Ground Wild Tobacco

muy chico sapote (Laughlin 1975: 245); finely ground wild tobacco (Holland 1963: 97)

Nichik'in: *Nichk'in* Flower Days; Dog Days; Dog Day Butterflies

nich k'in dog days; butterfly that appears in great numbers in the dog days (Laughlin 1975: 252)

nich flower; strength (Laughlin 1975: 252)

nichim flower; jasmine (Laughlin 1988: 272)

Ok'in Ahwal: *O' K'in Ahval* Water Days Lord

Ok'en Kahval: *Ok'-en K-Ahval* My Lord Has Wept

Ok'in: *O' K'in* Water Days

H'ok'en Ahwal: *H'ok'em' Ahval* Wailer Lord; *H'o' K'in Ahval* Waterman Days Lord. Variants: *Okinahual* [in the calendar of 1944/55]; *Okin-Ajual*

ho' canal, pond, rain, water (Laughlin 1988: 340)

ok' tongue; cry, weep (Laughlin 1975: 67)

o' water, spring (Laughlin 1975: 64)

k- my (Laughlin 1988: 85)

ahval master, person who afflicts or kills us (Laughlin 1988: 151), under *ojov*

Noting first that Bricker associates the *tun* or *teponastli* drum with rain, warfare, and the scaffold sacrifice and second that the Yukatek equivalent to H'ok'en Ahwal is Pax, whose activities involve warfare and sacrifice (and whose name denotes another kind of drum), Karl Taube (1988b: 335–37) suggests the *ok'* in the name, which signifies not only crying, weeping, or howling but also the sound of a musical instrument, might allude both to the sound of the instruments typically played during sacrificial rites and to the cries of the victims. A Late Classic vase (Coe 1973: 76, #33) depicts both a grotesque executioner disemboweling a victim tied to a scaffold and four musicians playing rattles, flutes, trumpets, and two kinds of drums.

The informant Manuel in Guiteras-Holmes (1961: 195) identifies Okinahual or Ok'en Ahwal with a three-inch-long edible caterpillar. Its head is black, with two teeth, little eyes on either side, and four tiny legs nearby. The creature has no hair or tail or wings. It is born and lives in only two kinds of tree. The larva becomes a butterfly, with spotted blue wings looking like silk.

Olalti: *Olal Ti'* Heart Mouth [Chest Wound Heart Bowl]; Heart of Woodpecker; *O'lal Ti'* Center Mouth [Cave]; *Ol Ti*: *Ol Ti'* Center Mouth [Cave] *olal* heart (Laughlin 1988: 154)

ol community, counsel (Laughlin 1988: 153) middle, center (Laughlin 1975: 64)

ti' to eat; big-lipped mouth (Laughlin 1988: 313); mouth, opening, entrance (Hurley-Delgaty and Ruiz Sánchez 1978: 197); hairy woodpecker (Laughlin 1975: 337)

"Heart Mouth" might denote the slit cut into the chest to rip out the heart or the mouth of an idol that held the bloody offering (Tozzer 1941: 110 n. 502, 118 n. 541, 119 n. 543, 162). Perhaps the bowl that the priest put the heart into, sometimes covered with another bowl (Tozzer 1941: 143 n. 684; Smith 1932a: pl. 5), functioned as a ritual mouth. "Heart of Woodpecker" recalls that the Yukatek Maya cut out and burned the hearts of many different birds and other animals in their month Mak (Tozzer 1941: 163).

The *o'lal* of this month name presents the Chamula Tzotzil term for "half" (Gossen 1974b: 235, 242). Analogy indicates it also means "center, middle" because *o'lol* (its counterpart in other dialects) has that meaning (Hurley-Delgaty and Ruiz Sánchez 1978: 95; Laughlin 1975: 64). The natives describe their community as the navel of the world, so any local cave could be the "mouth" near that center. If *o'lal/o'lol* refers to the center, middle, or bowels of the earth, the name could gloss as "Underworld Mouth" to denote any cave or water hole as an entrance to the nether realm.

Pom: *Pom* Honey, Incense

pom incense (Laughlin 1975: 282); honey (Laughlin 1988: 287)

Sbabinkil: *Sba Vinkil* First Score; One Score Ago

sba first (Hurley-Delgaty and Ruiz Sánchez 1978: 460)

ba first, best, top (Laughlin 1988: 161)

Schanibal Vinkil: *S-chanib'al Vinkil* Fourth Score; Four Score Ago

s-chan-ib-al fourth (Hurley-Delgaty and Ruiz Sánchez 1978: 460)

chaneb dialectal variant of *chanib* four (Laughlin 1988: 188). Some numbers take *-ib*, others *-eb* (Hurley-Delgaty and Ruiz Sánchez 1978: 458–61); *chan-* usually adds *-ib*.

chan four, compound form (Laughlin 1975: 110)

Schibalbinkil: *Schibal Vinkil* Second Score; Two Score Ago

Sisac: *Sisak, Sasak* White, White Corn; *Sij Sak, Sik Sak* Frost White Corn

sak For its sense as "white corn" see **Muctacac** above.

sik cool, cold (Laughlin 1988:1:297); frost (Hopkins 1967:23)

The rule relevant to *sik* reads [$k > h \sim 0/$ ___ C]: "cold" might designate a type of corn from the significantly cooler highlands. "Cold" and "frost" could also denote a race of maize notably resistant to cold and frost or a type whose planting or harvesting depends on the frost or cold weather.

Tsun: *Tz'un* To Plant with a Stick

tz'ui   ant, breed, stab; be planted (Laughlin 1975: 105); a sticking-into-the-ground (knife, pole) numerical classifier (Berlin 1968: 224)

Unenuch: *Unen Uch* Small Opossum; Small Drizzle; Small Yellowing (of Cornfield)
unen infant, child (Laughlin 1988: 160); small, young (Laughlin 1975: 74)

Yaxk'in: *Yax K'in* Green Days
yax blue, green, gray (Laughlin 1975: 384)

Yoshibalbinkil: *Y-Ox-ib-al Vink-il* Third Score

Ox-Binquil: *Ox Vinkil* Three Score
ox three, third (Laughlin 1988: 155)

In Yoxibinkil the form has deleted the *-al* of *yoxibal* then lost the final *b* to geminate consonant deletion. In Yoxchibal Uinicil the *ch* presents a lexemic error, perhaps a repetition of the *ch* in the immediately preceding name *xchibal uinicil*, an example of phonemic change via analogy or "contamination."

SUMMARY: Largely indebted directly to Tzeltal for most its names, Tzotzil distinguishes itself with the internal replacements, alternates, or modifications for many of them. Like Tzeltal, it has three or four different first months, depending on location, with some overlapping (table 3.10): B'atz'ul, Muk'ta Sak, Tz'un, and maybe Mok. It has developed no substitutes for the Five Days. The standard Elech as well as the rare Hoyoh, Muk, and Muy appear to spring straight from Tzotzil in form, though in sense this remains debatable. Also unique to this tradition is the 1–4 Vinkil series, four consecutive months each prefixed by a numeral, starting with 1 and ending with 4. This sequence focuses on something other than the Tzeltal series of the 7–3 Binkil set. Finally, the stemmas indicate that Tzotzil lends out or inspires so few names that it ranks almost last in lexical exports.

TABLE 3.10
Summary of Tzotzil month names

Names in source	Interpretations	Translations
Ala Uch	Alal Uch	Small Opossum
Bats'ul	Ba Ts'ul	Top of Amaranth
Bikit Uch	Bik'it Uch	Little Opossum
		Little Drizzle
		Little Cornfield-Searing
Elech	Elech	Small, Log-Dwelling Mammal
	El-ech	Greasy Back; "Go!"
Ghunviniquil	Hun Vinikil	One Score
Hoyoh	Hoyoh	Area or Object Being Enclosed
H'uch	H'uch	Male Opossum
		Opossum Hunter
		Drizzle Maker
Julol	Hul Ol	Pierce the Heart
		Pierce the Child/Corn
	Jul Ol	Arrival of Corn or Offspring
		Arrival at Midpoint
Me'el Uch	Me'el Uch	Old Woman Opossum
		Old Woman Drizzle
		Old Woman Cornfield Yellowing
Me'okinahual	Me' Ho' K'in Ahval	Mother Rain Days Lord
Moc	Mok	Wall
Mol Uch	Mol Uch	Old Opossum
		Old Man Drizzle
		Old Man Cornfield Yellowing
Mol H'uch	Mol H'uch	Old Opossum Hunter
		Old Man Drizzle Maker
		Cornfield Yellower
Muctacac	Muk'ta Sak	Great White Corn
Muctasái	Muk'ta Sakil	Great White Corn
Muk	Muk	Bury, Plant
	Muk'	Big
Muktauch	Muk'ta Uch	Big or Old Opossum
		Big Drizzle
		Big Cornfield Yellowing
Mush	Mux	Hominy
Muy	Muy	Chico Sapote Tree
		Finely Ground Wild Tobacco
Nichik'in	Nichk'in	Flower Days
		Dog Days
		Dog Day Butterflies

(continues)

Table 3.10

Summary of Tzotzil month names (*continued*)

Names in source	Interpretations	Translations
Ok'in Ahwal	O' K'in Ahval	Water Days Lord
Ok'en Kahval	Ok'-en K-ahval	My Lord Has Wept
Ok'in	O' K'in	Water Days
H'ok'en Ahwal	H'ok'em' Ahval	Wailer Lord
	H'o' K'in Ahval	Waterman Days Lord
Olalti	Olal Ti'	Heart Mouth
		Heart of Woodpecker
	O'lal Ti'	Center Mouth [Cave]
Ol Ti	Ol Ti'	Center Mouth [Cave]
Pom	Pom	Honey, Incense
Sbabinkil	Sb'a Vinkil	First Score
		One Score Ago
Schanibal Vinkil	S-chanib'al Vinkil	Fourth Score;
		Four Score Ago
Schibalbinkil	Schibal Vinkil	Second Score;
		Two Score Ago
Sisac	Sisak or Sasak	White, White Corn
Sij Sak	Sik Sak	Frost White-Corn
Tsun	Tz'un	To Plant with a Stick
Unenuch	Unen Uch	Small Opossum
		Small Drizzle
		Small Cornfield Yellowing
Yaxk'in	Yax K'in	Green Days
Yoshibalbinkil	Y-ox-ib-al Vinkil	Third Score
Ox-binquil	Ox Vinkil	Three Score
Ch'ay K'in	Ch'ay K'in	Fiesta of the End
		Days Lost
		Days of Destruction
		Days of Smoke

Yukatek Definitions

Itzaj, Lakandon, Mopán, and Yukatek together constitute Yukatekan. Of these only Lakandon does not enter into this study. Yukatek Maya vowels are articulated in four tones: neutral, *a* or *a'*, only with short vowels; low or falling, *àa*, with long vowels; high or rising, *áa*, also with long vowels; and broken or rising-falling or glottalized, *áa*, with *a'a* and most *a'* segments (McClaran Stefflre 1972: 61–62, 81–82, 180; *a* represents any vowel). In most sources a written *aa* can represent a vowel sequence with any of the four tones. Angle brackets enclose words whose tone is unmarked in the sources or uncertain with inflected or compound forms (table 3.11). Marking for

TABLE 3.11

Summary of Yukatek month names

	Names in source	Interpretations	Translations
1	Poop	Póop	Reed Mat
2	Woo	Wo'	Toad
		Wo'	Pitahaya Cactus
		Wooj	Glyph
2	Uol	Wo'ol	Ball
			Hole
			Heap
			Shrub
3	Siip	Sìip	Offense
			Spite
			Ripe
		<Siip>	Release; Miss
			Deer Guardian
3	Zip'	Sìip'	Swelling, Ripe
4	Sootz'	Sòotz'	Bat
4	Tzoz <Tzos>	Tz'òo's	Bat
5	Sec	Sek	Bat
5	Tzeec	<Tzek/Tz'ek>	Foundation
		Tzèek	Clearing
		Tze'ek	Splinter
		Tzek'	Skull
6	Xuul	Xùul	End
		Xúul	Digging Stick
		Xú'ul	Lance Pod Tree
		<Xuul>	Striped Mullet; Goby
		<Xul>	Maimed
7	Tz'e Yaxkin	<Tz'e'> Yáax K'iin	Little First Sun
			Little Dry Season
		<Tz'e'> Yá'ax K'iin	Little Green Days
			Little Reed Days
			Little Dry Season
			Little Green Grasshopper
8	Mool	Mòol	Pile; Harvest
		Mó'ol	Paw
9	Ch'een	Ch'é'en	Cave with Water
10	Yaax	Yá'ax	Blue-Green
			Reed
		Yáax	First; Great
11	Sac	Sak	White
			White Flower
			Small Silvery Fish

(continues)

TABLE 3.11

Summary of Yukatek month names (*continued*)

	Names in source	Interpretations	Translations
12	Ceeh	Kéeh	Deer
13	Maac	Màak	Cover
			Pit
			Giant Sea Turtle
			[Turtle] Shell
			Granary
		Mak	Cornfield Measure
14	K'ank'in	K'ank'ìn	Days of Ripening Harvest
			Red Bead Days
		K'àan K'ìin	Cord Days
			Field-Measuring Days
			Field-Clearing Days
		K'áan K'ìin	Net or Hammock Days
15	Muan	Muan	Hawk; Owl
16	Paax	Pàax	Upright Drum
17	Kayab	K'ayab'	Horizontal Drum
17	K'ay Yab	K'àay Yá'ab'	Song of Many
		K'ay Há'ab'	Announce the Year
18	Cumku	Kúumk'uh	Thunder of the God
			Pot God
		Kùum K'uh	Cloud-Thickening God
		Kum K'uh	Throne God
		Kúum K'ú'	Thunder Cave
			Potter's Kiln
		Kum K'ú'	Throne Cave
19	Uayab	<Way-ab'>	Poison
			Hex
			Bed
		Wàay Há'ab'	Poison of the Year
		Wáay Há'ab'	Animal Companion of the Year
		<Way> Há'ab'	Bed of the Year
19	Uayeb	Wayib'	Bed

tone appears only in a few recent sources. The typical entry below presents the term first in its original spelling then in the orthography of this study. The abbreviations are GTZ: Greater Tzeltalan, LL: Lowland Mayan, PM: proto-Mayan, and PY: proto-Yukatek. Unreferenced PY reconstructions are mine. The more frequently repeated parts of names are defined here.

-ab' instrument (Smailus 1989a: 121)

ah agent, male (Bricker 1986: 39, table [hereafter T.] 19)

-ah active gerund; nominals from transitive verbs (Blair 1964: 44, 96: #524)

h- agent; male (Bricker 1986: 39 T. 19). Victoria Bricker (ibid.) and Marlys McClaran Stefflre (1972: 186) note *h-* as the modern counterpart to the Classical Yukatek *ah*.

<*eb'*> stairs, steps (Ciudad Real 1984: 160v)

-eb' instrument. This represents *-ib'*, as borrowed by Yukatek. The true Yukatek cognate is *-ab'*.

-e' the, as for; often not translated; topicalizer or referential enclitic, refers to something in mind, under discussion, or marked as a topic for comment. In nominal stative constructions the topic must end with *-e'* if it precedes the comment (Blair 1964: 133).

há'ab' year (Fisher 1973: 230, item 183)

-il relational, possessive; abstractive (Blair 1964: #515, #542; Bricker 1986: 39 T. 19)

k'ìin sun, day (Fisher 1973: 267, item 365); a clear-green grasshopper or locust that usually flies at night (Pacheco Cruz 1939: 77)

u his, her, its 3sA (Bricker 1986: 21 T. 3)

u-y prevocalic variant of 3sA *u* possessive (Bricker 1986: 21 T. 3)

u-ADJ-*il* attributive frame (Smailus 1989a: 126–27)

u-ADJ-*ol* attested variant of attribute frame *u*-ADJ-*Vl* (Smailus 1989a: 126–27)

y- reduced form of *uy-* 3sA (Bricker 1986: 21)

yáax blue (Mengin 1972: 24v); green; tender (Bricker et al. 1998: 312); <*yax*> a little reed or liana [*bejuquillo*] that the horses and goats eat well (Mengin 1972: 27r)

yáax first; prior (Bricker et al. 1998: 312)

Yukatek Definitions

Ceeh: *Kéeh* Deer (Fisher 1973: 248, item 272)

Ch'een: *Ch'éen* Cave with Water

ch'éen cave of water (Barrera Vásquez et al. 1980: 131); hollow of cacao and other trees (Mengin 1972: 124v); tract of low-lying fertile land especially suited to the production of cacao; well, cistern (C. H. Berendt in Brinton 1882: 125 n. 4).

Cumku: *Kúumk'uh* Thunder of the God; Pot God; *Kùum K'uh* Cloud-Thickening God; *Kum K'uh* Throne God; *Kúum K'ú'* Thunder Cave; Potter's Kiln; *Kum K'ú'* Throne Cave

kúunk'uh This names the major *chàak* or rain god, father of the four directional rain gods. Living in the ground at the eastern horizon, he causes the great rumbling thunder heard in the east during July and climaxing in August, but he begins to thunder on 13 June, the date for the heliacal rising of the Pleiades (Sosa 1985: 379, 453–54).

The Maya at Chan Kom revere San Miguel Arcángel as the chief rain god. They rank Kunku-Chaac second and relate that he rides across the sky at the head of the others. Some say that on 2 June, the end of the dry season, he and all the others assemble at Coba to receive their orders from San Miguel Arcángel and that on 3 June they ride forth, old men on horses seen as clouds. The first thunder heard in the east announces their coming (Redfield and Villa Rojas 1934: 115–16). This rumbling they call *jumkú* or *kumkú*, the "sound or noise of God."

kúumk'uh <cumkú> "a thunderclap, or noise like the report of a cannon . . . heard in the woods while the marshes are drying" (Pío Pérez 1843 303); "the loud explosion as of a distant cannon report . . . heard, and produced at the beginning of the rainy season perhaps by the marshes that split as they dry, or by the blasts of lightning bolts in distant squalls. They are also called *humkú* the noise or sound of God" (Pío Pérez 1864: 380, 382).

kúum <kum, kuum> sound of kicking or stamping [*pateo*] (Michelon 1976: 74); noise of kicking a hollow thing or walking quickly (Pío Pérez 1866–77: 62)

kum Spanish: *abatanar* beat down (Po'ot Yah and Bricker 1981: 15)

Abatanar is also defined as "to full, to shrink and thicken cloth." "To full" means "to thicken by moistening, heating and pressing, as cloth"; its older usage glosses as "to beat down; to trample; to destroy." The nominal form would be *kùum* fulling; beating down; trampling; destroying (Fisher 1976: 30–31). As part of the rain god's name, *kum/kùum* suggests that this deity beats the clouds, perhaps made of cotton, with the flat of his machete or flint axe to thicken them or to produce thunder. If *kúum* "noise" represents a variant, then the methods of producing that ruckus indicate that the god might also kick or stomp on the clouds. The Ch'orti rainmaker deities beat the clouds but produce rain (Wisdom 1940: 382, 396–97), whereas the Yukatek Kúunk'uh "only thunders" (Sosa 1985: 454).

k'uh god (Fisher 1973: 270, item 383); temples, ancient pyramids, places of worship (Pío Pérez 1866–77: 183)

kúumk'u' = *kúum* + *k'ú'* potter's kiln (Ciudad Real 1984: 89v; Mengin 1972: 10v); literally "jug nest" (Ralph Roys in Barrera Vásquez et al. 1980: 352)

kúum pot (Fisher 1973: 256, item 317); loan from Nahua *koomi-tl* pot, jug, pitcher (Justeson et al. 1985: 25 T. 12); <cum> little or portable furnace [*hornillo*] for making ink (Ciudad Real n.d.: 131v); <cuun> furnace in which ink is made from smoke (Ciudad Real 1984: 90r)

<kum> supposed root of <kuman>, a thing that is seated or settled, laid down or fixed (Ciudad Real 1984: 89v)

k'ú' nest (Fisher 1973: 271, item 388); nest or pen of animals (Pío Pérez 1866–77: 183). Several sources denote the burrow of animals with *áaktun* "cave, cavern" (Ciudad Real 1984: 2v).

K'ú' "nest" might function as a metaphor, a ritual term for "cave," especially if sacred leaves, flowers, garlands, and boughs bedeck the grotto floor (MacLeod and

Puleston 1979: 72). Some (Pohl and Pohl 1983; Bassie Sweet 1991: 199, 238–40) conclude that kings performed seating rites in caves. *Kum k'u'* thus could mean "seat nest" or "throne cave." The many vessels extant in caves (MacLeod and Puleston 1979: 72) prompt the "pot nest" or "pot cave" interpretation.

<*kum ahaw*> demon (Pío Pérez 1898: 18)

<*kumhaw*> Lucifer, prince of demons (Ciudad Real 1984: 89v)

A literal interpretation reads "*kúum* lord," with "hell" an extension for *kúum* "small furnace for making ink from smoke." *Kúum* in its basic sense of "pot" might also signify "underworld," because Diego López de Cogolludo's illustration of the Maya cosmos (in Gates 1931: 4) depicts the world tree rooted in a large bowl, labeled <*cum*> by Berendt (Brinton 1895: 47, fig. 9) in his version of the drawing. The Itzaj *ch'en* "*olla*/pot" (Armas 1897: 30) and the Ch'orti *ch'een* "a low pottery bowl" (Fought 1972: 256 n. 16.91, 127 n. 1.45) reinforce the link because the literal sense of each is "cave."

With *kúum* as "thunderclap," *kúum ahaw* yields "thunder lord." If the original entry <*cumhau hum, hauhum ahau*> "Lucifer, prince of demons" (Mengin 1972: 138v) really records <*cumhau, humhau, hum ahau*>, then *hum(a)hau* probably identifies Lucifer with Landa's (1978: 95) "prince of all the demons . . . Hunhau." This name might translate as "1 Ahaw, 1 Lord," the principal day name for Venus as morning star (Thompson 1972: 62–67). A relation between the major *cháak* and the demon prince remains to be established.

The possible derivation of this name from its glyphic and Ch'ol counterparts *ol* and *olh* via the Ch'orti' or Ch'olan *ku'm'-k'u'* requires the definitions below. They suggest that *ol* "ball" becomes *kum* "egg," then *kum-k'u'* "egg in nest," and finally *kum-k'uj* "thunder god."

ku'm Ch'orti': egg; testicle; any large or ovoid fruit (Wisdom 1949: 604); avocado, kernel, bean; stone; egg (Girard 1949: 135)

k'u' Ch'ol: nest (Aulie and Aulie 1978: 43)

Kayab: *K'ayab'* Horizontal Drum, Tunk'ul

k'ayab' = *k'ay* + -*ab'* to sing it + instrument (Kaufman 1971: 16, 75, item 71.4)

k'ay for men to sing and for the birds, cicadas, etc., to warble and sing, and such a song or call (Ciudad Real 1984: 234v)

k'àay music, song (Bastarrachea Manzano et al. 1992: 99)

k'á'ay for those who want to marry to publish the banns and to proclaim and make public and divulge by proclaiming whatever thing (Ciudad Real 1984: 234v)

k'ayom Lakandon: drum (Bruce 1979: 179); *k'ay-o'* = <*cayó*> Ch'ol: drum (Becerra 1937: 275, item 9)

k'ayob' Tzeltal: drum (Ara 1986: 257; Slocum 1953: 31); large drum (Redfield and Villa Rojas 1939: 116); *k'ayob'* = *k'ay* + -*ob'* to sing it + instrument (Berlin et al. 1974: 135)

One Tzeltal source (Berlin et al. 1974: 135) defines this term at length: "The native drum is produced by hollowing out a short section of . . . the avocado trees . . . onto which *two* pieces of cowhide are stretched over the opening at each end. The skins are pulled taut by vine or cord lashing. A small cross-shaped opening is cut into the side of the drum."

Seler (1899: 110) correctly translates *kayab* as "with what one sings." Identifying the main sign of the compound as a turtle head, he concludes that the name denotes a turtle shell drum, its plastron beaten with a deer antler. Icons and glyphs with more detail demonstrate that the head refers rather to a parrot (Stuart 1985: 2–4; Justeson 1975). The term is listed in no Yukatekan source. *Kayab* immediately follows *páax* "upright drum," a pairing that hints at the meaning "horizontal drum" for *k'ayab*.

K'ank'in: *K'ank'ìn/ K'an K'ìin* Days of Ripening; Harvest; Red Bead Days; *K'àan K'ìin* Cord Days; Field-Measuring or Clearing Days; *Káan K'ìin* Net or Hammock Days

k'ank'ìin time of ripening or maturity (Justeson 1989a: 78); time of maturity (harvest) (Justeson 1988: 14, 15 T. 3)

k'an k'ìin yellow time, harvest (Justeson 1989b: 26 T. 3.1)

k'an yellow; ripe (Ciudad Real 1984: 238r); matured or ripened fruit; ripened ears of corn and beans; beads or stones that served the Indians as money and as adornment for the throat (Ciudad Real 1984: 238r)

k'áan string, cord, rope; hammock; net of fishers and hunters (Ciudad Real 1984: 237v); a measure of one cord with which the Indians measure their cornfields, called *mecate* (Ciudad Real 1984: 237v, 238r); a measure of 20 meters or 24 *varas*, a *mecate* (Tozzer 1941: 96 n. 427); *k'àan mecate*, a surface measure of 400 [20 by 20] square meters, and a lineal measure of 20 meters (Pérez Toro 1942: 51)

K'ay Yab: *K'àay Yáab'* Song of Many; *K'ay Háab'* Announce the Year

yáab' many (Fisher 1973: 353, item 775)

<*yaab'*> year in Chilam Balam texts (Bricker 1990a, 1990b); *háab'* year (Bricker et al. 1998: 92)

Maac: *Màak* Cover; Pit, Giant Sea Turtle; [Turtle] Shell; Granary; *Mak* Cornfield Measure

màak lid of a box, stopple of a vessel, or lock, and a door that opens and closes (Ciudad Real 1984: 281r); box-like trap (*cepo*) for catching deer or jaguars (Mengin 1972: 56v); pit, trap; giant sea turtle or shell (Mengin 1972: 226); to close up in traps, granaries, etc. (Pío Pérez 1866–77: 210)

máakan galerón, *enrededera*/trellis (McClaran Stefflre 1972: 243: 22.254); *enramada*/bower; to place logs in the form of a quadrilateral to a height of two meters, with a roof of grasses or palm leaves (Pacheco Cruz 1960: 111); Itzaj: granary

(Schumann 1971: 82); *tapanco, troje*/attic, storage space, storage for corn on cob, granary (Hofling 1997: 429)

<*mak*> *estado*, a general measure for cornfields; same as <*k'aan*> (Mengin 1972: 101v); *estadal de doce brazas*/a measure of twelve double arm spans (Pío Pérez 1866–77: 210)

Mool: *Mòol* Pile; Harvest; *Móol* Paw

mol to gather (Fisher 1973: 283, item 446)

mòol collect, join, group, couple (Bastarrachea Manzano et al. 1992: 105)

mòol a thing gathered or brought together, taken and collected (Ciudad Real 1984: 306v), collection, heap. "Harvest" derives from the Itzaj *mol-ik* to harvest (Schumann 1971: 82) or pick up, collect, gather (Hofling 1997: 449), the Lakandon *mal-ik* to gather, pile up, harvest (Bruce 1979: 191), and the Mopán *mol-ik* to harvest (Ulrich and Ulrich 1976: 130).

móol track or print and the paw and feet and hands of the cat, dog, lion, tiger, and other such animals (Ciudad Real 1984: 306v); burr plant (Barrera Marín et al. 1976: 113, item 784)

Muan: *Muan* Hawk; Owl

<*muan*> owl, hawk (Barrera Vásquez et al. 1980: 531, 532). Muwan is spelled several times in the inscriptions as **mu-wa-ni.**

Many scholars long considered the bird representing the month identical to the feathered figure often placed by glyphs or sky bands in the heavens, but some recently have declared them distinct, reading them respectively as *muan* "hawk" and *kuy* "owl" (Macri and Vail 2009: 74–75).

Paax: *Pàax* Upright Drum

pax to play (music) (Fisher 1973: 303, item 541)

pàax music (Fisher 1973: 301, item 532); music, musical instrument (Bastarrachea Manzano et al. 1992: 111); drum, tambour and to play these instruments (Ciudad Real 1984: 369r); erect wooden drum with a single head and carved feet, played by hand (Roys 1949: 175 n. 198); upright drum made from the hollowed out trunk of a tree (Rivera y Rivera 1977: 29, 31)

Poop: *Póop* Reed Mat

póop reed mat (Fisher 1973: 309, item 567); seat of principals (Ciudad Real n.d.: 29v); calendar, almanac (Solís Alcalá 1949: 32, 109). This recalls the mat design on Stela J of Copán as well as Stela H at Quiriguá (Looper 2003: figs. 3.19, 3.21), and the matrix for the day names on page 128 of the Chilam Balam of Kaua (Bricker and Miram 2002: 260).

Popp: *Pohp'* Reed Mat

<Popp> appears several times in Landa, with *pp* usually indicating *p'*, in a dialectal variant of *póop*. It parallels the East Central Tzeltal *pohp'* mat, the local version for the normal *pohp*. Applied to the transition from proto-Tzeltal-Tzotzil to East Central Tzeltal, the rule states that stops and affricates after *h* in the same morpheme glottalize; or more specifically that after **h* the consonants **p*, **t*, **ts*, **ch*, and **k* change to their glottal counterparts. **Pohp* thus becomes *pohp'* (Kaufman 1972: 55, items 3–4; Campbell 1988: 34–35).

Sac: *Sak* White; White Flower; Small Fish

sak white; cataract; diminished or imperfect (Ciudad Real 1984: 91v), white flower (Roys 1967: 85); small fish, sardine, sand smelt, or silversides (Bastarrachea Manzano et al. 1992: 116; Schoenhals 1988: 263, 274, 281)

Siip: *Sìip* Offense; Spite; Ripe; *<Siip>* Release, Miss; Deer Guardian; Shrub

sìip sin, offense, accusation (Fisher 1976: 30–31); *sip* to sin (Po'ot Yah and Bricker 1981: 27); ripe, swollen (Fisher 1973: 321, item 627)

sí'ip to sin, offend, commit an error or crime (Bastarrachea Manzano et al. 1992: 117)

<sip> slippery (Arzapalo Marín 1987: 560); to loosen the bow or snare (Michelon 1976: 441); tutelary god of deer in Yucatán described as a large stag that can never be caught and that appears before the hunter to make him miss his shots (Barrera Vásquez and Rendón 1948: 182 n. 6); supernaturals who guard the deer, making hunters miss; they resemble deer but are small, the size of a dog (Redfield and Villa Rojas 1934: 117–18); they have more developed antlers and bear dangerous winds (Villa Rojas 1986: 181–82); a tree (Pío Pérez 1843: 302)

Sootz': *Sòotz'* Bat (Fisher 1973: 322, item 630)

Tzoz <Tzos>: *Tz'òo's* Bat (1566; Landa's variant of **Sootz**)
On forms with plain *tz* in the original below, (+) marks sources that contrast *tz'* to *tz*, while (−) indicates those that do not.

tzos <tzoz> K'iche': bat; a certain paste (Ximénez 1985: 576–77) (−); same for *<tzotz>* (ibid.) and *<zotz>* (ibid.: 652–53)

<tzoz> Kaqchikel: bat; a certain paste (Santo Domingo 1693: cxix-r) (−)

tzootz' <tsoods'> Yukatek: bat (Schuller n.d.: 37)

tz'utz' <tz'u'tz> Ch'ol: bat (Stoll 1938: 62)

tzutz <tsuts> Chontal: bat (Blom and LaFarge 1927: :475) (−)

tz'otz' <tz'ò'tz> Tzeltal: bat (Sapper 1897: 425)

tzʼotz <*tzʼotz*> Kʼicheʼ: bat (Alvarado López 1975: 70)

tzotzʼ <*tzotz, zotz*> Kʼicheʼ: bat (Schultze Jena 1972: 279)

<*tzotz*> Kʼicheʼ: bat (Basseta 2005: 227r) (–)

<*tzotz*>, <*tzoz*> Kʼicheʼ: bat; a certain paste (Ximénez 1985: 574–75) (–)

(*tzotz*) Poqomchiʼ: bat (Fernández 1937: 195) (–)

Barring metathesis or a loan from Kʼichean, the source and tone of Landaʼs <*tzoz*> remain uncertain.

Tzeec: <*Tzʼek/Tzek*> Foundation; *Tzèek* Clearing; *Tzekʼ* Skull; *Tzeʼek* Itzaj Splinter

<*tzek*> foundations of old houses (Pío Pérez 1866–77: 362)

<*tzʼek*> foundation (Pío Pérez 1866–77: 429); front part of the house (Ciudad Real n.d.: 72r)

tsèek clearing (Bricker et al. 1998: 42)

tsekʼ skull (Ciudad Real 1984: 117r)

tzeʼek Itzaj: a splinter (Hofling 1997: 625)

Tzeec and Sec could descend from *tsekʼ* "skull" through two changes, one from *tz* to *s*, the other for *kʼ* to *k*. Data that support these shifts appear in the following variants of names for the same plants: *x-makʼulam/makulan* (Roys 1931: 263), *ahtokʼ/ahtok* (Brasseur de Bourbourg 1864: 480–81), *tukʼ/tuk* (Barrera Marín et al. 1976: 148–49, items 1186, 1188), *tsiʼtsim/siʼisim* (Barrera Marín et al. 1976: 205, item 96), *tzúulubʼ/súulubʼ* (Bricker et al. 1998: 45, 251), *tsum/sum* (Brasseur de Bourbourg 1864: 294, item 853), and *tsunikʼaax/sunilkʼaax* (Brasseur de Bourbourg 1864: 299, item 892).

Sec: *Sek* Bat (Lakandon). Sec might also derive from one of the words under **Tzeec**.

sek Lakandon: bat (Bruce 1979: 217, 285). Two kinds of vampire bats inhabit the Lakandon rain forest, both called *äh sek* or *mehen sek* (Rätsch and Kʼayum Maʼax 1984: 186 n. 1).

Tzʼe Yaxkin: <*Tzʼè*> *Yáax Kʼìin* Little First Sun; Little Dry Season; <*Tzʼè*> *Yáʼax Kʼìin* Little Green Days; Little Reed Days; Little Dry Season; Little Green Grasshopper

<*tzʼè*> little; bound form of <*tzʼetzʼ*> little, small (Ciudad Real 1984: 129r), from <*tzʼeʼ tzʼè*> (Fisher 1973: 37)

tzʼéeh chip, verbal noun from *tzʼeh*, transitive verb "chip, crack, break into pieces" (Bricker et al. 1998: 49)

<ꟼ*eyaxkin*> summer, the severest part of it, when the rains are about to come (Ciudad Real 1984: 128r)

yáax-kʼìin spring (March, April, May) (Bricker et al. 1998: 312); season of dry spells; summer (Bastarrachea Manzano et al. 1992: 133); <*yax kin*> the summer and autumn of this land during which it does not rain, and the fields dry up and parch;

dry season (Ciudad Real 1984: 217v); autumn, the time in which it does not rain and is cold (Mengin 1972: 154v); spring, from February to April (Ciudad Real n.d.: 185v); *yax* [*sic*] *k'iin* literally "first sun," a hot and dry time of the year in March (Sosa 1985: 72)

yáʾx = k'iin time of new growth [literally "green days"] (Justeson 1989a: 78)

yaʾx k'íin [*sic*] Yukatekan green time, spring (Justeson 1989b: 26 T. 3.1); etymologically: new growth time (Justeson 1988: 13). Thompson's extension of *yáax* from "green" to "new" appears questionable, as <*yax*> means either "first, prior" (*yáax*) or "green; tender" (*yaʾax*) (Bricker et al. 1998: 312), but it does not mean "new" (Bricker 1982b: 354).

Uayab: <*Way-Ab'*> Poison, Hex, Bed; *Wàay Háʾab'* Poison of the Year; *Wáay Háʾab'* Animal Companion of the Year; <*Way*> *Háʾab'* Bed of the Year

way to corrode, contaminate (Poʾot Yah and Bricker 1981: 34)

wàay poison, contagion nominal form of *way* (Fisher 1976: 30–31)

wáay apparition (Bricker 1978: 13); the familiar that the necromancers, witches, and sorcerers have, which is some animal, [into] which by the pact that they make with the devil they change themselves fantastically: and the evil that befalls such an animal also happens to the witch whose familiar it is (Ciudad Real 1984: 439v); to see visions, as when dozing, half asleep, or dreaming [*entre sueño*]; to transform oneself by enchantment; cell, chamber, room or alcove where one sleeps, and the bed itself (Ciudad Real 1984: 439r)

Uayeb: *Wayib'* Bed (Ch'olan); *wäyib'* bed (Ch'ol) (Aulie and Aulie 1978: 133); literally "sleep instrument"

Uol: *Woʾol* Ball, Hole, Heap

woʾol round thing (Fisher 1973: 352, item 771); ball (Mengin 1972: 28v); pellet, lump (Michelon 1976: 387); *uol* hole (Ciudad Real 1984: 451r); heap (McClaran Stefflre 1972: 238). As a month name this appears in no other original list, so it represents an innovation.

Woo: *Woʾ* Toad; *Woʾ Pitahaya* Cactus; *Wooj* Glyph

wóʾ giant toad, as large as one's hat; mud colored, pure fat; call of <*woʾ wóʾ*> (Harrington 1957: 270); poisonous "giant toad" (Schoenhals 1988: 333); certain frogs of much lard and fat, good to eat; they give sad cries (Ciudad Real 1984: 451r); small, nearly black frog with yellow line down its spine (Thompson 1971: 107); bullfrog, common and vocally active in henequen fields during August (Bolles 1990: 86); small, wood-colored frog or toad that booms at night after the first rains have fallen and abounds only in the rainy season (Pacheco Cruz 1958: 272); a frog appearing in large numbers in puddles during the rainy season (Baer and Merrifield 1971: 239) *wòob'/wo'* pitahaya cactus (Barrera Marín et al. 1976: 159, items 1316, 1318–19); red pitahaya or night-blooming cereus, with showy red flowers (Schoenhals 1988: 91)

<*wooh*> letter or character (Ciudad Real 1984: 451r). This interpretation seems likely, given the glyphic spelling *wo-ji* precisely for the month name on a Classic ceramic vase (fig. 7 Uo in appendix C)

Xuul: *Xùul* End; *Xúul* Digging Stick; *Xú'ul* Lance Pod Tree; <*Xuul*> Striped Mullet or Goby; <*Xul*> Maimed

xùul end, finish, extermination (Bricker et al. 1998: 264)

xúul coa or digging stick, sharpened, sometimes with an iron tip; point, tip (Barrera Marín et al. 1976: 43)

xú'ul lance pod tree (Schoenhals 1988: 150); a tree from which they take certain poles for the thatched huts (Ciudad Real 1984: 464r)

<*xul*> maimed, with cut off hand, foot or fingers (Ciudad Real 1984: 464r)

<*xul/xuul*> lisa, a fish (Ciudad Real 1984: 464r), striped mullet or white mullet (Schoenhals 1988: 268); goby, a freshwater fish (Pedro Beltrán, in Brasseur de Bourbourg 1872: 424)

Yaax: *Yá'ax* Blue-Green, Reed; *Yáax* First, Prior

<*yax*> in composition of nouns: first; in composition of verbs: first time (Ciudad Real 1984: 217r); *yáax* first, prior

<*yax*> the color green (Ciudad Real 1984: 217r); *ya'ax* green, tender (Bricker et al. 1998: 312)

Yaaxkin. See **Tz'e Yaxkin** above.

Zip': *Sìip'* Swelling, Ripe

sìip' ripe; swollen (Bastarrachea Manzano et al. 1992: 117)

<*çipp*> tumor, swelling, weals; ripe, swollen; fat, thick (Ciudad Real 1984: 105v). This appears only in Hun Batz' on Men's (1983) list as his innovation on Siip.

SUMMARY: Some but not all the Yukatek names appear in the glyphs. Ceeh, Ch'een, Cumku, Kayab, Poop, Siip, Tzoz, Tz'eyaxkin, and Xuul are never represented as they are pronounced, while Sac and Yaax are recorded, but only as prefixes. The glyphs do represent Mac, Mol, Muan, and Yaxkin or their Classic Ch'olti' counterparts. They also occasionally set down the Yukatek innovations Kankin, Uo, Pax, and Tzec, or Seec. The glyphs seem to record as well the one Yukatek innovation with the most impact, *Sihoom. Even after this, as compounds in the Codex Dresden and Landa's manuscript demonstrate, Yukatek continued to generate new names. Though it was one of the most innovative, resilient, best attested traditions for month names, Yukatek faded first, between 1850 and 1900.

Conclusion

Apparently scholars have discovered most of the native calendars by now; for decades they have recorded virtually no new names. Many of the premodern rolls employ ambiguous orthographies that allow more than one interpretation, as most tables in this chapter demonstrate. Scientific alphabets, dictionaries, and grammars expedite

TABLE 3.12

Definition of *muan* in most Mayan languages

Language	Form	Source term/translation	Source	Comment
Script	*muwan*	raptor		
Ch'olti'	*muhan*	*milano, gavilán/* hawk, falcon	Morán 1935: 32, 46	
Ch'orti'	*mujan*	*águila/*eagle	Girard 1940: 311	
Ch'orti'	*muan*	*gavilán/*hawk, any raptor	Girard 1949: 1:103	
Ch'orti'	*mwahn*	hawk	Wisdom 1940: ix	
Ch'orti'	*muahn*	*gavilán/*hawk, generic	Wisdom 1949: 531	
Ch'orti'	*mwan*	*gavilán grande/* large hawk	Pérez Martínez et al. 1996: 149	
Yukatek	*moan*	*ave negra parecida al pich'* (*Dives dives*), *en Cantamayek*	Refugio Vermont Salas, in Barrera Vásquez et al. 1980: 531–32	
Yukatek	*moan muan*	Yucatan screech owl (*Otus choliba thompsoni*)	Tozzer and Allen 1910: 337; Thompson 1972: 55	
Yukatek	*moan muan*	sparrow hawk	Roys 1940: 43	surname
Yukatek	*moan*	*buho o tecolote cornado;* horned owl *tecolote crescendo* (*Otus guatemalae*)	Garza 1995: 89	*Otus guatemalae:* vermiculated (Guatemalan) screech owl (Peterson and Chalif 1973: 83, pl. 16)
Yukatek	*muan*	crested fal-con (*Spizaetus tyrannus*)	C. H. Berendt, in Brinton 1895: 74 n. 1	still in use
Itzaj	*mujan*	*gavilán/*hawk	Hofling 1997: 454	
Mopan	*mujan*	*gavilán/*hawk	Hofling 2011: 312	

Table 3.13

Meanings of *mo'* in relevant Mayan languages

Language	Form	Gloss	Source
Ch'olti'	*mo'*	macaw	Morán 1935: 32
Ch'orti'	*mo'*	macaw, parrot	Wisdom 1949: 528
Q'eqchi'	*mo'*	macaw	Sam Juárez et al. 2001: 213
Yukatek	*mo'*	macaw	Ciudad Real 1984: 306r
Itzaj	*mo'*	macaw	Hofling 1997: 451
Mopan	*mo'*	macaw	Hofling 2011: 311
Kaqchikel	*mo'*	*milano*/raptor	Coto 1983: 349
Kaqchikel	*mo'*	*gavilán grande negro*/ big black hawk	Guzmán 1984: 45
K'iche'	*mo'*	*gavilán*/hawk	Ximénez 1985: 389
K'iche'	*mo'*	mosquito	Basseta 2005: 203r
Ixil	*mo'*	*azacuán, pájaro negro*/hawk	Chel and Ramírez 1999: 164; *azacuán* is Swainson's hawk (Schoenhals 1988: 366)

correct translations, in the language of each name. For flora and fauna, nonetheless, accuracy usually remains at the species level, with one term applying to several variants of one or more species. Table 3.12 illustrates this with Muan. The diversity of referents likely stems from the ambiguity of the generic glosses in the sources, such as *águila*, *gavilán*, *milano*, hawk, and owl as well as from the nearly total lack of descriptive details. The term remains unattested in Ch'ol, Chontal, Lakandon, Tzeltal, Tzotzil, Chuj, Ixil, Kaqchikel, K'iche', Pokomam, Pokomchi', Q'anjob'al, and Q'eqchi.'

Furthermore, a "word" or sequence of phonemes does not always refer to the same entity across languages or regions or even within them. Sometimes it denotes animals rather dissimilar in appearance. *Mo'* for instance means "macaw" in Ch'olti', Ch'orti', Q'eqchi', Yukatek, Itzaj, and Mopan, while it refers to some kind of hawk or a sort of mosquito in Kaqchikel, K'iche', and Ixil. Languages not listed do not use the term (table 3.13).

Over the past 150 years some traditions have allowed only a few slight modifications, reinterpretations, or innovations. Such changes entered Q'anjob'al and Tzeltal from other rosters, but for Ixil, Kaqchikel, K'iche', Poqom, Tzotzil, and Yukatek they arose internally and did not diffuse. In most instances the reasons for these developments remain opaque. Barring major preservation efforts, nearly every tradition faces eventual oblivion.

CHAPTER FOUR

The Maya Month Initial Dates

This chapter considers the initial dates for the Maya months. It lists the standard names alongside their dates, establishes the dates for the first month of the year and the period of five days, and considers correlations with the zenith passages, solstices, equinoxes, and patron saints. Finally, it scrutinizes the initial dates of months with seasonal names in calendars frozen to the European count.

Ch'olan

Two counts survive from Ch'ol, but just one from Ch'orti' (table 4.1); nothing has come from Chontal. A few scattered names are recorded for Ch'olti', none with dates. One Ch'ol calendar yields both names and dates, but the other has only dates. The Ch'orti' round has lost all its names and frozen its dates.

For the Lanquín roster *K'an Jal-ib' is reconstructed from cognates in the glyphs, Poqom and Q'anjob'al. *Sutz derives from the glyphs. The original initial dates for the first and the twelfth month, Chac, both fall one day early, insignificant errors. If the reversal of the lengths of Mahi and K'än Jäl-ib' was intentional, its purpose remains obscure. The gloss *holob cutan* "nauseated or banished five days" immediately after Mahi, supposedly twenty days long, indicates an unintentional inversion.

The 1631 Ch'ol calendar and the 1962 Ch'orti' roster preserve no names but have secure dates. These demonstrate the Ch'orti' count was frozen, while the Ch'ol still kept to the unfrozen Tikal calendar, even a century after the conquest. The Ch'ol equated 0 *K'an Jal-Ib' (= Pop) to 14 July in 1548. By 1580 their count had receded to 6 July. When the Gregorian reform came in 1584, they let the Christians advance ten days but refused to alter their own year, so 0 *K'an Jal-ib', not 10 *K'an Jal-ib', corresponded with 16 July. In 1589 0 *K'an Jal-ib' matched 14 July, as the Lanquín calendar reports. By 1631 this day had retreated to 4 July. None of the Lanquín saint names designate a person born or canonized after 1500, so the time of its composition remains conjectural.

Table 4.1

Ch'ol and Ch'orti' calendars

Language Town Date	Ch'ol Lanquín 1589	Ch'ol Lanquín revised	Ch'ol Manché 1631	Ch'orti' Jocotán 1584	Ch'orti' Jocotán 1962
*Kän Jäl-ib'	28[29]-VII	14-VII	4-VII	18-VII	18-VII
I Cat	3-VIII	3-VIII	24-VII	7-VIII	7-VIII
Chacât	23-VIII	23-VIII	13-VIII	27-VIII	27-VIII
[Sutz']	12-IX	12-IX	12-IX	16-IX	16-IX
Cazeu	2-X	2-X	22-IX	6-X	6-X
Chichin	22-X	22-X	12-X	26-X	26-X
Ianguca	11-XI	11-XI	1-XI	15-XI	15-XI
Mol	1-XII	1-XII	21-XI	5-XII	5-XII
Zihora	21-XII	21-XII	11-XII	25-XII	25-XII
Yax	10-I	10-I	31-XII	14-I	14-I
Zac	30-I	30-I	20-I	3-II	3-II
Chac	18[19]-II	19-II	19-II	23-II	8-II
Chantemac	11-III	11-III	1-III	15-III	28-II
Uniu	31-III	31-III	21-III	4-IV	20-III
Muhan	20-IV	20-IV	10-IV	24-IV	9-IV
Ah Qui Cou	10-V	10-V	30-IV	14-V	29-IV
Ccanazi	30-V	30-V	20-V	3-VI	19-V
Olh	19-VI	19-VI	9-VI	23-VI	8-VI
Mahi	9-VII	9-VII	29-VI	13-VII	28-VI

The fixed dates of the 1962 calendar have been rearranged into the hypothetical 1532 list. This was done on the assumption that the earlier Ch'orti' tradition more closely resembled the glyphic, Ch'ol, Poqom, and Yukatek, with the first month starting in mid-July and the five final days immediately before it. The correlation of 0 Pop/*K'an Jalib' to 14 July 1548 allows two explanations for the 1962 equivalence of that date to 18 July. Either 0 *K'an Jalib' equated to 18 July in 1532 and the count was frozen then or it was frozen in 1572 at 8 July. The Ch'orti' then integrated the Gregorian correction in 1584 by letting the ten days accrue to the Christian round only, keeping the equation fixed at 0 *K'an Jalib' = 18 July.

Chuj

Termer (1930: 391–94) gives eighteen names from the pueblo Santa Eulalia but doubts the accuracy of the sequence. He records no dates. Recent dates but not names come from the town of San Mateo Ixtatán, where in 1981 the year ended on

1 March or 5 Hoyeb' Ku, fifth of the Five Days (Judith Maxwell in Edmonson 1988: 95). The roster in table 4.2 presents the names from Santa Eulalia alongside the dates from San Mateo Ixtatán, on the assumption that the two pueblos shared the same calendar. Only the last seven names are likely in the correct order, by analogy to the Q'anjob'al sequence. Hoyeb' Ku "Five Days" represents the Q'anjob'al Oyeb K'u and the Chuj Oyebin.

Ixil

The first list (table 4.3) combines two rosters from Chajul; the second (table 4.4) compares three from Nebaj. Each of the nine Ixil series, whether entire or partial, offers several unique names (table 2.11). Altogether innovations account for forty-three of the fifty-six names (77 percent). Probably due to obsolescence, the ten bold names of the Chajul sets differ in location in the sequence.

The Ixil New Year did line up with the Gregorian New Year in the early 1720s, so the natives apparently made leap-year corrections until that time, perhaps to keep

TABLE 4.2
Chuj calendar (1981)

Names	Dates
[Oyebin?]	2-III
Bex	22-III
Sacmay	11-IV
Nabich	1-V
Mo	21-V
Bac	10-VI
Tam	30-VI
Huatziquin	20-VII
Kanal	9-VIII
Yaxaquil	29-VIII
Yaxul	18-IX
Savul	8-X
Xujim	28-X
Mol	17-XI
Meak	7-XII
Oneu	27-XI
Sivil	16-I
Tap	5-II
Oyebin	25-II
Hoyeb' Ku	25-II

TABLE 4.3
Two Chajul rosters combined

Names	Dates
Mol Tche	6-XI-1940
Och'ki	26-XI
Mek'aj	16-XII
Koj'ki	5-I-1941
Talcho	25-I
Nimcho	14-II
A'ki	11-III
Tchohtcho	31-III
Kucham	20-IV
Petzetz'ki	10-V
Xukul('ki)	30-V
Yowal	19-VI
O'ki	6-III
Muen	9-VII
Chentemak	29-VII
Pactzi	18-VIII
Nol'ki	7-IX
Zil'ki	27-IX
Zoj'ki	17-X

TABLE 4.4
Three Nebaj rosters

1973 Don de León	1973 Pap Tek	1973 Don Felipe	Month
Talchój	Talchój	Talcho	March
Masnimchó	Moxnimchó	Nimchó	April
Ne'chó	Ne'chó		May
Lem			May
Akmór	Akmór		June
K'olk'óy	K'olk'óy	K'olk'óy	July
Tsanakváy	Tsanakbáy	Tsanakbáy	August
Maháb	Maháb		August
Muentyín	Muentyín	Muanchím	September
Moch'ú	Molch'ú		October
Paksí	Paksí	Paksî	November
Onchív	Onchív	Onchíl	November
Mochó	Mochnimchó	Pekxitx	November
Sonchój	Sonchój		December
Chantemák	Chantemák		December
Kahabtsé	Kaháb		January
Sojnóy	Sojnóy	Sojnóy	February
Mamak'i			March
Ok'í	Ok'í	Ok'í	March
Okok'í	Okok'í	Okok'í	March
Oval K'í	Oval K'í	Oval K'í	March

the names related to the seasonal and agricultural activities that most seem to designate (Colby and Colby 1981: 47). Only at Nebaj does the calendar continue to run frozen or virtually so. There, depending on the year, New Year's falls somewhere in the first fifteen days of March, suggesting a tie to Shrove Tuesday or Carnival, which falls between 3 February and 9 March (Bond 1889: 142, 143). If the native year did begin on 1 January in the 1720s but then ceased to adjust for leap years, it would now be starting earlier. This indicates that a shift in the choice of first month, not a dismissal of the bissextile, likely moved New Year's to early March.

Outside Nebaj the calendar runs nonfrozen (Colby and Colby 1981: 47). The day-keepers ignore leap-year corrections, so their equation recedes one day every four years. The following fixes on New Year's Day confirm this: 16 March 1927 (LaFarge and Byers 1931: 174), 12 March 1939 (Lincoln n.d.: 112; Colby and Colby 1981: 47); 11 March 1940 (Edmonson 1988: 94) and 1941 (Lincoln 1942: 117); 5 March 1967, 4 March 1971, 3 March 1972 (Colby and Colby 1981: 47). Just after midnight, when a certain star in the constellation Orion dips below the horizon and at the moment when the Seven Marías (Pleiades) attain some position now forgotten, the Ixil New Year begins (Colby and Colby 1981: 45).

By 1942–45, when Lincoln reported the names and dates, the system appeared moribund. The names, dates, and sequence for most of the months were uncertain, unknown, or conflicting. The daykeepers at Nebaj even devised a Europeanized calendar. Its first month lasts 34 days; the penultimate month, beginning in February, has only 26 days, while the last one preserves the traditional five days. All the others total 30 days apiece.

In the years that followed, the Nebaj calendrists reverted to a more native count, but by 1973 even they seemed uncertain about the names and sequence of the months. The three lists from that period not only label the months differently but also assign the same names to different dates. Many names had never been reported before. The year was virtually frozen to the Gregorian round, with New Year's Day beginning between 1 and 15 March. By this time the daykeepers no longer met annually to decide the precise date (Nachtigall 1978: 307).

The 1940–41 Ixil Gregorian dates do not fit exactly with the opening of the native year; this suggests that a ladinoized informant and a few calendar priests attempted to freeze names to the Gregorian calendar for a "perfect" fit with the agricultural round, perhaps to regain an earlier paradigm, when the year was divided into rituals exclusively tied to sowing, cultivating, and harvesting maize (Lincoln n.d.: 111, 120–21). Benjamin Colby and Pierre L. van den Berghe (1969: 99–100) apparently allude to this earlier arrangement of days when they recount that, in addition to the 365-day round not corrected for leap year, the Ixil earlier employed "another calendar of eighteen months associated with agricultural activities which has fallen into disuse, but it is still remembered in part by some of the older men." This calendar must have corrected for leap year to keep in step with the seasons.

The variety and uncertainty between 1939 and 1973 indicate the absence of a unified calendar authority, exacerbated by the passing of the last experts and by the difficulty of communication among settlements isolated in mountainous terrain. Small wonder this tradition records the highest total of both alternates and distinct names.

K'iche'

The K'iche' calendars present identical names in the same order, differing only in the designation and location of the Five Days. If they placed this unit in the same slot, all their dates would agree. The paired names and some of the single ones find counterparts in meaning and relative position among those of the count at Teotitlán, associated with the eastern Nahuas of Puebla, northern Oaxaca, Nonoalco, and Veracruz (Edmonson 1988: 224, 244).

Kaqchikel

The Kaqchikel calendar employs a nonfrozen Vague Year. Its various rosters exhibit no real differences. The many pairs and several individual names find counterparts in position and meaning in the count of Teotitlán, employed in Puebla and northern Oaxaca as well as in Veracruz, home to the Kaqchikel elite's ancestors.

K'iche'-Kaqchikel

These two closely related traditions each produced innovations (table 4.5). The correspondences between the two demonstrate a common legacy from their elite forefathers, likely attributable to their conquest of the native Maya populations and assimilation into them during the period 1350–1450 CE. The K'iche's and Kaqchikels probably shared identical calendars before the arrival of these invaders.

Of the counterpart pairs, half are cognates and one consists of synonyms; the rest juxtapose words altogether unrelated. K'iche' preserves more terms derived from the glyphic list, while Kaqchikel invents more designations. Both share the same three to five names acquired from the Gulf Coast ancestors, but Kaqchikel accepted one from the Aztecs as well. Kaqchikel's eight innovations apparently pushed cognate counterparts out of alignment, perhaps a result of the Kaqchikel revolt in 1493 and the following thirty years of strife with the K'iche's.

Poqom

The four names from Morán's Poqomam and Poqomchi' dictionaries preserve almost unaltered the sounds of their glyphic counterparts. Mol and Muan occupy the same

TABLE 4.5

Kaqchikel and K'iche' calendars in standard order

Kaqchikel (1685)	K'iche' (1722)
Tacaxepual	Tequexepual
Nabeitumuzuz	4,ibapopp
Rucantumuzuz	Çac
CibixiΣ	4hab
Vchum	4,api Σih
Nabeimam	Nabe Mam
Rucabmam	Ucab Mam
LiΣinΣá	Nabe Liquin Ca
NabeitoΣiΣ	Ucab Liquin Ca
RucactoΣiΣ	
Nabeipach	Nabe Pach
Rucanpach	Ucab Pach
Tziquin Σih	4,iquin Σih
	4,içi Lakam
Cakan	Cakam
Ybota	Botam
Katic	Nabe Çih
Yzcal	U Cab Çih
	Urox Çih
Pariche	Chee
Tzapi Σih	

Note: Cognates are aligned and Kaqchikel innovations are in **bold**.

positions in the sequence as their inscriptional forebears, Petcat seems one slot off, while Canazi errs by one place, nudged out by the Five Days. Kanjalam, Tzikin-Kij, Yax, Sak, Chantemak, Uniu, Muguan, and Olj occupy the same niches as their glyphic antecedents. Kaseú is just two places out, but Kanasí a startling thirteen.

Q'anjob'al

The six lists from this tradition (table 2.25) share most of their names. Only Cubihl, Baktan, and Mak each appear on just one roster; all the other names appear in two or more. Q'anjob'al lent most of its calendar to Chuj or imposed it.

List f in table 2.25 resembles the glyphic sequence more closely than the others, with the inversion of Yac (= Yax) Sihom and Sax (= Saq) Sihom being the only discrepancy within its Sihom segment. This color series is correctly framed between Mol and Mak and preceded by Yacakil (= Yax Ak'il) and Gwatsikin (= Watz' Tz'ikin), the counterparts to Yaxkin and Tz'ikin. Though synonymous with Muan "Hawk" of the inscriptions, Mo' appears not in the fifteenth position but the fourth, opposite the Ch'olan winged predator, Sutz' "Bat."

Q'eqchi'

The most remarkable aspect of the Q'eqchi' tradition is its recently devised fixed calendar. This frames the year in terms of an ancient sacred number, thirteen, a unique format.

Tzeltal

This tradition allows little variation in its names, tolerating only the alternates Muk'/Muk'ul and Alal/Alaj/Yalaj/Yaj/B'ik'it/Ch'in before Uch, and these are synonyms. The core of a name never changes. But its position within the sequence sometimes does, though with no apparent significance.

Two features characterize Tzeltal and Tzotzil calendars: First Month and Type. One designates the score of days beginning the native count; its initial day is New Year's. This day's Gregorian equivalent determines the Type. In Tzeltal nearly all the calendars are now nonfrozen, so for comparability they are all set to the same New Year's, here 1 Joken Ajau, in the nonleap year of 1900 (table 4.6). With the earliest date as Type+0, the rest are labeled by the number of days between them and this point. For example, in 1900 Oxchuc started 1 Joken Ajau on 9 May; with the earliest equivalent, the Type for this roster is +0. Lists that convert this date to 10 May are +1. Tzeltal has in all six rounds, consisting of combinations of three First Months (A, B, O) and four Types (+0, +1, +2, +4). The different First Months in these two traditions seem to reflect contacts with Nahua speakers (table 4.7).

In sum, B occurs only with +1 calendars, while O is found with all four Types. Some Tzeltals shifted their First Month to Batzul, probably due to mounting pressure from Aztec merchant-spies (Edmonson 1988: 105, 221–22) and the military advances against the Tzotzil in 1498 and 1511. In 1548 1 Batzul corresponded to 17 January, just like the 1 Izcalli of the Aztecs at Tenochtitlán (Edmonson 1988: 76, 105, 142–44, 221–22, 257–58; Caso 1967: table 10, notes 4, 5). Ajelch'ak, too, has a counterpart in a Nahua tradition that may have come into contact with the Tzeltal. That is Tlacaxipehualiztli in the Teotitlán calendar (Edmonson 1988: 221, 224, 257) of the same K'ichean warrior ancestors who impacted the Tzotzil count between 1250 and 1350 CE. Edmonson (1988: 218, 222, 224) suspects that Joken Ajau was adopted from the Nahuatl Tecuilhuitontli. As it is not a Nahuatl First Month like the other two, Joken Ajau might be the original Tzeltal initial score, as suggested by its position at the head of the list in one source (Schulz 1942: 11) and the prefixed "1" in another (Becerra 1933: table 8).

TABLE 4.6

Tzeltal Types (from 1 Joken Ajau in 1900, with First Months, towns, and sources)

Date in source	1-I-1900	Type	First Month	Town and source
6-V-1912	9-V	+0	O	O; Chanal? Becerra 1933: 45–46 T. 8
6-V-1988	9-V	+0	A	O; Gómez Ramírez 1991: 215–16; frozen to 1915
7-V-1912	10-V	+1	O	C; Becerra 1933: 45–46 T. 8
6-V-1917	10-V	+1	B	C; Schulz 1953: 114–15 T. IIb
30-IV-1941	10-V	+1	O	C, O, T; Schulz 1942: 11
30-IV-1941	10-V	+1	B	C, O, T; Schulz 1953: 114 T. Ia., 1b;
30-IV-1942	10-V	+1	B	Yochib at Oxchuc; León Portilla 1973: 146–49
30-IV-1942	10-V	+1	B	Dzajalchen at Oxchuc; Villa Rojas 1946: 573–75
29-IV-1944	10-V	+1	?	T; Barbachano 1945b: 116; Villa Rojas 1946: 577
27-IV-1952	10-V	+1	B	C, O, T: Schulz 1953: 114 T. 1b; O: Slocum 1953: 87
24-IV-1966	10-V	+1	B	T; Berlin et al. 1974: 119 T. 5.16, 120–24
8-V-1915	11-V	+2	O	T; Becerra 1933: 45–46 T. 8
24-IV-1978	13-V	+4	O	San Cristóbal: Edmonson 1988: 257–58, 281

Note: First month: A: Ajelch'ak; B: Batz'ul; O: Joken Ahau. Town: C: Cancuc; O: Oxchuc; T: Tenejapa. In sources, T. stands for table.

TABLE 4.7

Tzeltal First Months by calendar Types and towns

Town	+0	+1	+2	+4
Cancuc		B, O		
Casas				O
Chanal	O			
Oxchuc	A, O	B, O		
Tenejapa		B, O	O	

Note: A: Ajelch'ak; B: Batz'ul; O: Joken Ahau.

Only two communities use more than one Type. Oxchuc has +0 and +1, while Tenejapa is familiar with +1 and +2. If not due simply to a delay in adopting the Gregorian reform, the reason behind the four Types still needs explaining.

Only Canuc, Oxchuc, and Tenejapa use Batzul as the first month, and all three also report people who start the year with Joken Ajau. The dichotomy suggests a rivalry rooted in foreign domination, because Batzul corresponds to the first month of the Aztecs, who achieved considerable influence in the region during the early 1500s. This implies that Joken Ajau served as the first month among the native Maya. The fourth column in table 4.8 demonstrates the nonfrozen status of the Tzeltal calendars by comparing the number of leap year days due (d) since the previous date to the actual difference (Δ) between the two dates.

Tzotzil

This tradition permits more variation than Tzeltal, granting synonymous modifiers to a few names and alternates to others (table 4.9). Like Tzeltal it adopted and translated one Nahuatl name then shifted the start of the first month to the New Year's Day in three different Nahuatl counts. The sequence of names seems well established, as exceptions are rare.

The 1901 variant repeats the names and sequence of the 1917 standard, but it abolishes Ch'ay K'in distributing its members among the other units. These are seven months of 21 days; ten of 20, and one of merely 18. The shortest includes most of the briefest European month, February. In each Western month of 31 days a winal of 21 days begins; not one begins in a month of 30 days. This system links the month of 30 days to the native unit of 20 days, then applies the deviation above or below 30 to 20, so every month of 31 days corresponds to a winal of 21, while February,

Table 4.8

Dates indicative of nonfrozen status in Tzeltal

Town	1 Joken Ajau	Type	d/Δ	Source
Cancuc	7-V-1912	+1	0/0	Becerra 1933: 45–46 T. 8
Cancuc	6-V-1917	+1	1/1	Schulz 1953: 114–15 T. IIb
Cancuc	30-IV-1941	+1	6/6	Schulz 1942: 11, 1953: 114 T. Ia, 1b
Oxchuc	30-IV-1941	+1	0/0	Schulz 1942: 11, 1953: 114 T. Ia, 1b
Oxchuc	27-IV-1952	+1	3/3	Slocum 1953: 87
Tenejapa	30-IV-1941	+1	0/0	Schulz 1942: 11, 1953: 114 T. Ia, 1b
Tenejapa	29-IV-1944	+1	1/1	Barbachano 1945b: 116; Villa Rojas 1946: 577
Tenejapa	24-IV-1966	+1	5/5	Berlin et al. 1974: 119 T. 5.16, 120–24

Note: T. in sources stands for table.

TABLE 4.9

Usual Tzotzil dates and Cancuc variants

1917 Mitontic	Start date	1910 Cancuc	Cancuc total
Tzun	28-XII	1-I	[20]
Batzul	17-I	21-I	[21]
Sisac	6-II	11-II	[18]
Chaiquin	26-II	abolished	
Muctasac	3-III	1-III	[20]
Moc	23-III	21-III	[21]
Olalti	12-IV	11-IV	[20]
Ulol	2-V	1-V	[20]
Oquinaghual	22-V	21-V	[21]
Uch	11-VI	11-VI	[20]
Elech	1-VII	1-VII	[20]
Nichilquin	21-VII	21-VII	[21]
Ghunvinquil	10-VIII	11-VIII	[21]
Xchibalvinquil	30-VIII	1-IX	[20]
Yoxibalvinquil	19-IX	21-IX	[20]
Xchanibalvinquil	9-X	11-X	[21]
Pom	29-X	1-XI	[20]
Yaxquin	18-XI	21-XI	[20]
Mux	8-XII	11-XII	[21]

usually two days shy of 30, has 18 days. A February of 29 days probably generates a winal of 19 days (table 4.10).

Frozen, the Tzotzil calendars derive their Type from the Gregorian equivalent to New Year's, 1 Tzun. The variants range from earliest to latest. This yields seven Types; combined with three or four First Months, these give the attested twelve to fifteen calendars. Not only does Tzotzil have more Types than Tzeltal, but they all start later, ranging from five to nine days instead of one, two, or four. Some communities use two Types separated by merely one day (table 4.11). This could result from two freezings, one to the Julian calendar in 1548 and one to the Gregorian in 1584, with the result of an overall one-day advance because the leap years between the freezings had canceled nine of the ten days' difference between the calendars (Edmonson 1988: 136). This does not explain why one community would keep both systems or why the Tzotzil developed more than two Types.

The Tzotzil calendars use altogether four First Months. Batzul and Tzun probably did not qualify, however, until assigned that role by the encroaching Aztec. In 1498 the hosts of Ahuitzotl subjugated Comitán (Campbell 1988: 277); the armies under Moteuczomah Xocoyotl seized the highly desirable commercial center Zinacantán in 1511 or 1512 (Hassig 1988: 216–18, 228, 231). It seems that Comitán abruptly shifted the start of the year to Batzul, the equivalent for the invaders' initial month

TABLE 4.10

Tzotzil calendars: frozen (with Types, First Months, towns, and sources)

1 Tz'un in source	Type	First Month	Town	Source
20-XII-1947	+0	B	CN	Berlin 1951: 157
25-XII-1912	+5	M	CC, CN, M	Becerra 1933: T. 7
26-XII-1844	+6	m	IL	Edmonson 1988: 183–84
26-XII-1914	+6	m	IL	Becerra 1933: 39 T. 7
26-XII-1943	+6	T	M	Barbachano 1945a: 16
26-XII-1947	+6	B/m	IL, YU	Berlin 1951: 157
26-XII-1967	+6	T	CH	Gossen 1974b: 242
27-XII-1941	+7	M	CH, CN, IL, A, YU	Schulz 1942: 9
27-XII-1944/55	+7	M	CN	Guiteras-Holmes 1946: 188
27-XII-1944/55	+7	T	CN	Guiteras-Holmes 1961: 32–33
27-XII-1957	+7	M	IL	Holland 1963: 295–96 n. 14
27-XII-1965	+7	M	CH	Whelan 1967
27-XII-1967	+7	T	CH	Gossen 1974b: 242
28-XII-1688	+8	B	CM	Juan de Rodaz, in Ruz 1989: 143–44
28-XII-1688	+8	M	G	Rodaz in Vogt 1969: 603
28-XII-1912	+8	M	TA, YU	Becerra 1933: 40 T. 7
28-XII-1917	+8	T	M	Schulz 1953: 114–15
28-XII-1941	+8	M/m	M, P	Schulz 1942: 9, 11–12; Berlin 1951: 158
28-XII-1968	+8	T	CH	Gossen 1974b: 231
29-XII-1967	+9	M/m	CH	V. Bricker in Edmonson 1988: 260
1-I-1901	+12	T	C	Starr 1902: 72

Note: Most Tzotzil calendars function as frozen. The letters for the first months are B: B'atz'ul, m: Mok, M: Muk'ta Sak, T: Tz'un. The abbreviations for the towns are as follows: C: Cancuc; CC: Chalchihuitán; CH: Chamula; CM: Comitán; CN: Chenalhó; G: Guitiupa; IL: Istacostoc-Larraínzar; M: Mitontic; P: Pantelhó; TA: Tanhobel-Aldama; YU: Yolotepec-Utrilla. T. in sources stands for table.

Izcalli (Caso 1967: table 10 nn. 4–5; Caso 1971: 339–40); Chenalhó, Istacostoc, and Utrilla followed suit, suggesting that they also shouldered Aztec tribute demands, synchronizing themselves with the imperial count to ensure timely delivery. Though apparently placed first in the Aztec capital, Izcalli figured last in the counts of virtually all other Nahua calendars (Caso 1967: table 10).

Frozen to the European round in 1548, the Aztec New Year's day corresponded to 17 January Julian (Edmonson 1988: 136, 144), the identical date for Comitán's counterpart 1 Batzul when that Tzotzil roster along with others was also tied to the Western round that year (Dionicio Pereira in Ruz 1989: 143–44). For reasons

TABLE 4.11

Tzotzil towns with Types and First Months

				Type			
	+0	+5	+6	+7	+8	+9	+12
Cancuc							T
Chalchihuitán		M					
Chamula			T	M, T	T	M, m	
Chenalhó	B	M		M, T			
Comitán					B		
Guitiupa					M		
Istacostoc-L.			B, m	M			
Mitontic		M	T		M, m, T		
Pantelho					M, m		
Tanhobel-A.				M	M		
Yolotepec-U.			B, m	M	M		

Note: The letters refer to first months: B = B'atz'ul, M = Muk'ta Sak, m = Mok, and T = Tz'un.

not yet apparent, however, the daykeepers at Istacostoc and Utrilla hesitated to fix their count to the Western year until sometime in the span 1556 to 1559, for their 1 Batzul equals 15 January. The calendrists of Chenalhó temporized even longer, up until 1580–83, because their 1 Batzul corresponds to 9 January (table 4.13). Only Comitán has just one First Month, and that is Batzul.

As Batzul mirrored Izcalli (the initial score of days at Tenochtitlan), so Tzun (the First Month in other Tzotzil towns) corresponded to Tititl, the start at Texcoco (Edmonson 1988: 222, 258–59). Close neighbor to the Aztec capital, this city and the smaller Tacuba nearby joined Tenochtitlán to form the Triple Alliance, a pact for empire building. Expeditions for trade and tribute were joint ventures. Though superior in cultural endeavors, Texcoco conceded military and tributary primacy to Tenochtitlán. Its share in tribute fully equaled that of the capital, but all imperial lucre flowed first to the seat of power, whose lord then redistributed it to the other two members of the league (Davies 1987: 42–45, 267). The shift to Batzul and Tzun implies that the Aztecs divvied up the conquered or cowed Tzotzil towns with Texcoco then moved the First Months to integrate them into tributary schedules.

The third Tzotzil roster begins with Muk'ta Sak, counterpart to the First Month on the Teotitlán calendar, Tlacaxipehualiztl (Edmonson 1988: 224, 259). This count entered Tzotzil country with the forebears of the K'iche'-Kaqchikel elite sometime between 1250 and 1350 CE, during the first stage in their trek to northwest Guatemala. The Kaqchikels in particular always used this month as their first, while the K'iche's may have already chosen another by the time of their encounter with the Tzotzils. Perhaps this usage survives as a relic of the prestige if not overlordship of some lineages that stayed behind. Muk'ta Sak is the only First Month at Chalchi-huitan, Guitiupa, and Tanjobel. It alone corresponds to Type +5.

The fourth Tzotzil month used to begin the year is Mok. Whenever it occupies the initial position, it immediately follows the five days of Chaykin, a pattern not followed by the other first months. Tequexepual, its K'iche' equivalent, appears on that list in the fifteenth position, not the first (Edmonson 1988: 237, 259). The cognate Teotitlán Tlacaxipehualiztli and Kaqchikel Tacaxepual are First Months but correspond to Mukta Sak. As no other Nahuatl tradition significantly impacted the Tzotzil, they themselves evidently installed Mok as the first month, for reasons still unclear.

Tzeltal-Tzotzil

Tzeltal represents the earlier, more conservative slate as well as the source for many of the Tzotzil names. In their several First Months and in a few names both record the impact of three Nahuatl traditions between 1250 and 1520 CE. With their signature ordinal series, the Tzeltals seem to count down to an event, but the Tzotzil away from one. Though frozen in 1584, virtually all the Tzeltal counts have been unfrozen since 1848–51, while most Tzotzil rounds remain fixed.

Yukatek

The names exhibit virtually no variation; only Yaxkin changes, to Tz'eyaxkin. A small number of lists assign incorrect lengths, but slips of the pen likely account for these. Landa's list presents the standard (table 4.12).

All Calendars: First Months and Five Days

In most frozen calendars the First Months fall in December, January, February, March, and July. The Five Days occur in February, March, May, July, and December, many of them not immediately before their First Months (table 4.13). Apparent explanations involve sun worship or encounters with the Nahuas. The data are too deficient to correlate with patron saints.

The Tzotzil calendars present eight "nonsequential" rosters: their First Months do not immediately follow their Five Days. Two of the four First Months, Batzul and Tzun, appear exclusively in nonsequential rounds, accounting for seven of the sets. The four headed by Batzul start in mid-January, while the three under Tzun begin toward the end of December. Neither may have been first, however, until the Aztecs began to encroach on the region shortly after 1500; by 1512 they had conquered Zina-cantan (Hassig 1988: 231). Batzul paralleled Izcalli, the lead score in Tenochtitlán, the

TABLE 4.12

Landa's Yukatek calendar (1566)

Names	Dates
Pop	16-VII
Uo	5-VIII
Zip	25-VIII
Zodz	14-IX
Tzec	4-X
Xul	24-X
Yaxkin	13-XI
Mol	3-XII
Chen	23-XII
Yax	12-I
Zac	1-II
Ceh	21-II
Mac	13-III
Kankin	2-IV
Muan	22-IV
Pax	12-V
Kayab	1-VI
Cumku	21-VI
Uayeb	11-VII

TABLE 4.13

Names and initial dates of First Months and Five Days (with possible correlations to Nahua counterparts and solar events)

Tradition/Site	Five Days	Date	Event	First Month	Date	Event
GLYPHIC						
Tikal, Palenque	Way-ab'/Hab'	nf		K'än Jäl-ab'	nf	
Caracol	Lok Ajaw	nf		K'än Jäl-ab'	nf	
CH'OL						
Lanquín	Mahi	nf		*K'an Jal-ib'	nf	
CH'ORTI						
Jocotán	?	3-II	fm-5, c fm	*K'an Jar-ib'?	8-II	c
CHUJ						
Santa Eulalia	Hoyeb' Ku?	nf		no name	nf	
Solomá	Oya Cwal	15-III	ve-5	no name 20-II	ve	
IXIL						
Chajul 1942bc	O'ki	nf		A'ki	nf	
Nebaj 1973abc	Ok'í	nf		Tâlchój	nf	
Nebaj 1942F	O'ki	6-III	ve-15	Muenchin	11-III	

(*continues*)

TABLE 4.13 (*continued*)

Tradition/Site	Five Days	Date	Event	First Month	Date	Event
K'ICHE'						
unknown: 1698	Hunobixquih	nf		**Nabe Tzih**	nf	
Momostenango	4,api Σih	nf		Nabe Mam	nf	
Chichicastenango	Tzapiquih	nf		Tequexegual	nf	
KAQCHIKEL						
unknown	Tzapi Σih	nf		Tacaxepual	nf	
POQOM						
Amatitlan/ Cahcoh	unknown	30-V	ss-20?	***K'an Jar-ib'**	14-VII	
Cahcoh	Kaxik-laj-kij	23-XII	ws	**Kan Jal-am'**	16-VII	
Q'ANJOB'AL						
Santa Eulalia	Oyeb' K'u	nf		Wex?	nf	
Q'EQCHI'						
Cahabón	Mayejick	31-XII	yr-end	**Pop**	16-VII	
TZELTAL						
C-I-C-CH-O-T	Ch'ay K'in	nf		Jok'en Ajaw	nf	
C-O-T	Ch'ay K'in	nf		**B'atz'ul**	nf	ac
Oxchuc	Ch'ay K'in	1-V	zn-5	**Ajelch'ak**	10-II	ffm, c; gc
TZOTZIL						
Y	Ch'ay K'in	16-III	ve-5	**B'atz'ul**	15-I	ffm; ac
I-Y	Ch'ay K'in	16-III	ve-5	Mok	21-III	ve
CN	Ch'ay K'in	18-II	c	**B'atz'ul**	9-I	ffm; ac
CM	Ch'ay K'in	18-III	ve-5	**Muk'ta Sak**	27-II	c; gc
CM-M-P	Ch'ay K'in	18-III	ve-5	Mok	23-III	ve; e
CH-CN-M	Ch'ay K'in	23-II	c	Muk'ta Sak	28-II	c; gc
CM-M	Ch'ay K'in	24-II	c	**Tz'un**	26-XII	ws+5, fm; tc
I-Y	Ch'ay K'in	24-II	c	B'atz'ul	15-I	ffm; ac
T-CM-CN-I-Y	Ch'ay K'in	25-II	ve-25; c	Muk'ta Sak	2-III	ve-20; c; gc
CM-CN-IP	Ch'ay K'in	25-II	ve-25; c	**Tz'un**	27-XII	ws+5, fm; tc
C	Ch'ay K'in	26-II	c	B'atz'ul	17-I	ffm, ac
CM-M IP	Ch'ay K'in	26-II	c	**Tz'un**	28-XII	ws+5, fm; tc
G-M-P-T-Y	Ch'ay K'in	26-II	c	Muk'ta Sak	3-III	ve-20; c; gc
YUKATEK						
Campeche	Way-ab'	nf		Póop	nf	
Mayapán, Valladolid	Way-eb'	11-VII		Póop	16-VII	

Note: **Boldface** marks nonsequential rosters, whose Five Days do not immediately precede their first month. All calendars with specific dates for "5d" and "1m" are frozen. Key: 1m: First Month; 5d: Five Days; ac: Aztec counterpart; c: Carnival; e: Easter; ffm: fictive first movement (out of winter solstice); fm: first movement; gc: Gulf Coast Nahua (Teotitlán) counterpart; nf: not frozen; ss: summer solstice; tc: Texcoco counterpart; ws: winter solstice; zn: zenith passage north; zs: zenith passage south; zst: zenith passage south, Teotihuacán. Tzeltal towns: C: Cancuc; CH: Chanal; LC: Las Casas; O: Oxchuc; T: Tenejapa. Tzotzil towns: C: Comitán; CH: Chalchihuitán; CM: Chamula; CN: Chenalhó; G: Guitiupa; I: Istacostoc; IP: Istacostoc 1845; M: Mitontic; P: Pantelhó; T: Tanhobel; Y: Yolotepec.

Aztec capital (Caso 1967: table 10 nn. 4–5). Tzun corresponded to Tititl, the head month in Texcoco (Edmonson 1988: 222, 258–59), partner of the Aztec partner. Mok, another First Month, parallels Tozoztontli, roster head in Nahuatl-speaking Cuitlahuac (Edmonson 1988: 259, 224, 221), who had been Aztec subjects for some eighty years by this time (Hassig 1988: 150). Whenever initial, Mok immediately follows the five last days. The final nonsequential score is Muk'ta Sak, equivalent to Tlacaxipehualiztli, first month in the Teotitlán calendar (Edmonson 1988: 224, 259; Caso 1967: table 10 n. 15). This count entered the Tzotzil ken between 1250 and 1350 CE via the ancestral K'iche'-Kaqchikel elite migrating to Guatemala.

Likely due to pressure from the Aztecs around 1500–10 (Hassig 1988: 231), some of the nearby Tzeltals shifted the start of their year to Batzul, counterpart to Izcalli, the first score in the year for the interlopers from Tenochtitlán, at least according to one source (Edmonson 1988: 257, 221–22; Caso 1967: table 10 n. 5). In each roster that it heads Batzul does not follow right after the final five days. The next Tzeltal First Month, Joken Ajau, finds its Nahuatl counterpart in Tecuilhuitontli, head of the list in the town of Chalco (Edmonson 1988: 257, 224, 222), tributary to the Aztecs since 1453 (Davies 1987: 56). The last nonsequential month, Ajelch'ak, has a Nahua not Nahuatl counterpart, Tlacaxipehualiztli. As the first score in the Teotitlán calendar (Edmonson 1988: 221, 224, 244, 257), it points to the 1250–1350 CE intrusion by the warrior ancestors of the migrating K'ichean elites.

Only the unique K'iche' variant recorded by Domingo de Basseta in 1698 but referent to 1588–91 ran nonfrozen. Its First Month, Nabe Tzih, corresponds to that of the Texcoco calendar, Tititl (Edmonson 1988: 221, 237).

In sum, it comes as something of a surprise that all the nonsequential rosters point to ties with the central Mexican highlands. The Tzeltal, Tzotzil, and K'iche' peoples each begin their count with a month that corresponds to the First Month in a Nahua tradition with which they were or once had been in direct contact (the Aztecs, the Texcocans, or the Putuns, the Mexicanized Gulf Coast ancestors of the K'iche's and Kaqchikels). Imposed tribute schedules might account for the shifts. Only the First Months record significant encounters with Nahua traditions. The role for the Five Days remains unclear. Before the Spanish conquest this unit changed position, perhaps to synchronize Maya First Months with Nahua counterparts; but in the colonial era it aligned itself to the brief dangerous period of Carnival or Holy Week.

The First Months and their Five Days demonstrate unexpected variation within and among themselves, as table 4.13 records. Of the fifteen traditions only five present one name for the First Month: the glyphs, Kaqchikel, Q'anjob'al, Q'eqchi', and Yukatek. Ch'ol, Ch'orti', and Chuj each have two initial months, while Ixil, K'iche', Poqom, and Tzeltal use three. Tzotzil has the most, with four. Almost all uniquely designated Five Days occupy their own distinct slot in a list, but in Tzeltal and Tzotzil each such name occurs in two places. In six traditions the Five Days immediately precede their First Month: glyphs, Chuj, Ixil, Kaqchikel, Q'anjob'al, and Yukatek. K'iche', Poqom, and Q'eqchi', however, position their Five Days once before a month other than the First Month, while Tzeltal does this twice and Tzotzil three times.

All Dates

As table 4.14 indicates, not all the calendars started the months on the same days, even in 1548–49, when the first counts were frozen. Ch'ol (1631), K'iche', and Kaqchikel divide the year into many identical segments, as do Poqom, Yukatek, Tzeltal, and Tzotzil. Chuj, Q'anjob'al, and Ixil each section the annual round at points quite unlike those in other traditions. The difference between any two calendars likely results from shifts of the Five Days or freezes imposed on whole counts.

Foci for the Frozen Calendars

With "frozen" or fixed calendars attention turns to the dates of First Months and Five Days. The data indicate explanations in annual markers such as astronomical events, church festivals, and the seasons.

TABLE 4.14

Dates in July 1548, with Five Day periods in known positions

Ch'ol	Ch'ol	Ch'orti'	Chuj	Ixil	K'iche'
1932	1631	1962	1981	1942bc	1698
1548	1548	f1532	1973	1548	1548
nf	nf	f	f	±f	nf
9-VII	14-VII	18-VII	20-VII	9-VII	14-VII
29-VII	3-VIII	7-VIII	9-VIII	29-VII	3-VIII
3-VIII	23-VIII	27-VIII	29-VIII	18-VIII	23-VIII
23-VIII	12-IX	16-IX	18-IX	7-IX	12-IX
12-IX	2-X	6-X	8-X	27-IX	2-X
2-X	22-X	26-X	28-X	17-X	22-X
22-X	11-XI	15-XI	17-XI	6-XI	11-XI
11-XI	1-XII	5-XII	7-XII	26-XI	1-XII
1-XII	21-XII	25-XII	27-XII	16-XII	21-XII
21-XII	10-I	14-I	16-I	5-I	10-I
10-I	30-I	**3-II**	5-II	25-I	30-I
30-I	19-II	8-II	**25-II**	14-II	19-II
19-II	11-III	28-II	2-III	**6-III**	11-III
11-III	31-III	20-III	22-III	11-III	31-III
31-III	20-IV	9-IV	11-IV	31-III	20-IV
20-IV	10-V	29-IV	1-V	20-IV	10-V
10-V	30-V	19-V	21-V	10-V	**30-V**
30-V	19-VI	8-VI	10-VI	30-V	4-VI
19-VI	**9-VII**	28-VI	30-VI	19-VI	24-VI

(continues)

TABLE 4.14

Dates in July 1548, with Five Day periods in known positions (*continued*)

Kaq	Pc	Q'eq	Tzeltal	Tzotzil	Yuk
1703	1914	1932	*1548	1688	1553
f1548	f1932	f1548	f1548	f1548	f1548
nf	f	f	f	f	f
14-VII	16-VII	12-VII	21-VII	21-VII	16-VII
3-VIII	5-VIII	1-VIII	10-VIII	10-VIII	5-VIII
23-VIII	25-VIII	21-VIII	30-VIII	30-VIII	25-VIII
12-IX	14-IX	10-IX	19-IX	19-IX	14-IX
2-X	4-X	30-IX	9-X	9-X	4-X
22-X	24-X	20-X	29-X	29-X	24-X
11-XI	13-XI	9-XI	18-XI	18-XI	13-XI
1-XII	3-XII	29-XII	8-XII	8-XII	3-XII
21-XII	**23-XII**	19-XII	28-XII	28-XII	23-XII
10-I	28-XII	8-I	17-I	17-I	12-I
30-I	17-I	28-I	6-II	6-II	1-II
19-II	6-II	17-II	26-II	**26-II**	21-II
24-II	26-II	**9-III**	18-III	3-III	13-III
16-III	18-III	14-III	7-IV	23-III	2-IV
5-IV	7-IV	3-IV	27-IV	12-IV	22-IV
25-IV	27-IV	23-IV	**17-V**	2-V	12-V
15-V	17-V	13-V	22-V	22-V	1-VI
4-VI	6-VI	2-VI	11-VI	11-VI	21-VI
24-VI	26-VI	22-VI	1-VII	1-VI	**11-VI**

Note: The upper dates designate each calendar's source year; the lower dates indicate referent years. The "f" means frozen, "nf" never frozen, and "<+−>" mostly frozen. **Boldface** marks the Five Days.

FIRST MONTHS AND FIVE DAYS

In regard to First Months or Five Days, no one name or date is shared among all the traditions. Sometime between the end of the Classic (900 CE) and the first year of ethnographic record, several counts shifted the Five Days from the end of the year to other points, most in February and March.

Solar Movements: Equinoxes, Solstices, and Zenith Passages

Six annual solar events qualify as likely candidates for the initial dates of First Months and Five Days: the two solstices, the two equinoxes, and the two zenith passages. Table 4.15 gives the dates for the solstices and equinoxes, while appendix G contains the dates for the local zenith passages. The summer solstice marks the start of summer and the sun's northernmost rise or set position along the horizon. The winter solstice begins winter, with the sun at its southernmost rise or set position. The equinoxes define the onset of spring and fall, when the sun rises and sets about midway between the solstice points. The zenith passages take place when the sun, heading north or south along the horizon, stands so exactly overhead at noon that an upright gnomon casts no shadow (Aveni 1980: 60–66, 118). These zenith dates vary by latitude, with the northern passage (first solar zenith) occurring between 29 April and 28 May, and the southern (second solar zenith) from 16 July to 14 August.

All the sources allow some variation for the events but do not specify limits. As Anthony Aveni (1980: 62–63) observes, the daily shift in the sun's position on the horizon varies most noticeably around the equinoxes, almost one full sun's width a

TABLE 4.15

Range of dates for equinoxes and solstices

Spring equinox	
19-III	Kluepfel 1986 in Bricker 1997: 8
20-III	Kluepfel 1986 in Bricker 1997: 8
ca. 21 III	Aveni 1980: 61, fig. 24; 99; Carlson 1990: 89; Kluepfel 1986 in Bricker 1997:8
Summer solstice	
ca. 21-VI	Aveni 1980: 61, gig. 24, 62; 100; Kluepfel 1986 in Bricker 1997: 8; Krupp 1991: 84; Vogt 1997: 111
ca. 22-VI	Harris and Levey 1975: 2559; Kluepfel 1986 in Bricker 1997:8; Sosa 1985: 421; Vogt 1997: 111
23-VI	Kluepfel 1986 in Bricker 1997: 8
Fall equinox	
ca. 20-IX	Aveni 1980: 61, fig. 24, 62; 99
22/23-IX	Kluepfel 1986 in Bricker 1997: 8; Krupp 1991: 84
ca. 23-IX	Harris and Levey 1975: 884; Newman 1967: 417; Ridpath 1979: 70
24-IX	Kluepfel 1986 in Bricker 1997: 8
Winter solstice	
ca. 20-XII	Aveni 1980: 100; Kluepfel 1986 in Bricker 1997: 8
ca. 21-XII	Aveni 1989: 290, 294
21/22-XII	Kluepfel 1986 in Bricker 1997: 8; Krupp 1991: 81
ca. 22-XII	Aveni 1980: 61, fig. 24; 62; Harris and Levey 1975: 2559
23-XII (= 13-XII Julian)	Julian: Landa 1978: 96, 173 n. 60

day in the Maya latitudes, but only very slightly about the solstices, hardly changing at all in a week. Perhaps each event runs for a range of two or three days in a cycle of four consecutive years. Brief extensions to the event dates could accommodate observations to features of the local topography such as peaks, notches, and holy spots, or to the vicissitudes of politics, or to the schedule of the local syncretism of native beliefs with Christianity.

As none of the Five Days or First Month initial dates fall between 25 July and 25 December, the autumnal equinox cannot be a referent. The vernal equinox, however, if reckoned as 22 and 23 March, might serve as the referent for all six March dates: 21 and 23 March would serve as alternates for the equinox itself, while 2 and 3 March or 16 and 18 March would herald its arrival, twenty or five days in advance. The Poqom of Amatitlán start their Five Days on 30 May, a possible countdown to 4 June, the beginning of the month that contains the summer solstice.

Several initial dates apparently focus on the winter solstice. The relation between the winter solstice and its associated dates needs some clarification. Vogt (1997: 111) observes that while the Zinacantán Tzotzils

> are aware of the solstices, the times of the solstices have been conceptually absorbed into the saints' days in the Catholic calendar. The winter solstice is conceived of as the time of the ceremonies of the birth of the Christ child and the major winter festival of San Sebastian [20–25 January]. This is the six-week period of the most intensive and complex ceremonial activity . . . and it appears to symbolize the end of the old year and the beginning of the new year (Vogt 1976; Vogt and Bricker 1996).

A few Tzotzils at Chamula realize that by 1 Tz'un (26, 27, or 28 December), their New Year's Day, *"the sun's apparent position at sunrise has moved ever so slightly to the north . . .* these few careful observers see the beginning of tz'un as the passing of the winter solstice" (Gossen 1974b: 231, emphasis added). Another Tzotzil, listing the sunrise points along the horizon, states that *the sun is in motion even through January 6* [Feast of Three Kings]; it then "stays over the hamlet of P'ih most of January and *starts coming back* at the end of the Fiesta of San Sebastian [January 25]" (Vogt 1992 in Vogt 1997: 111; emphasis added). The Tzotzils of San Pedro Chenalhó begin their month B'atz'ul on *16 January, when "the sun begins to move to the north, and the days become longer"* (Guiteras-Holmes 1961: 36 n. 1, emphasis added). In Jocotán the Ch'orti's equate the first day in their calendars of 260, 360, and 365 days to *8 February, the canonical day of the sun's first perceptible northward movement along the horizon,* out of its winter solstice extreme in the south (Girard 1962: 3–4, 7, 29–30, 148, 325; 1966: 4, 7, 20, emphasis added). For Evon Vogt, as the real motion begins much earlier, the winter solstice is better defined as a period than as an astronomical event. An alternative explanation recasts the apparently Fictive First Motions as former Real First Motions shifted one or two positions due to changes in First Months, plus perhaps moves of Five Days. The Fictive First Motion of 16 January, for example, was originally a Real First Motion on 27 December; 8 February really indicated the sun's first perceptible move on 30 or 25 December.

With 21, 22, or 23 December as the solstice, the Tzotzil First Months beginning on 26, 27, and 28 December mark the start of the year, five days after the solar standstill. Given the native concepts, 26, 27, and 28 December can be interpreted not only as the beginning of the year but also as the day of the sun's first detectable movement. Given the range assigned, from 26 December to 8 February, other dates fit in with this phenomenon, specifically 30 December; 9, 15, and 17 January; and 3 February. Events involving tribute to the Aztecs might explain the three Tzotzil First Months in January.

The crossing over of the sun from the southern skies to the northern skies or from the northern to the southern is called the zenith passage. All tropical sites experience both passages, but for each the dates depend on the local latitude. In the Maya area the northern passage occurs between 29 April and 28 May and the southern between 16 July and 14 August. Within these limits fall four dates related to New Year's: 1 May, 16 July, 19 July, and 24 July. The first day of May would alert the Tzeltals of Oxchuc to the arrival of the northern zenith passage there five days later. The crossover occurs throughout Tzeltal country between 6 and 8 May, during the month Hulol "arrive at the middle" (27 April–16 May). The neighboring Tzotzils observe the same phenomenon between 1 and 9 May, during their own month Hulol (2–21 May). Perhaps the southern zenith passage serves as the starting point of the Tzotzil series of months named 1–4 Score. At Zinacantán the passage occurs during the feast of the solar god San Lorenzo, which runs from 7 to 11 August (Hunt 1977: 222–24); virtually every Tzotzil calendar starts 1 Score somewhere in 7–11 August. Perhaps the variation depends on a town's location or its date for the winter solstice.

In summary several astronomical events tie into the initial dates of some First Months. These are the northern zenith passage, the vernal equinox, and particularly the winter solstice as both a standstill and a period lasting until first perceived or fictive movement.

THE CHURCH CALENDAR

Some motivations for First Month and Five Days start dates lie with the church calendar. These involve the patron saint's feast and alternate dates for Shrove Tuesday.

Only one town relates the year to the day of its patron saint. Jocotán, a Ch'orti' community in eastern Guatemala, lives under the aegis of Santiago. His feast on 24 July begins the year, but it also falls exactly twenty days before the local southern zenith passage.

Carnival or Shrove Tuesday falls anywhere from 3 February to 9 March (Bond 1889: 142, 143). This encompasses the six initial dates for both First Month and Five Days, a dozen dates from 3 February through 3 March. These appear in table 4.16. It is not yet possible to assign a specific year for each date.

TABLE 4.16

First Month and Five Days initial dates versus Carnival

Start:	Years corresponding to start dates
Five Days	
3-II	1598, 1693, 1761, 1818
18-II	1586, 1597, 1608, 1670, 1676, 1681, 1692, 1738, 1744
23-II	1599, 1610, 1621, 1632, 1694, 1700, 1751, 1762, 1773
24-II	1626, 1632, 1637, 1648, 1705, 1716, 1784, 1789, 1846, 1852, 1857, 1868, 1903, 1914, 1925
25-II	1648, 1653, 1659, 1664, 1716, 1721, 1727, 1732, 1800, 1868, 1873, 1879, 1884
26-II	1591, 1596, 1664, 1675, 1686, 1732, 1743, 1748, 1754, 1805, 1811, 1816, 1884, 1895
First Month	
8-II	1622, 1633, 1644, 1701, 1712, 1780, 1785, 1796, 1842, 1853, 1864, 1910, 1921
10-II	1587, 1592, 1660, 1671, 1682, 1728, 1739, 1750, 1807, 1812, 1880, 1891
27-II	1596, 1607, 1618, 1629, 1691, 1748, 1759, 1770, 1781, 1816, 1827, 1838, 1900, 1906
28-II	1623, 1634, 1645, 1656, 1702, 1713, 1724, 1775, 1786, 1797, 1843, 1854, 1865, 1876, 1911, 1922
2-III	1588, 1593, 1604, 1672, 1677, 1683, 1688, 1740, 1745, 1756, 1802, 1808, 1813, 1824, 1892, 1897
3-III	1604, 1609, 1615, 1620, 1688, 1699, 1756, 1767, 1772, 1778, 1824, 1829, 1835, 1840, 1908

Note: Years for Carnival come from Bond (1889: 138–43).

Seasonality

Frozen Calendars and Commensurates

A month name is "seasonal" if it denotes an activity in the agricultural round or a prominent point for an event in the annual meteorological cycle, like onset, peak, or end. Such a name serves as an annual marker only if it is also "commensurate," frozen to the Gregorian year, linked by its dates to the task or phenomenon indicated by the name. The fifteen traditions present fifty-five seasonals: forty-two innovated and thirteen derived from glyphic antecedents. Of these seasonals thirty-five are commensurate.

Two or three commensurates in Tzeltal, Tzotzil, and Ixil as well as quite a few in Poqom and Yukatek clearly come from the glyphic roster. Glyphic seasonals probably never were commensurates, as that calendar did not adjust for leap year; they likely became colonial commensurates just by coincidence, due principally to their multiple possible referents. For instance, Yaxk'in as "dry season" correlates to a period lasting three to six months, depending on location, so it is likely to qualify as a commensurate due to chance. Mol "Pile" in the first century of the glyphic count might have denoted the heaps of dried out slash burned from March to May, with the peak activity in March, or it could refer to the *mol che'*, the collection of chunks of wood left over after these burnings for use as charcoal and firewood. By the time several calendars were frozen in 1548 and 1584, however, Mol could only refer to the gathering of maize, a useful coincidence.

The earliest glyphic Mak perhaps indicated the "enclosure" or cessation of the rains about 550 BCE (Bricker 1982a: 102, 103 T. 3). This would refer not to the end of the rainy season, however, but to the first general decrease of rainfall in late July and early August, a period of two or three weeks, The rains really stop in late October or November, after a second peak in September or October. When some calendars became fixed to the Western round, Mak fell in the very dry days of middle or late March and could no longer point to the diminution of the rains. Instead it might denote "enclose in granaries" (Pío Pérez 1866–77: 210), because at least in Yucatán all the maize rests secure in the bins by the middle or end of March (Redfield and Villa Rojas 1962: 83–84; Pérez Toro 1942: 43). Even if Mak as "Granary" does indicate the days immediately after harvest, in view of the considerable variation in other Maya lands in the races of corn, the types of soil, the altitudes and temperatures, the dependence on irrigation or rainfall, and the quantity and timing of the rain, the name may well fit several harvests, widely spaced across the year. Commensurates should apply to specific areas.

Nonfrozen Calendars and Seasonals

In nonfrozen calendars a date recedes from the true position of the sun about one day every four years. These counts, then, cannot have commensurates, only seasonals. Without the constraints of dates the status of a term as a seasonal depends on its definition. The totals for seasonals are eleven for Ch'ol, ten for Chuj, fourteen for K'iche', twelve for Kaqchikel, and sixteen for Q'anjob'al. The sum for apparent seasonals among the nonfrozen calendars exceeds that for all the commensurates and seasonals of the frozen traditions combined. This contrasts sharply with the expectation that in nonfixed counts neither commensuration nor seasonality would motivate the choice of names. Perhaps the reasons behind name selection here will help explain the retention of so many nonseasonal names in the fixed rounds and why almost every commensurate appears first among the drifting Vague Year of the glyphs. These reasons challenge the assumptions behind commensuration: that a calendar is frozen to track with the solar year and that each commensurate has a correlate.

The multivalence and generality of seasonals help explain their presence in fixed counts, as well as the purposes of both frozen and nonfrozen calendars. Their numerous referents allow them to perform as symbols, as prisms for content. Detached from dates, seasonals focus on meaning. As names for annual phenomena, they represent the recurring, perennial, fundamental realities, natural or cultural: they are mythic. Together the seasonals and other month names summarize and symbolize major features of the culture.

COMMENSURATES OR SEASONALS: WEATHER OR MYTH?

With millennia of collective experience, competent farmers keenly appreciated the environment, the soil, the weather, and the seasons; they knew when to clear, sow, weed, double the stalks, and harvest the ears. Farmers could not know from one year to the next, however, precisely which of the days allotted to an operation fell under the benign or malevolent auspices of various gods. The combinations of divine, natural, personal, and unusual elements guaranteed that no two days with the same names, numerals, or both would show the same face. The interpretation of these puzzles required the esoteric learning of the daykeeper. This specialist procured answers via prayers, crystals, and sortilege. Above all the daykeeper relied on the Sacred Round, the calendar of 260 days that preserved the knowledge indispensable to ascertaining the character of any one day. This count rather than the annual round guided or determined agricultural decisions (Edmonson 1988: 214; Tozzer 1941: 28 n. 154, 112 n. 509). Together the experience of the farmer and the nature of the tzolk'in rendered commensuration redundant if not optional.

Commensuration and the bissextile do not prove essential to the success of farming. For example, intensely agricultural ancient Egypt did not devise the leap year until about 240 BCE, but even then the reform did not gain acceptance until decades later (Quirke and Spencer 1992: 50). The farmers and dynasts had managed splendidly without the correction for 3,000 years. Originally divided into seasons designated Inundation, Sowing, and Harvesting, the calendar quickly slipped out of alignment with the sun; but still the Nile receded and farmers sowed. At first each season consisted of four months bearing only its name and numbered 1–4, but between 1300 and 1100 BCE each of the twelve months received a distinct name derived from its principal festival (Quirke and Spencer 1992: 28). These names made evident the religious, mythic nature of the calendar. Similarly, Mesoamericans named some months after their prominent rituals (Edmonson 1988: 214–15). For the Aztecs in particular, "ceremonies which mark the 18 20-day 'months' of the solar year reenact the myths" (Graulich 1981: 48). Seasonal as well as nonseasonal names can operate as summaries of myth simply by denoting a ceremony or festival, its focus deity, or some item of the paraphernalia.

Various tropical year anniversaries of dynastic events in texts of Palenque, Yaxchilán, Copán, and other sites demonstrate that the Maya knew the true length of the solar year but saw no need to impose corrections. The inventors of the calendar of Kaminaljuyu had achieved modern accuracy already in 433 BCE with their value of 365.2422 days for the tropical year (Edmonson 1988: 117–18). This too points up the symbolic rather than utilitarian nature of the year.

Some daykeepers froze their counts in 1548 or 1584, doubtless under pressure to expedite coordination with the Spanish ecclesiastical and civil authorities. The Tzeltal, Tzotzil, and Yukatek peoples may even have invented several commensurate names. The possible reality of commensurates in the fixed counts, though, merits the consideration below. The observations relate the names, through their translations and initial dates, to seasonally appropriate activities or events, in Gregorian terms.

Seasonal Names

Ixil

The dates for the 1973 calendars from Nebaj are virtually frozen, so four names there likely connect to seasonal referents. The synonyms Moochóo and Pekxitx appear to denote a large brown hawk, while the months with these names start in early, mid-, or late November (Nachtigall 1978: 307) These two names and their dates could indicate the southward migration in October and early November of huge flocks of Swainson's hawks (B. Tedlock 1985: 80, 83; León 1945: 180; Land 1970: 66). Chantemak as "granary on tall posts" falls in December, appropriately, as in this month the Ixil harvest and store seven varieties of corn (Nachtigall 1978: 96–98); one calendar glosses this name: "Little is lacking before the harvest." This contrasts with the only totally frozen Ixil count, the 1942f roll from Nebaj: its Chantemak, Molchu "Flock of Grackles" and Tzikin'ki "Hawk Days" could be commensurate but are not. Even in fixed counts, exactly where they could serve as annual indicators, seasonal names might perform other functions.

Poqom

The Poqoms froze their calendar to the Julian count in 1548 (Edmonson 1988: 235). These Julian dates became Gregorian in 1584, when the count jumped ahead ten places in both systems, preserving the day-by-day correlation. Several month names then corresponded to modern seasonal remarks about them.

For example, Mol "Gather, Heap" begins in early December (3-XII). Its definitions suggest that workers began later in the month to cut and pile up underbrush. This sounds plausible, because in the next period, Yax "Green" (28-XII), peasants clear and sow fields for summer corn by just the tenth day (6-I); the first green sprouts likely poke through before the last day and so supply the name a referent. Next the farmers sow small corn in forested areas. Analogy with the Tzotzil Muk'ta Sak "Large White," the name for both a month and a variety of maize, suggests that the designation of the period as Sak "White" (17-I) denotes this corn. The same applies to Kchip (26-II), when for the second time they sow kernels of the diminutive species, because its name means "Little." As "Granary on Tall Posts" Chantemac (18-III) relates directly to its major activity, the first corn harvest.

During the spring in Guatemala some thirty-eight species of birds migrate north (Land 1970: 18), among them Swainson's hawks, passing through in flocks several thousand strong in late March and into April. Muwan "Hawk" could refer to these, though belatedly, because it spans 27 April–16 May. Their passage north would attract attention not only as a spectacle but also as a sign of imminent rain (Reina 1966: 190). The first weeding around the large yellow maize planted earlier dominates the next month, Kcham "Weeds" (17-V). Caseu (25-VIII) means *pacaya* and so denotes several kinds of palm tree and their shoots. From September into April the trees put out shoots commonly consumed in the dry season (Standley and Steyermark 1958: 251; Wisdom 1940: 64, 476). Kanajal "Yellow, Ripe Corn" (4-X) starts in early October, aptly named for its focal event, the ripening of the maize fields. Finally, the month Tzikin Kij commences on 24 October. Its name can be translated as either "Bird Days" or "Hummingbird Time." Both senses prove apropos: at the end of October the ruby-throated hummingbird, the sole migrant of the thirty-seven species of hummingbirds in Guatemala (Johnsgard 1983: 179; Land 1970: 152; Peterson and Chalif 1973: 106–7), begins to return from the north, while from mid-October through early November swarms of Swainson's hawks barrel through, headed south (B. Tedlock 1985: 80–85).

Q'anjob'al

According to Oliver LaFarge (1947: 168 T. 2), the Q'anjob'al of Santa Eulalia in 1932 began the month Tap on 17 February. The period of five days, called Oyeb K'u, started next, on 9 March. As 1932 was a leap year, these dates mean that Tap lasted not twenty but twenty-one days, thirteen in February plus eight in March. In other words, the Q'anjob'al people decided to freeze their calendar starting with that very year. If 29 February had counted, the five days would have begun on 8 March. The frozen status could mean several commensurates, but the locations of the names are not at all secure and so preclude any reliable assignment to dates.

Q'eqchi'

The only complete Q'eqchi' roster, apparently frozen, dates to 1979. It consists of thirteen months of 28 days each, plus 1 day at year's end, 31 December, and another day after that every leap year. Half the names likely refer to agricultural activities or events, but the lack of local data precludes confirmation. The seasonals are Rakol "Mark Off Fields" (13-VIII), Gkalec "Clear and Weed Fields" (10-IX), Gkatoc "Set on Fire [the cut underbrush]" (8-X), Auuck "Sow"(5-XI), Raxgkim (29-I) as a native copy of the Ch'olan yaxk'in "green days; dry season," a correctly placed reference to the rainless time of year, as this extends from November through February in Guatemala (Edmonson 1988: 90), Gkan "Yellow, Ripe" (26-II), Gkoloc "Harvest Maize" (26-III), and Moloc "Collect [into granaries?]" (23-IV). Though they lack specific dates, two more names from another subtradition may qualify as seasonals. Glossed as "Autumn," Chantemac "Granary on Tall Posts" might refer to the storage of the ears of some variety of maize, while Muhan, glossed as "Spring" but really meaning "Hawk," could refer to the spectacular annual migrations of Swainson's hawks from late March through early April (B. Tedlock 1985: 80–85).

Tzeltal

The Tzeltal people froze their calendar in 1585, with 1 Batzul at 17 January (Edmonson 1988: 257–58). Various dates from several communities in the first half of the twentieth century indicate that many unfroze the count in 1848–51. The dates discussed below correspond to 1585–1848.

Tzun "Sowing [with a planter stick]" (28-XII) could correspond with the sowing of a particular strain of maize, as this takes place between 1 January and 9 February (Gómez Ramírez 1991: 216).

Mak "Wall, Enclosure, Lid" (18-III) corresponds ultimately to "Granary, Bin." The month could refer to a harvest of maize later in March on Lowland riverside plots (Breedlove and Laughhlin 1993: 234).

Hulol (27-IV) might involve two events. As "Piercing" it could indicate the dibble stick sowing finished by 5 May in the high, cold plots (Gómez Ramírez 1991: 216) or carried out in the hot, low country at exactly this time (Redfield and Villa Rojas 1939: 107, 117; Villa Rojas 1946: 573–75) or the final stage of sowing the *sijomal* "maize fields" in the temperate zone between 1 and 16 May (Gómez Ramírez 1991: 216). Glosses from two sources seem appropriate: "Best Time to Sow" and "Spring." As "Arrive at the Middle" Hulol describes the northern zenith passage; this happens throughout Tzeltal country between 6 and 8 May.

Hokinahaw as "Wailer Lord" or "Wailing Days Lord" (22-V) seems to mark the first full month of rains. Among the neighboring Tzotzil the cognate denotes either a demon whose wails and tears explain the thunder and cloudbursts of the season or the crying of the sun-Christ deity, who permits the Earth Lords to send the storms (Gossen 1974a: 296, tale 83, 1974b: 236). The rains start in late April or early May.

The next two months are Alal Uch (11-VI) and Muk Uch (1-VII). If the Tzeltal *uch* "opossum" also means "drizzle" as it does in Tzotzil, then these names as "Little Drizzle" and "Big Drizzle" refer to the rains of the season.

Huk Binkil "Seven Score" (21-VII) heads a seven-score countdown to the start of the month (8-XII) that includes the winter solstice (21-XII). This may serve as the anchor of the series because toward its end, between 5 and 7 August, the sun passes through the zenith on its way south. Four more months complete the downward progression, from "six" to "three"; the assumed "two" and "one" disappeared long ago. The final score, Yaxk'in "Green Days; Dry Season" (18-XI) qualifies as a commensurate, for the dry season does start in November (Gossen 1974a: 28).

Tzotzil

When the Tzotzils froze their calendar to the Gregorian count in 1584, they perhaps replaced several names with newly minted commensurates, but none was recorded until a friar set the first list to paper in 1688. The people of Chamula and most other Tzotzil towns start the year with the winter solstice: "In this order, the meanings of the names of months fall strikingly close to specific agricultural, weather, cosmological and ritual facts about their respective periods" (Gossen 1974b: 243). The highland maize cycle follows it rather closely (ibid., 243, 247).

The head of the year, Tzun "Sowing [with a planter stick]" (28-XII), might allude to three activities. First, it likely indicates the planting of black beans and fava beans in the hot country (Gossen 1974b: 248). Second, it could denote the sowing of some species of maize in Tzotzil land similar to that around the Tzeltal community Oxchuc, where the men sow corn between 1 January and 9 February (Gómez Ramírez 1991: 216). Third, if the definition included not only planter sticks with fire-hardened tips but also those with broadleaf blades called *coas de hoja*, then the name could refer to the use of these sticks in breaking up the ground for the cornfields; sowing starts in earnest in the next month (Gossen 1974b: 248).

Moc as "Wall, Enclosure, Cover" (23-III) might derive from an earlier "Granary, Bin." This sense could relate it to a corn harvest in temperate lands in early March or to a harvest later in March on Lowland riverside plots (Breedlove and Laughhlin 1993: 234).

Hulol "Arrive at the Middle" begins on 2 May and likely indicates the northern zenith passage, because this occurs throughout Tzotzil country between 1 and 9 May. If instead this name denotes "Piercing," it recalls the late April to mid-May dibble stick sowing of maize in the Highlands and temperate zones (Breedlove and Laughlin 1993: 234) as well as in the Lowlands, with the first rains of early May (Stauder 1966: 149, 157; Vogt 1969: 36, 45–46).

The first full month of rains finds acknowledgment in the very next period, Okinahual "Water Days Lord" or "Weeping-Days Lord" (22-V). The second translation recalls the story of a demon whose wails and tears are thunder and downpours as well as the belief that these come from the crying of the sun-Christ deity, who grants the Earth Lords leave to send storms (Gossen 1974a: 296 tale 83; 1974b: 236).

Though slightly off due to Hulol, the next two months focus on the rains. The first, Uch (11-VI) means not only "Opossum" but "Drizzle" (Laughlin 1975: 72); its alternates signify "Small Opossum, Small Drizzle." The second, Elech (1-VII), defies definition, but its equivalents on other Tzotzil rosters clearly denote "Big Opossum, Big Drizzle." May brings the first full month of rains in the Chiapas Highlands (Gossen 1974b: 236) and Lowlands (Gossen 1974a: 33; Stauder 1966: 157). In June the rainfall reaches its first maximum (Vogt 1969: 6). By mid-July the showers taper off into a brief dry spell (Gossen 1974b: 225 T. 1).

As "Flower Days" Nichikin (20-VII) might well correlate with the flowers or tassels that adorn the maize in mid-July (Guiteras-Holmes 1961: 44; Gossen 1974b: 238; Vogt 1976: 56). If the name means instead "Flowers of the Sun," it likely denotes the swarms of brown butterflies that arrive in July and August (Gossen 1974a: 317, tale 122; Laughlin 1975: 252). The sense of *nich* as "strength" apparently applies to the sun, for *nich k'in* denotes the canicula or dog days, an especially hot, dry interval of several weeks in July and August between the heaviest parts of the rainy season (Gossen 1974a: 28, 1974b: 225 T. 1; Laughlin 1975: 252).

The next seasonal name is Hun Vinkil "First Score." Beginning on 10 August, this is the first month after the southern zenith passage of 4–8 August (appendix G). It heads a series of four, with the last one, closing on 28 October, marking the end of the rains.

Yaxk'in "Green Days; Dry Season" (29-X) designates the final Tzotzil seasonal. It fits with the tropical year because the dry season begins in November (Gossen 1974a: 28).

Ch'olan

The one frozen Ch'olan calendar survives in the almost nameless Ch'orti' roster from Jocotán, reported in 1962 but probably frozen in 1532 or 1572. Together with the names and dates of the Lanquín (1589) and Manché (1631) Ch'ol calendars the sparse data available allow the reconstruction of three Ch'orti' names.

Tzikin now denotes the Days of the Dead, 31 October and 1 November, but originally designated the days 21 October–9 November. As *tz'ikin*, the name means "bird, hawk, hummingbird," depending on the area of Guatemala. It perhaps came to identify the days of the dead because the Ch'orti's equate the souls of the departed with birds (Girard 1949: 1:217); possibly the festival assumed the name of its month. Its seasonality rests on three annual events. One is the passage of thirty-eight species of birds through Guatemala every autumn (Land 1970: 18). Another is the start of the southward fly-through or residency of the ruby-throated hummingbird in late October (Johnsgard 1983: 179; Land 1970: 152; Peterson and Chalif 1973: 106–7). The last is the spectacular southward migration in late October of thousands of Swainson's hawks (B. Tedlock 1985: 80, 83; León 1945: 180; Land 1970: 66).

The other reconstructible months are the next two in the series. *Yaxk'in as "Green or Clear Days, Dry Season" (Wisdom 1949: 766) would coincide with the start of the "spring" or near-dry season in early November (Wisdom 1940: 12–13, 468). *Mor as "A Heap, Gather" would correlate with either the second harvest of maize in the Lowlands (Wisdom 1940: 446, 468) or the piles of husked ears.

Yukatek

Of all the traditions, Yukatek has the most remarks on apparent commensurate names, many with details. In 1548 the Yukatek stopped regarding bissextiles as common days and fixed their calendar to the Julian count, so their initial dates lagged ten days behind the seasonal year. When the Gregorian reform eliminated the difference in 1583, both systems simply ignored ten days that year, locking in their day-to-day correlation. As some 250 years of month lists in the Chilam Balam books demonstrate, 1 Pop still corresponded to 16 July, but from then on the two calendars would keep abreast of the sun. Seasonal names would remain relevant.

Martín de Palomar and the scholarly native noble Gaspar Antonio Chi recorded their apparent support for this scenario in their *Relación de la ciudad de Mérida* (1579, in Garza et al. 1983): that at their latitude of 20°20′ north the sun passed through the local zenith on its way north on 10 May Julian (Garza et al. 1983: 65, 73). In one of the rosters of the Chilam Balam of Tizimín the entry opposite 20 May reads *u xocol y oc kin*, "the counting of movement of the sun" (Edmonson 1982: 122, line 3274), more literally "the counting of the leg or staff of the sun." The same date and phrase appear in the Chilam Balam of Ixil (Miram 1988a: 43) and the Codex Pérez (Miram 1988b: 30, 107). The expression evokes the use of a pole as a gnomon to register zenith passages, while the date is the Gregorian equivalent of 10 May Julian. If the note does document native observation of this zenith passage, it links Landa's dates in sources after 1583 to the Gregorian count. Before then Landa's dates are Julian, ten days behind the true position of the sun. For instance, 1 Kumk'u falls on 21 June. If this day is Gregorian, the month begins on the summer solstice; if it is Julian, the month starts ten days after the solstice. The Yukateks froze their calendar to the Julian count in 1548 (Edmonson 1988: 76, 202); support for this assertion comes from Landa, who clearly states that the natives corrected for leap year and so always started their round on 16 July, the first day of Pop (Pagden 1975: 96–97, 107). On adopting the Gregorian correction in 1583, both the Europeans and the Maya must have advanced their counts ten days, as the correlation between them remained identical before and after the reform. Even if the Yukateks did remain on Julian time, the commensurates still related to their events or activities. The original Julian dates appear in the discussion below.

With the meaning "Ripe, Swollen" and the initial date of 25-VIII, Sip could correspond to a period when the milpas or cornfields ripen. In Yucatán this happens in late August and early September (Redfield and Villa Rojas 1962: 83–85). As the term for a small deer or a supernatural protector of deer (Redfield and Villa Rojas 1934: 117–18), the name goes well with the month that brings the start of the deer season, celebrated by hunters through bloodletting as well as a dance with an arrow and a deer skull (Tozzer 1941: 155; Pearse 1977: 115; Redfield and Villa Rojas 1962: 208).

Xul (24-X) might involve several events. As "End" it would coincide with the cessation of the rains in late October (Bolles 1990: 85–87; Page 1933: 418–21) and the termination of the harvest of a variety of maize in November (Harrington 1957: 270–71; Thompson 1925: 122–23). If originally a Q'eqchi' loan for "Bird," the name can refer to either a number of the thirty-eight bird species migrating out of Guatemala in autumn (Land 1970: 18) or to the southbound fly-through of the ruby-throated hummingbird in late October (Johnsgard 1983: 179; Land 1970: 152; Peterson and Chalif 1973: 106–7).

Yaxk'in "Green Days; Dry Season" (13-XI) has as its colonial counterpart Tzeyaxkin "Little Green Days; Little Dry Season." One sense correlates with the onset of the dry season all over Yucatán in mid- and late November (Bolles 1990: 86).

In late December the men finish the harvest of a particular variety of corn, *maíz menudo* "fine-grained [?] maize" (Ciudad Real 1984: 246v; Pérez Toro 1942: 44; Bolles 1990: 85–87). An apt name for this activity designates the month: Mol "Harvest; Pile" (3-XII).

Sak "White" (1-II) seems to denote a white flower in bloom at this time; white flowers of the ceiba tree blossom in late January or early February (Schele and Mathews 1998: 113).

Ceh "Deer" (21-II) might relate to hunting deer. In March, during the dry season, many trees shed their leaves. This makes the deer more readily visible, bringing out the hunters (Bolles 1990: 85–87). The lack of rainfall also drives the deer to more reliable sources of water, known all too well among their pursuers.

Two seasonal meanings can apply to Mak (13-III). One should be "Enclosure, Granary," from *mak* "enclose in granaries" (Pío Pérez 1866–77: 210), because by the middle or end of March the maize ears rest safely in the granaries (Redfield and Villa Rojas 1962: 83–88; Pérez Toro 1942: 43). The other sense, "Turtle, Turtle Shell," can also render the name a commensurate. First, the native author of the Chilam Balam of Chumayel notes in the gloss for Mak that in this month turtles deposit their eggs (Roys 1967: 85). Second, scholars confirm that one or two species do start laying eggs in March (Ernst and Barbour 1989: 125; Harrington 1957: 270–71; Schoenhals 1988: 320, 331–32). Finally, metaphor and iconography link the definitions. On many vessels a supernatural turtle actually is a granary, because from its carapace cleft open by lightning emerges the youthful corn god representing the maize as it sprouts out of the earth (Miller and Taube 1993: 175; Taube 1988a: 195).

Muan "Hawk, Screech Owl" (22-IV) might designate a bird heralding the first rains in late April or early May (Bolles 1990: 85–87). The occasional variant Moan, however, relates to the season more directly: *moan kin* means "cloudy and drizzly day" (Pío Pérez 1866–77: 221), while *moan* itself connotes "overcast and mizzly."

The last score starts on 21 June. Kumk'u "Boom God" designates a major thunder deity of eastern coastal Yucatán, the source of the rains. These associations coincide with the season, as thunder rumbles most often in June and July (Bolles 1990: 87; Sosa 1985: 379).

Two names were not commensurates originally, but arguably the natives reinterpreted them upon yoking their count to the Spaniards' count in 1548. The first, either Wo' "a kind of frog [toad]" (Bricker et al. 1998: 306) or Wo "Bullfrog and Its Call," seems relevant to its span (5-VIII), because these creatures do boom out during August (Bolles 1990: 87; Harrington 1957: 270–71). This event, however, did not generate the name, as this debuted at Chichén Itzá in 880 CE, when it covered 19 January–7 February. An earlier variant spells out the name as *wo-ji* or *WOJ* "glyph." The second noncommensurate, K'ank'in "Yellow Days" (2-IV), could refer to the yellow skies of April produced with the burning of undergrowth hacked away to clear land for the maize fields (Bolles 1990: 87); when the name first appeared in 743 CE at Xcalumkin, though, it fell in the rainy days of 21 October to 9 November. Seasonal names do not necessarily coincide in time with their referents. Some commensurates seem due to reinterpretation.

NAMES AND DATES

The compounds of the scribes apparently focus on meaning, while the glyphs of Landa likely stress dates. No matter how well the seasonal names might fit, the regression inherent in the nonfrozen inscriptional calendar pulls them out of kilter with annual activities and events. Only after 1,507 years does an alignment return. Seasonal names seem pointless in such a system, so their meanings should count for more, as suggesting meanings beyond the seasons. The Yukatek system recorded by Landa fits the seasons well, with twelve commensurates, but four of these turn out to be innovations (Sip, Xul, Keh, and Kumk'u) while two others (Wo and K'ank'in) become commensurates via reinterpretation. The many fits suggest the priests invented new seasonals specifically for the freezing of the count in 1548. The alternative, linguistic origins for these apparent commensurates as well as for the new, nonseasonal Pop, Chen, and K'ayab' are episodes in the Early Postclassic (900–1250 CE) history of Yukatekan.

The invention of commensurates might also have occurred in Tzeltal and Tzotzil, as several focus on visible solar events rendered useful as reference points by the 1548 freezing of those calendars. The Tzeltals counted to the month of the winter solstice, while the Tzotzils counted the months after the southern zenith passage and perhaps invented a name to mark the northern zenith passage. In all instances the ties to the dates did nothing to neutralize the mythic content of the old names or to preclude its permeation of the new ones.

Conclusions

Of the 522 senses permitted for the names, by far most concern agriculture (154) and animals (97) then ritual and religion (73). Under agriculture the largest totals accrue for seasons and weather (76) then activities (61). Mammals (34) and birds (30) constitute the bulk of the animal category. No doubt many of the terms for animals as well as rites and their paraphernalia also relate to agriculture. Though the majority speak to rural concerns, many recall foci of the Late Classic and Postclassic court culture such as accession, bloodletting, and human sacrifice. In some cases too many data preclude the confident recovery of a referent; in others, too few.

Some fixed seasonal terms tie in to dates for specific yearly events but correlate less reliably when allowed two or more referents or several translations. Even in frozen calendars some seasonals are not commensurates, so symbolism and ritual might determine their presence on the lists more than agriculture and the weather do. In a fixed tradition with many commensurates, seasonal translations seem more likely than alternatives.

THE TRADITIONS

The Ch'ol people stayed on the Tikal calendar, never freezing it. The calendar from Lanquín, the sole Ch'ol roster with names and dates, refers to 1548 or 1588. The correlation of 0 *K'an Halib' with 14 July in both 1548 (Julian) and 1588 (Gregorian) is confirmed via the recession from those dates to the Manche Ch'ol New Year's of 4 July 1631.

The Ch'orti' people of Jocotán did freeze their count but over time lost all the names. The dates indicate either 1532 or 1572 for the immobilization. If this occurred in 1572, then only the European count added the ten days of the Gregorian reform in 1583. If the Five Days period was fixed to an actual Carnival or Shrove Tuesday, this correspondence rules out 1532, as the festival cannot fall on 3 February in a leap year; the only possibilities are 1598, 1693, 1761, and 1818. Whenever it happened, by the late 1940s the Five Days were no longer the last unit in the sequence. Perhaps one lone Ch'orti' commensurate survived, Tzikin. As "Bird" it would refer to the migration of numerous avian species during its span of 26 October–14 November.

The Chuj and Q'anjob'al calendars are identical in their dates. The first of the Five Days for Chuj fell on 25 February in 1980–83; recession demonstrates this coincided with 9 March in 1932, the very date recorded for the same day in Q'anjob'al. In that same year Q'anjob'al apparently was frozen; no information on it has come down since. Chuj receded until 1983. Then, with the death of the last daykeeper, it was frozen, perhaps for the first time. In both counts the Five Days always immediately precede the first month. The seasonal names in these two calendars cannot become commensurates: first, because the counts were not frozen; second, because the positions of the names come down as too uncertain or contradictory to allow any alignment.

For the Ixil in 1941 the first of the Five Days fell on 6 March. The noninsertion of leap year days equates this to 8 March 1932. Ixil was exactly where it should be, starting the same unit one day before Chuj and Q'anjob'al. Perhaps when these two immobilized their calendar in 1983, the Ixil followed suit; until then their round had operated as nonfrozen. At Nebaj in 1941 somebody recast the year into twelve months and five days, a variation on the European format never reported again. By 1973 the calendar had become nearly frozen, restricting New Year's to 1–15 March; unfixed, it would have fallen on 3 March. Maybe this window was framed with an eye to Shrove Tuesday, as this festival falls between 3 February and 9 March. The Five Days end the year, but the identity of the first month remains debatable due to the marked uncertainty about the positions of the months. For the same reason as well as because of the earlier nonfrozen status of the count, the many seasonals cannot qualify as commensurates.

As dates contemporary with their sources indicate that the K'iche' calendar always ran nonfrozen, its seasonals cannot serve as commensurates. The names always remain in the same sequence. Without changing location three distinct months each begin the year, one in a dictionary of 1698, another in a lexicon from 1703, and the third in all the rest. Each member of the trio seems to result from a major political event. The label of the Five Days does vary, but just once. Hunobixquih, the only other designation known for the unit, was recorded solely in the 1698 roll. The 1703 list substitutes Tzapiquih, and all subsequent calendars use this name.

In most essentials the Kaqchikel count mirrors the K'iche' count. It runs nonfrozen and its names remain unchanged in form or position; many with cognates in K'iche' coincide with their dates. As in K'iche', several names derive from Nahua, reflecting 300 years of interaction with its speakers. By contrast, Kaqchikel always starts the year with the one same month and ends it with the Five Days; these units both have only one name.

The daykeepers of the Poqoms developed two variant counts and apparently tied them to the Julian round in 1548. Anchored to dates, many seasonals can operate as commensurates. A number of months reflect Ch'ol cognates in form and location.

Tzeltal sun watchers bound the count to the Julian round in 1548, then the Gregorian in 1584. Reasons not yet appreciated led them to release their calendar during 1848–51. Most now still run unfrozen, though in four communities some timekeepers refroze it, in 1917 or 1965–66. While they operated as fixed for over 250 years, numerous versions reinterpreted seasonals into commensurates or even invented them. The daykeepers apparently innovated a single name that marks the northern zenith passage as well as a series that counts down from the southern zenith passage to the winter solstice. The Five Days stay in the same site even though this tradition assigns three months to begin the year. These evoke foreign domination, as they correspond to the First Months of several Nahua communities.

Like the Tzeltals, the Tzotzil people fixed their annual round to both European counts in turn; but in contrast they have kept them frozen into the present. This immobility has converted more than a handful of seasonals into commensurates. Tzotzil sun priests probably devised a series of months to track time after the southern zenith crossover. From their Tzeltal neighbors they accepted the newly minted or reinterpreted name that alludes to the northern zenith passage. The Tzotzils maintain all the names in their places, including the Five Days. The communities altogether employ four distinct First Months; three correlate to Nahua First Months, a likely vestige of foreign overlordship. The initial dates of these four vary between five and nine days, perhaps symptomatic of civil or religious strife.

In 1548 the Yukateks joined their calendar to the European count and never altered the correlation. An apparent allusion in several colonial native books to the occurrence of the northern zenith passage on 20 May instead of 10 May suggests that with the Gregorian reform of 1583 they advanced both the Christian dates and their own by ten days to maintain the correlation. In either case, the freezing converted many seasonals into commensurates. Among these appear four names not attested in the glyphs, possible innovations due to the freeze. Two others debuted in the inscriptions centuries earlier but could have served as commensurates after one freeze or the other. All the Yukatek names but one remain unchanged, and that one simply adds an optional prefix. They also stay in their places. The first month never changes. The Five Days always end the year.

The Foci for Frozen Initial Dates

Whether for the first month or the Five Days, many of the initial dates in the frozen calendars fall on dates significant in annual solar observations or in the church's season of Lent. Fifteen of the twenty fixed rounds commence between late December and late March. The other five start up in the middle of July or toward its end. Most of the Five Days now occur in February and March, but in eight of the twenty frozen rosters they immediately precede their First Months.

The initial dates in March might all refer to the vernal equinox. The one at the end of May could herald the month of the summer solstice. Those of late December probably relate to the winter solstice. Three in January from Tzotzil might find explanation in that people's history with the Aztecs or, along with 3, 8, and 10 February, in "localized" solstices: in the ritual or fictive first movement of the sun in various communities. One initial date seems to presage the local northern zenith passage. Others fall within the range of this event, 29 April–28 May, or the southern zenith passage, 16 July–14 August, but none corresponds to either event in its own region. Thus these initial dates still present a puzzle.

Shrove Tuesday or Carnival could serve as the alternative reference point for the dates of February and March because it falls between 3 February and 9 March. Only 23 March could tie in with Easter, occurring between 22 March and 25 April, but it might also initiate a count based on the vernal equinox.

Seasonality

The daykeepers never intended their seasonal month names to operate as commensurates because they never devised any form of leap year. The very presence of seasonals in such a year indicates a symbolic character. Unattached to dates, the names focus on the meaning behind the annual phenomena; they represent the perennial basic realities of nature and culture.

The retention or addition of noncommensurate seasonals in a frozen calendar underlines their mythic role, as does the operation there of nonseasonals, where every name could be a commensurate. The peasants did not need pragmatic reminders of the familiar facts about crops, weather, and pests. Rather, they wanted advice to make agricultural decisions specific to the day, and for this they looked to the divinatory calendar of 260 days. The round of 365 served symbolic purposes.

The names in the annual cycle recapitulate the fundamental realities of nature, agriculture, and ritual. They articulate the round of ceremonies, festivals, and myths to integrate nature and culture as well as to bind together all the levels of a society through shared beliefs and activities. They relate people not to dates but to time, not to the weather but to the natural world, not to literal interpretations but to complexes of meanings. This year does not track the rains or follow the sun; instead it intimates the myths and distills the rituals. Before contact the months changed their dates, but they kept their names.

CHAPTER FIVE

Continuity in Sound and Sense

The degree of linguistic relatedness among the numerous Maya calendrical month names is rather surprising (see appendixes D and E). Without exception, most or even all the units in a tradition find a synonym, if not a full or partial cognate, in at least one other roster. Even in the case of Ixil, which has the lowest relatedness percentage, forty-one out of fifty-eight names (71 percent) are connected to others. Most striking is the number of names and traditions, beyond Yukatek and Poqom, with cognates and synonyms in Ch'olan. The incredible relatedness of the month names indicates a long-lived, high regard for Classic culture.

Themes: Agriculture, Animals, and Religion

In appendix F every name appears in one or more of the following categories, with subcategories listed here in parentheses: agriculture (activities, plants, seasons and weather); animals (amphibians, birds, crustaceans, fish, insects, mammals, worms); architecture; artifacts; astronomy; body parts; colors; drink and food; geography; gods; hunting; illnesses; kinship; plants (general, trees); ritual and religion; society; time; and trade. Table 5.1 presents the tallies for each theme, by tradition.

Allowing multiple definitions for some names, the total of meanings comes to 522. By far the largest sums accrue for agriculture (152) and animals (101), then ritual and religion (71). Negligible totals typify the remaining subsets. Overall, the emphasis falls on farming, nature, and worship. While the majority of the calendar month name meanings reflect rural concerns, many names under the ritual and religion category preserve features of the court culture attested in Late Classic and Postclassic iconography. Examples include accession, bloodletting, heart excision, arrow sacrifice, and dance.

In terms of translations of other names, including multiple definitions, the totals are Yukatek (51), K'iche' (81), and Ixil (69). Ixil owes its total to its many lists, each with inventions. The lowest scores belong to Chuj (23) and Q'eqchi' (16).

TABLE 5.1
Themes by linguistic tradition

	Theme	Tradition											**Total**
		g	cj	cn	ix	k	kq	p	q	qq	te	to	y	
1	**Agriculture**	**15**	**8**	**9**	**24**	**16**	**17**	**12**	**10**	**8**	**7**	**17**	**9**	**152**
1a	Activities	2	5	4	12	7	5	5	5	6	2	4	3	**60**
1b	Plants	2	2	1	1	0	3	2	3	0	0	3	0	**17**
1c	Seasons and weather	11	1	4	11	9	9	5	2	2	5	10	6	**75**
2	**Animals**	**5**	**6**	**7**	**27**	**4**	**9**	**3**	**6**	**4**	**9**	**11**	**10**	**101**
2a	Amphibians	0	0	0	0	0	0	0	0	0	0	0	2	**2**
2b	Birds	3	3	5	8	3	2	1	3	1	0	1	1	**31**
2c	Crustaceans	0	1	0	1	0	0	0	1	2	1	0	0	**6**
2d	Fish	0	0	0	0	0	0	0	0	0	0	0	2	**2**
2e	Insects	0	1	1	3	1	4	1	1	1	2	1	1	**17**
2f	Mammals	2	0	1	11	0	1	1	0	0	6	9	4	**35**
2g	Worms	0	1	0	4	0	2	0	1	0	0	0	0	**8**
3	Architecture	0	0	0	1	1	1	0	0	0	1	1	2	**7**
4	Artifacts	7	2	4	1	10	3	1	4	1	4	4	7	**48**
5	Astronomy	0	1	1	0	7	1	0	1	0	0	1	1	**13**
6	Body Parts	0	1	0	0	1	0	0	1	0	2	1	1	**7**
7	Colors	0	0	3	0	2	1	2	0	2	0	0	2	**12**
8	Drink and Food	2	1	1	3	0	0	0	0	0	1	2	0	**10**
9	Geography	0	0	1	0	0	0	0	0	0	1	0	2	**4**
10	Gods	0	0	0	1	2	1	2	0	0	2	6	2	**16**
11	Hunting	0	0	0	2	1	0	1	2	1	1	3	2	**13**
12	Illnesses	0	1	2	3	2	2	1	0	0	0	0	2	**13**
13	Kinship	0	0	0	0	2	4	0	0	0	0	0	0	**6**
14	**Plants**	**5**	**1**	**1**	**1**	**5**	**3**	**3**	**5**	**0**	**0**	**1**	**6**	**31**
14a	General	4	1	0	0	3	3	1	5	0	0	0	5	**22**
14b	Trees	1	0	1	1	2	0	2	0	0	0	1	1	**9**
15	Ritual and religion	4	1	5	3	26	10	3	3	0	10	3	3	**71**
16	Society	0	0	1	2	0	0	1	0	0	0	0	0	**4**
17	Time	1	1	0	1	2	2	0	0	0	1	1	2	**11**
18	Trade	0	0	1	0	0	1	1	0	0	0	0	0	**3**
	Totals	**39**	**23**	**36**	**69**	**81**	**55**	**30**	**32**	**16**	**39**	**51**	**51**	**522**

Key: g: Glyphic; cj: Chuj; cn: Ch'olan; ix: Ixil; k: K'iche'; kq: Kaqchikel; p: Poqom (= Poqomchi', Poqomam); q: Q'anjob'al; qq: Q'eqchi'; te: Tzeltal; to: Tzotzil; y: Yukatek. **Boldface** marks totals. These include multiple names for one definition, multiple definitions for one name, and multiple occurrences of the same name.

Counterparts among Traditions: The Stemmas

A "counterpart" denotes a linguistic equivalent in sound, sense, or both; a cognate or synonym, full or partial; or a homonym. Across all the traditions, more than four out of five names (82 percent) relate to one another as cognates, homonyms, or synonyms, while nearly three out of four (71 percent) derive from the glyphs. The descent lines or "stemmas" in appendixes D and E trace the development of these names and characterize each transfer of sound or meaning. The following sections discuss the results from the construction of the stemmas.

The Late Classic, 600–900 CE

Two developments characterize the Late Classic period, 600–900 CE. The first involves the flow of names from the southern Lowlands west and south to Tzeltal-, Q'anjob'al-, Ixil-, K'iche'-, Poqomam-, and Poqomchi'-speaking regions and also north to the Yucatán. The second entails the appearance of the Yukatek innovation Sihom and its extensive impact on the southern Lowlands and the Highlands.

DIFFUSION OF THE GLYPHS

Ixil

Conservative traits in their post-Conquest dress as well as their Postclassic architecture have undergone only slight change among the Ixil since the Late Classic, when they lived under the general influence of the neighboring Classic Lowland Maya (Fox 1978: 121, 279; Colby 1976). Among the relics of this contact belong several names. These include **MOL** "Group," expanded later to Molch'ú "Flock of Grackles" and Mol Tche "Herd of Deer." The first half of Yax'ki represents the Ch'olan "green" or the Ixil "crab" (Kaufman 1974: 972), as *cha'x*/*tcha'x* is Ixil for "green" (Rodríguez Sánchez et al. 1995: 39); the second half translates the glyphic **K'IN** "sun, day." Tz'ikin Ki reflects the ***TZ'IKIN** of the texts, Ki or "Days" being an addition common in this tradition. Association suggests that this name is the source for Soj Ki as *sok q'ih* "nest days," found only in Ixil subtraditions that do not have Tz'ikin Ki or its apparent equivalent Tzunun Ki "Hummingbird Days." As *sok* also denotes *yagual* or "head-ring," Soj Ki seems the likely source for Yowal as *yawal* "head-ring." A Ki "Water Days" represents a modified version of the glyphic ***HA'** "water, rain," the core of the four color months. For **MUWAAN** "hawk," Ixil devised counterparts in Moochó "Rat Hawk," denoting a large brown hawk, and in Pekxítx "Cacao Wing," another term for the same bird. Assuming that the K'ichean equation of sprouts and shoot tips with sewing needles and bloodletters (compare K'iche' Tz'izil Lacam) also served the Ixil, Hui'ki "Tip Days," might have descended from the glyphic **OOL** "sprout, shoot," as did perhaps Muen (Chin) "Sprout or Seedling (Bag)." From ***KAN JAL-AB'** "yellow weave-instrument, loom" the Ixil could derive Tzu'ki "Loom Bar Days." Finally, the inscriptional **CANAZI** "yellow woodchopper" likely fashioned Zil'ki "Firewood Grove Days" and Tzil'ki "Split-the-Firewood Days."

Kaqchikel and K'iche'

If the word-final variation of *b'* with *m* known in Poqom, Ch'ol, and Tzotzil provided the alternate ***HAM** for ***HAB'**, a base of the glyphic color series, then the Kaqchikel and K'iche' Cakam, along with its variants, might preserve a K'ichean ***KAQ-AM** "Red Spider or Bubo" as a reinterpretation of the glyphic ***CHAK HAM** "red rain-device, red season."

Poqom

The number of Classic loans in Poqom confirms significant contact with the Late Classic Maya cultures. Canazi, Kasew, Mol, Muan, Ojl, and Uniw enter unaltered. Once in Poqom, several other forms undergo modification, resulting in Kanjalam, Petcat, Tam, Chab, and Tzikin Kij. One of the color months inspired the *Ah Qui Cou appearing later only in Ch'olan.

With its final -*m* Kanjalam dates itself. Unknown in the inscriptions, where the name reads K'an Jal-<u>ab'</u>, with a terminal -*b'*, Kanjalam must result from the *b'-to-m* shift that typified first Poqomam and afterward the western dialects of Poqomchi' sometime after 1000 CE, when these languages or dialect groups became distinct (Kaufman 1978: 103, 105, 111). K'an Jal-ab', then, appeared earlier in Poqom in the three centuries prior to the shift.

Glyphic precursors to Petcat are found on four "Calakmul Dynasty" vases (Vessels 127:G6, 128:J3, and 129:L3 in Robicsek and Hales 1982: 99–100; and K6751:H4 in Coe and Kerr 1998: pl. 114). The compounds each read 8 Yax, with the value for the second sign determined by the subfix: **ta** for **K'AT** "wicker," **na** for **TAN** "middle," and **la** for **YOL** "center." The vase K6751 dates by style to the eighth century (Coe and Kerr 1998: plates 114, 115), as must the very similar Vessels 127, 128, and 129. The Emblem Glyphs designate Calakmul. The invariable **ka-** prefix ensures that the main sign represents ***KAN**, "serpent" in proto-Yukatek, not the proto-Ch'olan cognate ***CHAN**. If the language were Yukatekan, the clear superfix would yield ***YÁ'AX** "green" or ***YÁAX** "great/first," synonymous with the usual superfix **CHAK** "red/great." Poqom calendrists chose "first," translating it with their unambiguous Pet. Synonymy and homophony together transformed proto-Ch'olan *Chäk K'ät "Red Wicker" to proto-Yukatek *Chäk K'ät "Red/Great Wicker" to proto-Yukatek *Yáax K'ät "Great/First Wicker" to Poqom *Pet K'aht "First Burn," *kaat* having in Poqom only the homophone *k'aht* "burn." *Yax K'at arose in Calakmul probably after 761 CE.

Phonetic complements on the glyphic versions of Chacc^cat and Icat indicate two values for the main sign, **K'AT** and **TAM** ($c^c = k'$). Apparently both survive in

Poqom: one in Petcat, the other in Tam. Via the **ta** subfix the first confirmation of **K'AT** for Icat dates to 727 CE (Dos Pilas, Stela 8:G13) and to 761 for Chacc^cat (Sacul Stela 1:C4). The more common **na** subfix indicates **TAN** for the main sign as early as 615 (Naranjo Stela 25:B8) or 633 (Naranjo Lintel 1:B1); attested by 692 (Palenque, Temple of the Sun, north *alfarda* [balustrade]: I2), the **ma** subfix occasionally points to **TAM**.

The inscriptional **TZ'IKIN** inspired Tzikin Kij in Poqom. This counterpart consists of the cognate for "bird" plus the innovation *q'iij* "day." From Poqom the form entered Kaqchikel and K'iche'.

The second glyphic color month, understood as the proto-Yukatek *Ya'ax Ha' "Green Chocolate" or *Yáax Ha' "First/Great Chocolate," likely generated Ah Qui Cou "Cacao Merchant" and "Royal or Roasting-Ear Chocolate." This Ch'olan winal designation derives from Poqom, where it is unattested but implied as a month name. To either form of the first half the Poqom would most directly link their own *rax*, a word denoting not only "green in color, unripe, fresh" (Fernández 1892: 26; McArthur and McArthur 1995: 139; Morán 1725: 475v) but also "*elote* or roasting ear of tender maize" (Sapper 1897: 418). Perhaps to preclude confusion with *rax ha* "blow fly" (Fernández 1937: 188; Mayers 1956: 48) or "*agua cruda*/raw [?] water" (Morán 1725: 475v), they replaced *rax* "*elote*/maize" with the synonym *aj* "tender maize, roasting ear" (Fernández 1937: 53; Mayers 1960: 294:76), and *ha* with *kikow* "cacao, chocolate." By this time *HA' "water" should also mean "cacao, chocolate," as attested in various Yukatek entries (Barrera Vázques et al. 1980: 78; Ciudad Real 1984: 17r, 134r; Roys 1931: 222, 242). Homophony, plus the association of chocolate "the drink of the gods" with royalty (Miller and Taube 1993: 48), would facilitate interpreting *aj* either as "reed mat" (Teletor 1959: 127) or as a prefix denoting high status (Reina 1966: 308). Altogether the parts could render "royal cacao."

Q'anjob'al and Chuj

Throughout the Late Classic, Q'anjob'alan accepted names from Ch'olan then shared them with nearby Chuj. Several names remained recognizable, but others transmogrified into synonyms.

Mol remains pristine in both languages. Q'anjob'al, moreover, preserves Mak, apparently reinterpreted in Chuj to Meak, an altogether different word.

In Baktan "bone ashes," Q'anjob'al preserves the **TAN** "ashes" value for the main sign of the glyphic predecessors to Chacc^cat and Icat. This name seems to fuse two consecutive designations recorded separately in Chuj as Bac and Tam.

By 603 CE (Caracol Stela 6:B14) the sign for **TZ'IKIN** acquired its constant **ni** appendage that resulted in the spelling **tz'ik-(ki)-ni**. The name entered Q'anjob'al then Chuj, perhaps modified by *watz'* "squeak," though this arguably came later.

In its color + Sihom series Q'anjob'al preserves the sequence of the Classic color months. It offers the same colors, translated, in the same order, but includes "yellow" as an alternate to "green." It received Sihom from the language of the engraved texts, which apparently had accepted the term from its inventor, Yukatekan. Chuj did not adopt this set.

The **YAXK'IN** of the inscriptions persists in Q'anjob'al's Yaxakil and Chuj's Yaxaquil. The glyphic **YAX** denotes "blue, green; grass for provender or thatch." The Yax of the two neighbors means just "Blue, Green," while Ak/Aq provides the "Grass for Provender or Thatch." The -*il* as "place or time of, abundance of" replaces **K'IN** "days, time." Both names yield virtually the same translation, with Yaxk'in as "Green Grass Days" and Yaxakil/Yaxaquil as "Green Grass Time."

Native to Q'anjob'al but lent into Chuj, Oneu presents the 250–650 CE version of a pre-proto-Mixe-Zoque loan into proto-Mayan from many centuries before. By the middle of the Early Classic, proto-Ch'olan developed this into **uniw* (Fox and Justeson 1984: 210–14; Justeson 1988: 14 T. 3), attested in the glyphs perhaps as early as 416 CE (Tikal, Ballcourt Marker:E6b) and 584 (Altun Ha jade plaque:B5; Mathews and Pendergast 1979: 204), as suggested by the **wa** complement, but certainly by 727 and 736, as confirmed with fully phonetic spellings, the first at Dos Pilas (Stela 8:I13), the next at Yaxchilán (HS3 Step 1, tread:D1a). If **UNIW** motivated the use of Oneu, Ch'olan loans into Q'anjob'alan began about 400 CE.

With Mo "Hawk, Owl-like Hawk" Q'anjob'al supplied a synonym for the inscriptional **MUWAAN** "hawk, owl," lacking as it did a cognate. It then passed the term to Chuj.

As ***HA'** "water, rain" or ***HA-AB'** "rains-device," the glyphic main sign of the color compounds inspired the Q'anjob'al Nabitc "Rainy Months, Winter," source for the Chuj Nabich.

Q'anjob'al invented and shared with Chuj one name apparently equated to a hieroglyphic counterpart, borrowing a lexeme from K'iche'. Sivil "Woodchopper" corresponds to Ch'olan's Ccanazi as K'an Ah Si' "Yellow Woodchopper," precluding confusion with its Kan-a(l) Si' or Q'anjob'al's Q'an-a Si', both "Yellow/Dry Firewood."

Assuming that the K'ichean equation of sprouts and shoot tips with sewing needles and bloodletters (compare K'iche' Tz'izil Lacam) mirrored an ancient widespread metaphor, the glyphic **OOL** "sprout, shoot" might have developed into the Q'anjob'al (Kaq) Xuhem "(Red) Threader, Bloodletter." This entered Chuj as Xujim.

Q'eqchi'

The rolls of the Q'eqchi' also included several Classic compounds. Gkatok "Set on Fire" recast the glyphic **KAT** "wicker," while Moloc merely verbalized **MOL**. Muan represents a direct loan. Raxgkim consists of two words, *rax* "green" and *gkim* "thatch-grass." The lead member of this pair translates the first word of the inscriptional **YAX K'IN**, while the other mimics **K'IN**.

Tzeltal and Tzotzil

The unaltered glyphic contribution to Tzeltalan survives in two names, the Tzeltal Mac and Yashkin, the Tzotzil Moc and Yaxkin. The *o* of Moc is a clue to an approximate date for this contact. The glyphic **MAK** entered Tzeltalan before Tzotzil grew distinct enough from Tzeltal to change *mak* to *mok*, a process that started about 500–600 CE (Kaufman 1978: 111; 2006: 585, fig. 13.2; Campbell 1984b: 2, fig. 1).

Slightly modified relics from the inscriptions seem preserved in the Tzeltal variants Sakil Ha' and Sakilab. Sakil Ha,' understood as Sak-il Ha' "White Water," preserves its ancestor ***SAK HA'** "white water." This designation, synonym for a white maize gruel (Ciudad Real 1984: 92r; Houston et al. 1989: 722, 723, fig. 2), must have ultimately given rise to the Tzeltal-Tzotzil Mux "A Kind of Atole; Hominy."

Sakilab can be parsed as either Sak-il Ajb' "White Reed" or Sak Il-ab' "White Seer"; it stems from the earlier Tzeltal *Sak-il Hab' "White Year" (Justeson et al. 1985: 77 n. 43), itself rooted in the glyphic ***SAK HAB'** "white time." No variants of either the optional superfix or the frequent suffix represent -**IL**, -**Vl** or **lV** in the color months, so the attributive suffix probably debuted in the Tzeltal version. Sak-il Ajb' "White Reed" gave rise perhaps to one of Tzotzil's counterparts, Mucta Sac "Large White [maize]" or Si Sac "Frost White [maize]."

The glyphic **OL** seems to survive in the Hul or the Ol of the Tzeltal Hulol. The direct analysis casts **OL** as "heart," yielding Hul-ol "Pierce the Heart." If the Ol represents the verbal noun suffix -*ol*, however, the inscriptional name might have transmuted into Hul instead, assuming that the K'ichean equation linking tips of shoots with needles and bloodletting circulated also in Ch'olan and Tzeltalan (compare K'iche' Tz'izil Lacam). **OL** as "sprout, shoot" could produce "pierce, puncture," the essence of Hul. Tzeltal passed the term to Tzotzil. As "Arrival at the Middle" Hulol might instead represent an innovation from when the Tzeltal froze their count to the Gregorian calendar in 1584, created to highlight the score of days that included the northern zenith passage between 1 and 9 May. If so, **OL** as both "heart" and "sprout up, pierce" could have engendered the Tzeltal Olalti or "Heart Mouth," an apt reference to either the bowl for holding the excised organ or the knife that tears it out or the gaping wound it creates. The name passed into Tzotzil with the identical meaning. Tzeltal later innovated Alalti "Small Mouth."

Altogether unrecognizable as a further legacy of the Classic period, two of the color months might live on as Small Uch and Large Uch. The usual sense for *uch* is "opossum," but a different meaning, "drizzle," connects the Tzeltal and Tzotzil names to the glyphic ***HA'** "water, rain." The Tzotzil definition "drizzle" should also apply to Tzeltal, as this language is geographically closer to Ch'olan. Among the Tzeltal the big dark marsupial called *muk'ul uch* "large opossum" also goes by *yax uch* "gray [green] opossum." People designate his small, pale counterpart not only as *tzail uch* "small opossum" but as *sakil uch* "white opossum" (Hunn 1977: 202–3). The stemmas, then, proceed from the glyphic ***YAX HA'** "Green Rain" to the early Tzeltal *Yax Uch "Green Drizzle," then "Green/Large Opossum," and finally Muc(ul) Uch or Muk'(ul) Uch "Large Opossum." The glyphic ***SAK HA'** "white rain" becomes the early Tzeltal *Sakil Uch "White Drizzle," then "White/Small Opossum," and finally "Small Opossum" in its variants Alaj Uch or Yaj Uch, Bikit Uch, and Chin Uch.

Again the Tzeltals share these names with the Tzotzils, who render them as Unen Uch or Bikit Uch "Small Opossum" and Muctauch "Large Opossum."

Yukatek

The Ch'olan linguistic groups shared Maac, Mool, Muan, and Yaxkin directly with Yukatek speakers. The Yukatek also took on the color series but modified the main sign to **SIHOOM**. The Ch'olans lent the glyphs for three additional names, but the Yukatek occasionally assigned their own designations or even altered their appearance. By 743 CE at Xcalumkin they transmuted Uniw into Kankin; a scribe in Palenque (Temple of the Cross K9b) might have anticipated them in 690 if he really did enclose within **UNIW** the sign **K'AN** "yellow." The usual split-drum sign, representing perhaps **PA-WAB'** "split-drum," is replaced with **pa-xi** at a handful of sites, starting with Naranjo in 615 (Stela 25:D1; fig. 76 this book). **Pa-xa** becomes another alternate by about AD 650 (Tikal, Monument 25: B; Haviland 1992: 76, fig. 3.7c; Montgomery 2001: 130–33). Both innovations spell **PAAX** "upright drum," a term that survives and makes sense only in Yukatek. Finally, by 880 CE the Yukatek at Chichén Itzá *IK' K'AT/TAN as **WO**, indicated with the complement -**wo** (fig. 8 in appendix C).

The leap from *IK' TAN to **WO** proceeds from homonyms to synonyms. Proto-Ch'olan *Ik' Tan "Black Ashes" became proto-Yukatek *Éek' Ta'an "Black Ashes." The optional -**ki** indicates the transformation into *Éek' Säb'äk "Soot, Black Ink." This led to *tz'iib' "write; writing." The term reentered the texts with its prefix, perhaps as *Éek' Tz'ib' "Black Writing" but certainly as **WOJ** "glyphs."

For **UNIW** "avocado," the transmutation to **KAN K'IN** "yellow days" starts with the early variant **UN** "avocado," then adds *K'IN "days, time." Entrance into proto-Yukatek weakens the final **n** next to a glottalized consonant, yielding *UH K'ÌIN "bead days." Synonymy then produces **K'AN K'ÌIN** "bead/yellow days." This emerged at Xcalumkin in 743 CE.

The Appearance of **SIHOOM**

The main sign of the color months, topped with a multistranded knot, represents the phonetic **hi**. In a Mixe dialect *jip* denotes "scrape, sharpen, polish" (Schoenhals and Schoenhals 1965: 42, 263). In Yukatek, with several day names and maybe one month name (**TSEK**) from this Mixe source, *ha'* has precisely these Mixe senses and could function as a logographic synonym. One homonym means "water." This could have produced *hab'* "year, season," as Ch'ol *ha'* "water, rain" is echoed in *ha'* "time," a reduction of *hab'* "year, time" (Attinasi 1973: 266, 268)or *ha'b'*, itself shortened from *ha'-ab'/-ib'* "rain-device," apparently the year as a reckoning tool centered on the annual downpours (Houston et al. 2001: 51 n. 7; Justeson 1984: 340; Kaufman and Norman 1984: 120, item 142). In Yukatek *ha'ab'/hab'* means not just "year" but "file, whetstone" (Barrera Vásquez et al. 1980: 165–66; Bricker et al. 1998: 92). Another

Yukatek *ha'* means "soapberry" (*Sapindus saponaria* L.?) or "an herb whose berries or fruit serves as soap, with which the Indian women wash their heads" (Ciudad Real 1984: 170r; Barrera Vásquez et al. 1980: 165). This leads to the virtual synonym *sihom*, "a tree that bears a little fruit that serves as soap" (Ciudad Real 1984: 103r), at least for washing clothes (Mengin 1972: 18v, under "árbol jabonera"). The **ma** suffix suggests a widespread value of **SIHOOM** "soapberry" (Fox and Justeson 1984: 48–51, 52 n. 30) or, unexplained, "flower" (Houston et al. 1998: 282, 285), with the earliest instance at 593 CE. No phonetic complement confirms *HAB' or *HA'. The Yukatek then shared **SIHOOM** with the Ch'olans, who passed it to the nearby Q'anjob'al.

The chain leading to "flower" began with the Q'anjob'al, who took over both the form and senses of the glyphs. They shared the name with their neighbors the Mam. These heard in the *q'eq'*, *kaq*, and *saq* ("black," "red," and "white") of Q'anjob'al their own cognates *q'aq*, *kaq*, and *saq*, but they apparently did not know the foreign *yax* "green" and could relate only to *yaaxh* "lover," not a color term like the others in the series. Instead, they linked *yax* as a cognate to *yoox*, their own second term for "red," ignored it as tautologous and so reduced the members of the color set from four to three. This perhaps explains why the K'iche' counterparts total just three. In regard to **SIHOOM**, the Mam effected the one change that accounts for the semantic shift vital to the next step: they replaced the term with their cognate *siijan* "*jaboncillo*/soapberry" (Maldonado Andrés et al. 1983: 294). Next they exported this to the nearby speakers of K'ichean Proper, ancestral to Kaqchikel, K'iche', and Tz'utujil. To these people *siijan* sounded like their own *si'jan* "white flowers of the *siij* tree" (Santo Domingo 1693: 137). This they abbreviated to *si'j* "tree of white flowers" (Brasseur de Bourbourg 1961: 255), "certain white flowers" (Ximénez 1985: 644), or "white flowers of large trees" (Anonymous 1787: 132r). This form entered the K'iche' roster as the 1/2/3 Zih series. The "white" of the header *zih* "white flower" likely rendered the color prefixes confusing, so the calendrists replaced them with ordinals.

Among the Ixil **SIHOOM** sounded somewhat like their own noncalendrical *siom/sijom* "zenzontle" (Kaufman 1974: 631), a species of mockingbird (Schoenhals 1988: 379). This might provide the source for their *pactzi* "a kind of bird."

Summary

During the Late Classic, between 600 and 900 CE, the Maya glyphic tradition shared at least one name with every tradition; in no instance did it cede any variants of its names. In a very few cases there is no direct trace among the other rosters; these loners are *IK' HA', **CHAK TAN**, and **PAX**[-**AB'**]. In this same period Yukatek innovated **KANKIN**, **PAX**, **UO**, and **SIHOOM**. This last designation spread to the glyphs, Ixil, and Q'anjob'al; with changes in form and sense, it eventually reached Mam, Kiché, Kaqchikel, Poqomam, and Poqomchi'.

The Early Postclassic, 900–1250 CE: Reduction and Highland Regionalism

Reduction characterizes the linguistic developments during these three and a half centuries. This reduction manifests in the deletion of **SIHOOM**, the drop in the number of innovations or reinterpretations, and the diminished, virtually one-way contact from the Lowlands to the Highlands. These reductions apparently resulted from the decimation of widespread elite relations created by the decline of Classic civilization in the southern Lowlands, where most cities lay abandoned by 900 CE. In the northern Lowlands, by contrast, sites continued to prosper and network, trade and spar, under a series of dominant polities, through 1100 CE. Preeminent between 850 and 1100 was Chichén Itzá, a cosmopolitan hub of such military, commercial, and cultural impact that for some four centuries after its demise ruling houses throughout most of the Maya area as far away as the lands of the K'iche' and Kaqchikel peoples relied on real or fictive ties with it to bolster their authority. These included even its nearby successor, Mayapán, the last capital for much of the northern lowlands; about 1450 CE this town toppled to rebels, who then formed eighteen independent petty kingdoms (Sharer 2006: 580–87, 626–30).

REDUCTION IN THE LOWLANDS AND ONE-WAY DIFFUSION

The Yukatek and Ch'olan peoples maintained contact with one another and with most of the Highland traditions. The quantity of exchanges or exports, however, indicates tenuous, sporadic communication. The Highland traditions each accepted one or two names from the Lowland calendars but sent them virtually nothing, turning instead to each other.

Yukatek: Deletion of *SIHOOM*, New Names to and from the Outside

Yukatek names arose from the reinterpretation of glyphic forebears and counterparts from outside traditions. Several of these Yukatek names were then exported to other linguistic traditions.

While **SIHOOM** "soapberry" transmogrified into "white flower" among the Mam and Kaqchikel-K'iche', in Yukatekan it acquired other meanings, turned ambiguous, and eventually vanished. Besides "soapberry," *sihom* shares meanings with *sihum* and *sihnal* such as "native, natural, creature" (Ciudad Real 1984: 103r; Barrera Vásquez et al. 1980: 727–28). With the essential sense "one that is born" extended to "person, human," its reading changed to *wíinik* "person, human being." This seems likely, as a text from Ek' B'alam, dated to 794 CE, replaces **YAX SIHOOM** with a clear **YAX WINIK** "green person/month" (fig. 50 Yax). When the sense "month" prevailed, Yukatek speakers regarded it as redundant, soon dropping the new spoken term *winik* from the color series while retaining the old glyph. **SIHOOM** disappeared from every color set except Q'anjob'al.

First in the resultant color series, proto-Yukatek *Éek' "Black" led apparently to the near-homonym *Èek' "Spot" (Barrera Vásquez et al. 1980: 150). A colonial manuscript defines <*kax ek*> as "a stand of water . . . to which one cannot find a bottom" (Avendaño y Loyola 1996: 63); the term literally translates as "forest spot." As *k'áax* denotes "forest," **èek'* means "deep pool." **Èek'* became **CH'E'EN** "well," so "black" vanished.

YAX "blue-green" and **SAK** "white" remained on the Yukatek roll. Two examples of **SAK** from the cave of Naj Tunich display **ka** as a subfix, the sole phonetic complement to any color logograph. The location and prominence of **ka** suggest the compound reads just **SAK**. The date, 744 CE, would mark the onset of the deletion of **SIHOOM**. Whether the texts at this cavern represent a form of Yukatekan or Ch'olan, the color months in both language groups offer no counterpart to the glyph for "stone."

CHAK as its proto-Ch'olan reflex ***CHÄK** likely denoted, as in Ch'ol, not just "red" but also "to hunt, capture" (Attinasi 1973: 250) and "bare; without grass" (Aulie and Aulie 1978: 51). Here the Yukatek *ah céeh*, "deer slayer" and "baldy" (Barrera Vásquez et al. 1980: 308; Solís Alcalá 1949: 110, 124) offers a transition to the Yukatek month **CEEH** "Deer." Another route from ***CHÄK** to **CEEH** starts with the proto-Ch'olan *chik as a variant for "Red" and leads via the Ch'olti' homophone *chik* "Beast, Deer" (Morán 1935: 11, 33, 59) to **CEEH**. Some support for *chik* "red" appears in the Yokotan Chontal *chik* "red" (Blom and LaFarge 1927: 2:477), along with the Ch'ol *chik-ix* "it is red" (Attinasi 1973: 253).

The *Chäk–Chik equation facilitates the derivation of the Yukatek **ZIP**. Proto-Ch'olan *Chäk K'ät as "Deer Idol" finds a close match in **ZIP** as "deer protector."

The only phonetic evidence for a glyphic **POP** consists of the **po-po** spelling prefixed to the standard sign in Landa. As this innovation appears in Yukatek, Q'eqchi', and K'iche', it might have originated with any of them. If diffusion more often proceeds by proximity, Pop should appear in the Ch'olan tradition, too, which lies between Yukatek and the other two. Its meaning "woven mat" fits well with the inscriptional **K'AN JAL-AB'** "yellow weaving-instrument, loom" as well as the presumed Yukatekan predecessor ***K'ÄN HA'L-AB'** "yellow command-instrument" (Barrera Vásquez et al. 1980: 174). Maya everywhere associated the mat with power. ***POP** likely passed to the Ch'olans, then to the Q'eqchi's; next the K'iche's accepted the name but qualified it as Tz'iba Pop "Painted or Inscribed Mat." Finally, their neighbors the Ixils apparently translated it into their equivalent Mu "Shade; Canopy, Throne."

Only proto-Ch'olan and proto-Yukatek in tandem connect the glyphic **OL** to the Yukatek **KUMK'U**. Sparse Ch'olan data intimate that "Ball," a proto-Yukatek definition for Ol, extended to proto-Ch'olan as well. This facilitated the putative proto-Ch'olan *Ku'um K'u' "Nest Ball, Egg," composed of *ku'um* "ovoid, rock" as a cover term for eggs, stones, beans, testicles, avocados, and kernels (Girard 1949: 135; Wisdom 1949: 604), and *k'u'* "nest." This link generated perhaps the Ixil Soj Ki "nest days." Homophony led straight to the Yukatek name as *kúum* plus *k'uh*, for *kúumk'uh* "thunder god." **KUMK'U** in turn might have inspired the Ixil name Kchip "Last, Youngest Child," epithet for their smallest, most powerful thunder-lightning deity.

The Yukatek **XUUL** might stem from either of two hypothetical Q'eqchi' names. Some of the Ixils not only preserved the inscriptional ***TZ'IKIN** "bird" in their own Tzikin'ki "Bird Days" but also expanded its meaning to "partridge, quail" (Nachtigall 1978: 304, item 15). Next door the Q'eqchi's apparently shifted this to Chichin "A Certain Bird" (Haeserijn 1979: 144) or to *Xulul "Partridge or Quail" (Curley García 1967: 187; Schoenhals 1988: 381) then to *Xul "Bird" (Pinkerton 1976: 141). Arguably, Yukatek next borrowed *Xul to replace an obsolete *Tz'ikin "Bird," recasting it over time as Xúul "Planting Stick" (Bricker et al. 1998: 264).

Whether the Tzeltal Jok'en Ajaw represents *h ok'-en ahaw* "drummer lord" or the unique 1887 variant Hoquén-Hajab encodes *ok'-en ahaw* "music-instrument/drum lord," either hints at a source for the Yukatek **K'AYAB'** "sing-instrument, drum." Ok'-em "Call-Device" denotes a decoy-shaped whistle and its call. Ok' applies to cries, howls, birdsong and the beat of a drum. If Jok'en Ajaw "Drummer Lord" and its variant reinterpret Hoken Ajaw (*hok'-en ahaw*) "Water Hole Lord," they stem from its inspiration, the Ch'olan Sihom or Zihora "Origin Hole of Water." As these both represent proto-Yukatek loans, the ultimate source for all these names, including **K'AYAB'**, lies in that tradition.

Of the newly fashioned, exported Yukatekan names, most gained currency as loans or homophones, while the rest begat synonyms. In the first instance *Sihom and *Zihora took root in Ch'olan with form and sense intact, while their issue Che'en (*ch'é'en*) "well" lodged in Q'eqchi' as Che'en (*ch'en*) "mosquito." **POOP** "woven mat" passed as Pop probably to Ch'olan, then to Q'eqchi', and finally to K'iche' which qualified it to Tz'iba Pop "Painted or Inscribed Mat." **YAX** and **ZAK** also entered Poqom as loans, retaining their form, meaning, location, and sequence. These traits argue for an origin in the Yukatek tradition, after the deletion of ***SIHOOM** and the other two color terms. So do the Ch'olan months Yax, Zak, and Chak. Only the first half of **K'ANK'IN** "yellow (corn) days" endured: in Q'eqchi' it took as Gkan "Yellow"; as Kanal "Yellow Corn Ear" in Q'anjob'al and Chuj; and as Kanajal "Yellow Corn Ear" in Poqom.

The second group, the names that developed synonym counterparts, includes ***SIHOM** "origin hole," *Zihora "Origin Hole of Water," and **CHE'EN** "well," each a possible source for the Tzeltal Hoken Ahaw or Hokin Ahaw "Water Hole Lord." Xuul "Planter Stick" apparently inspired Tik-che-ik in Poqom and Tsun in Tzeltal and Tzotzil, which both mean "planting with a stick."

Ch'olan: Significant Retention, Minor Innovation, and Some Borrowing

As successors to the Classic culture, Ch'olan speakers preserved in several subtraditions nearly all the inscriptional month names, most in their original forms. Canhalib, Cazeu, Ccanazi, Chacc^cat, Icat, Mol, Muhan, Oḻh, Tz'ikin, Uniu, and Yazquin continued complete and unaltered. Chac, Yax, Zac, and Sihom recorded reduced glyphic compounds. Two came down only slightly altered, as Chantemac and Zihora. Three entered as loans: Ah Qui Cou, Ianguca, and Chichin. Six migrated out: Ah Qui Cou, Chac, Chantemac, Sihom, Yax, and *Ku'um K'u'. Lost to this tradition but scattered among others are ***HÁ'**, ***HAB'**, ***IK'**, **PAX-AB'**, **SUTZ'**, and **WAYAB'**.

Three of these names survived in a separate Ch'olan subtradition, but nowhere as the designation for a score of days. Canhalib "Yellow Shuttle/Mat" is part of a Ch'olti' name. Tz'ikin "Bird" among the Chorti' designates the "days of the saints," probably 31 October–1 November. Yazquin "Green Days" comes down in 1631 as a Manche Ch'ol term for the "summer" or dry season, lasting from November to February.

Four of the above names persisted from the color series. The Late Classic set consisted of four compounds, each a different color, **IHK'**, **YAX**, **ZAC**, and **CHAC**, prefixed to **SIHOOM**. Proto-Ch'olan followed proto-Yukatek in its deletion of the base **SIHOOM** and in its conversion of "black" into "well," here Sihom "Deep Origin Hole" then Zihora "Origin Hole of Water." These last two names each continued in a different Ch'olan community. Chantemac, a modification of its predecessor, not only did service in Ch'olan, however, but proved useful for Poqom, Q'eqchi', and Ixil.

Three loans into proto-Ch'olan replaced native month names, each from a source where it did not function as a month name. Ah Qui Cou "Royal Cacao" from Poqom substituted for the original "Split-Drum" glyph **PAX-AB'** "slap-instrument, drum" or **PAS-AB** "split-instrument"; several variants spell out **PAAX** "upright drum." The second import, Ianguca "Pour Divine Honey," entered from Q'eqchi' to stand in for the seventh month, preserved in Ch'olan only as Yazquin "Green Days." Ch'olan subtraditions preserved the third loan, also from Q'eqchi': Chichin "A Certain Bird." It replaced the glyphic **TZ'IKIN** "bird," which has survived only among the Ch'orti's; they no longer have any inkling of its meaning and so in many locales have reinterpreted it as *si k'in* "a series of days" to designate festivals honoring the dead from late October through early November.

Six Ch'olan names developed counterparts elsewhere. Ah Qui Cou "Cacao Merchant," itself a loan from Poqom, occasioned a virtual Q'eqchi' synonym in Rakol "Merchant." Chac as *Chäk "Hunt, Capture" could by itself bring out in Q'eqchi' Tzaack "Game Animal," but the transformation seems more plausible if Chac first developed into *Chik, which denotes both "Red" and "Deer." Never attested in the glyphs, Chantemac "Granary on Tall Posts" took hold unchanged in sound or sense among the Poqoms, Q'eqchi's, and Ixils. Its root meaning "Enclosure, Granary" and that of Te as "Log" make it a likely source for the K'iche' Chee "Prison, Granary; Log" as well as the Kaqchikel Pariche "In the Granary." In the Ch'olan Yax "Blue, Green" the Q'anjob'als apparently heard their own *yax* "blue, green" and *yax* "crab"; for clarity they dubbed their counterpart Tap "Crab." The Yax–Tap equation was drawn sometime after 750 CE, when the main sign of the color months had lost its readings, leaving just the colors. In the Ch'ol-Q'eqchi' calendar from Lanquín Yax perhaps performed a double role, first as the Ch'olan *Yax "Green," second as the Q'eqchi' *Yax "Crab's Pincers." The reconstructed Ch'olan *Ku'um K'u' "Nest Ball, Egg" accounted for the Ixil Soj'ki "Nest Days."

Finally, Sihom apparently migrated into Poqom under the reinterpretation *siij-johm "offering gourd-bowl" (*siij* "gift": McArthur and McArthur 1995: 155; *johm* "gourd": Campbell 1977: 53, item 110; bottle gourd bowl: Stoll 1888: 169). This form contracted to *sijohm* then was recast as *saq-johm* or *saqohm* "white gourd-bowl, white gourd" (gourd bowl: Stoll 1888 :187; Sapper 1897: 411, item 22; gourd: Mayers 1960: 294, item 77). This approximates the putative successor Sac-gojk (*saq q'ojq'*) "White Chilacayote," a gourd-like squash. As *q'ojq'* "chilacayote" denotes a white squash (Campbell 1977: 51, item 73) or a green gourd-like squash (Schoenhals 1988: 31, 41, 141), *sac* was added to clarify. An alternative derives Sac-gojk from **UNIW**; the chain would proceed from **UNIW** "wild avocado" to *Ooj "Avocado" (Morán 1725: 440r; Sapper 1907: 448) to *Oo "Chilacayote" (Fernández 1937: 185) to Sac-gojk "White Chilacayote."

Lost in the transition to Ch'olan between 800 and 900 CE, several glyphic names still linger in four other traditions. The ***HÁ'** and ***HAB'** of the color series glimmer on in the Sakilab and Sakil Ha of Tzeltal; the possible variant ***HAM'** might continue in the Kaqchikel and K'iche' Cakam. The ***IK'** of the same set persists as late as 1566 in the **i-ki** spelling prefixed to the main sign in Landa's portrayal of the **CHEN** compound. **PAAX**, **PAXAB'**, **SUTZ'**, and **WAYAB'** survive in the Yukatek **PAAX**, **SOOTZ**, and **WAYEB**.

Summary

In this period Yukatek deleted **SIHOOM** from its roster then altered beyond recognition two of the four resulting color names, ***ÉEK'** and ***CHÄK**; the remaining two it shipped off to the Highlands. It exported eight other names. Some served fully or partially intact as loans or homophones; the rest were transformed into synonyms. Yukatek provided Ch'olan with two or maybe three words (*Sihom, *Zihora, *Póop); Poqom with four (Yäx, Säk, *K'än Kiin, Xúul), and Q'eqchi' with two or three (*Ch'é'en, *K'än Kìin, *Póop), while Q'anjob'al and Tzeltal each accepted one (*K'än Kìin, Xúul), sharing it with their neighbors, Chuj and Tzotzil. Altogether, ten different month names accounted for fourteen counterparts among seven traditions. Each recipient either bordered the Yukatek-speaking region or lay immediately adjacent to a bordering linguistic region. Only Ixil, Kaqchikel, and K'iche' evidence no direct contact with the Yukatek-speaking area.

During these centuries (900–1250 CE) Ch'olan paralleled Yukatek by maintaining contact not only with its major Lowland neighbor but also with most of the Highland traditions. In sum, six of its month names propagated thirteen counterparts among nine traditions. The loans and their recipients are Chac (Yukatek); Chantemac, *Ku'um K'u' (Ixil); Chantemac (Kaqchikel and K'iche'); Chantemac, Sihom, Uniu

(Poqom); Yax (Q'anjob'al and Chuj); Chac, Chantemac, Ah Qui Cou (Q'eqchi'); and Sihom or Zihora (Tzeltal). Except for Chantemac, these exports are all converted into synonyms. Tzotzil alone preserves no evidence of direct communication with Ch'olan. Of all the calendars affected only Chuj, Kaqchikel, and K'iche' might not have bordered Ch'olan.

The Regionalization of the Highlands

During the period 900–1250 CE the Highland traditions maintained contact with Lowland Ch'olan and Yukatek, but the exchanges proved to be rather unequal. The lowlanders sent up sixteen names that propagated twenty-seven counterparts among the highland calendars, while merely two highlanders, Q'eqchi' and Tzeltal, exported a total of five names to the Lowlands, which each became just one counterpart. In contrast, the highlanders managed a much more lively trade among themselves, sharing a total of thirty-four names that generated forty-four counterparts; only Chuj made no contribution. These gross counts suggest, first, that Yukatek and Ch'olan retained a prestige status among the highlanders and, second, that the depopulation of the southern and central Lowland centers accelerated the development of the Highlands as a separate region with brisk internal dealings.

Diffusion from Ixil

From this calendar's homeland several month names migrated east and west. Awax Ki "Corn-Sowing Days" corresponds to the Auuck "To Sow" of Q'eqchi', as Koj Ki "Clear-the-Land Days" does to Gkalec "Clear and Weed the Field." Hui Ki "Tip Days" perhaps called up Batsul "Tip of Amaranth" in Tzeltal and Tzotzil, while Kahab "Honey" inspired the Tzotzil Pom "Honey; Incense," which entered Tzeltal reduced to just "Honey." Tzunun Ki "Hummingbird Days" perhaps modified the hypothetical Q'anjob'al and Chuj *Tz'ikin "Bird" to Wa Tsikin and Guatziquin, "Squeak Bird." Finally, the "large" and "small" Cho variations might have generated the pair of Mam months in K'iche' and Kaqchikel. If *cho* "mouse" functioned in Ixil as it did in Q'eqchi' and Yukatek as a term of affection for newborns (Haeserijn 1979: 146) or very young infants still suckling or barely walking (Ciudad Real 1984: 157r, 228r), the contrast of "big baby" to "little baby" should eventually have replaced Cho with Mam: precisely this term denotes both "grandfather" and "grandson," due to the ritual kinship relation called *k'ex* "exchange" that requires the old man to regard the infant as his replacement and so treat him in many ways as his equal. If "large" and "small" denoted not just size but generation or rank, the contrast would yield First Mam "Grandfather" and Second Mam "Grandson." The K'iche'-Kaqchikel names inspired Ixil's Mama'ki "Elder/Prayermaker Days."

Diffusion from K'iche'

Many names derive from the K'iche' roster of month names. Botam "Rolled Up" and its variants Batam and Botan "Rolled Up," Ibota "Large Roll of the Year," and Obota

"Five Rolls of the Year," inspired Kaqchikel's Botam "Rolled Up, Doubled Over," Bota "Roll of the Year," Ibota "Large Roll of the Year," and Obota "Five Rolls of the Year." Chee as "Tree, Log" parallels Ixil Kucham "Tree Trunk." As "Prison, Log" it compares to the Ixil Lajab "Ancient Trap, Deadfall" and Kahabtsé "Log Beehive," which within its own tradition became Kahab "Honey(comb), Bee," then Cajab('Ki) "Honey(comb) or Bee (Days)." Lik'in Ka "Stretched Out Sky" found itself in the Kaqchikel LiΣinΣa of the same meaning. Tziij "Set Afire" recalls the Ixil Xukul(Σih) "Piece of Firewood (Days)." Zih "White Flowers" makes a likely candidate for the origin of Utzumckigh "Flower Days," the term for "summer" among the next-door Poqoms, perhaps a lost month name. Zih may have inspired nearby Tzotzil's Nichik'in "Flower Days" as well. Mam "Grandfather, Grandson" generated Ixil's Mama'ki "Elder or Prayermaker Days."

Diffusion from Kaqchikel

This tradition distributed five month names. ToΣ (*toq'*) "Bloodletting," a variant of ToΣic, closely resembled ToΣ (*to'q*) "Loincloth"; this sense migrated into Q'anbjob'al to bring out Wex "Loincloth," itself the source for the Chuj Bex "Loincloth." ToΣic "Piercing" crossed into Poqom as *Tokgüik (*toq'-wik*), glossed not only as "Piercer; Bite, Perforation" but also as "Dog" (Teletor 1959: 167; Fernández 1937: 193); to disambiguate, the Poqom replaced this with Tsi (*tz'i'*) "Dog." ToΣic as "Bloodletting" apparently inspired the close-by K'iche' Tz'izil Lacam "Bloodletter Banner." K'iche' also received 4,iquin Σih "Bird Days." In Kaqchikel this came to mean "Hummingbird Days" (Fernández 1892: 22, item 19), source for the Ixil Tzunun Ki "Hummingbird Days." If this meaning developed exclusively in Kaqchikel, the proposed transfer into Ixil took place after Kaqchikel became distinct from K'iche', after 1000 CE. B'otam "Doubled Over" perhaps engendered in Ixil Mekaj "Bent-Over Stalks" and Paksi "Bend-Over Firewood." K'atic "Burning the Underbrush" entered K'iche' as Tziij "To Set Afire." Of all the Kaqchikel names Tumuzuz "Winged Ant; Bean Beetle; Maize Bug" found its way most often into outside rosters: in Ixil it is Leem "Bean Beetle, Maize Worm"; in Q'anjob'al and Chuj Bak "Corncob Worm"; and in Poqom Mox-Kij "Dung Beetle Days."

Diffusion from Poqom

This group passed Tzikin Kij "Bird Days" to the immediately adjacent Kaqchikel. Given the K'ichean equation of sprouts and shoot tips with sewing needles and bloodletters, the Poqom Ojl "Bud, Sprout" apparently gave rise to the Kaqchikel pair 1/2 ToΣic "First/Second Bloodletting" if not the K'iche' Tz'izil Lacam "Sprout/Bloodletter Banner" as well. Finally, the Poqoms transmitted Sak "White" and Chab "Arrow" to Kaqchikel and K'iche'.

Diffusion from Q'anjob'al

Uchum "opossum" entered the Kaqchikel count from the language and likely the calendar of the Q'anjob'als. The motive for the borrowing remains a puzzle.

Diffusion from Q'eqchi'

The Q'eqchi's shared one month name, Gkatoc "Set on Fire," with their neighbors the Kaqchikels, who shaped it into K'atic "Burning the Underbrush." The unattested but inevitable *Xul "Bird" they imparted to the Q'anjob'als and via them to the Chuj people, as Yaxul "Blue Jay," apparently *yax* "blue" plus *xul* "bird." The strengthening of word-initial *w-* to *gw-* so frequent in Q'eqchi' hints that its speakers first elaborated the Ch'olan or glyphic **TZ'IKIN** "bird" into Watz' Tz'ikin "Squeak Bird, Hummingbird," elided this by geminate deletion to Wa-Tz'ikin, then strengthened it to Gwa-Tz'ikin. Both forms after this entered Q'anjob'al and only survived there.

Diffusion from Tzeltal

Several month names from this calendar developed counterparts. Tsun "To Plant with a Stick" corresponds to Ixil Awax Ki "Corn-Sowing Days." Either Sakil Ha "White Water, Atole" or Mux "Hominy" can account for the Chuj Savul "White Maize Kernels, Hominy." The "large" and "small" Uch "Opossum" months of Tzeltal and Tzotzil correspond to the "large" and "small" Cho "Mouse, Rat" months in Ixil. Uch might relate to the K'iche' and Kaqchikel Mam "Grandfather," too, because on pages 24–28 of the Codex Dresden the authors depict the New Year's hierophants as opossums but label them **ma-ma** or **MAM**.

Diffusion from Tzotzil

Muy instead of Moy for "finely ground wild tobacco" seems to have been restricted to the dialect of San Andrés Istacostoc Larraínzar (Holland 1963: 97, 129). It was perhaps the model for the Chuj Sac May "White Tobacco," a name that passed to the nearby Q'anjob'al as Sacmay "White Danger."

Summary

The exchange of month names between the Highland and Lowland traditions remained rather lopsided. Ch'olan and Yukatek distributed among the Highland calendars sixteen names that generated twenty-seven counterparts. Of the highlanders only Q'eqchi' and Tzeltal shared names with the Lowlands, five in all; four of these lodged unchanged as loans in Ch'olan, while the fifth perhaps became a synonym in Yukatek. Among themselves the Highland traditions conducted a far brisker commerce, exchanging thirty-four names that altogether produced forty-four counterparts.

The Late Postclassic, 1250–1520 CE: Nahua in the Highlands, Innovation in the Lowlands

Two recorded developments define this period. One, attested in the chronicles and archaeology reports written long after the events, is the arrival of Nahua among several traditions of the Highlands. As the Gulf Coast dialect, it accompanied the Chontal Maya forefathers of the K'iche's and Kaqchikels who trekked in from the plains of Tabasco and southern Veracruz. The other, preserved in the Codex Dresden and the Landa treatise, is the advent of still more innovations in the Yukatek calendar.

Nahua in the Highlands

Diffusion to Ixil

In the Late Classic, **MOL** "Group" entered Ixil from the glyphic tradition, changed to Mol Tche "Herd of Deer" in the Early Postclassic, and finally transformed itself into the hybrid Mol Masat in the Late Postclassic, using the Nahuat *masat* "deer." The contact with Nahua came through either the Epi-Toltec K'iche'-Kaqchikel elite forefathers, speakers of Gulf Coast Nahua, who passed through the area perhaps between 1250–1350 CE, or their descendants, the K'iche' king Gukumatz and his nobles, whose armies subjugated the Ixil between 1400 and 1425. This one name entered the rolls sometime after 1300.

Diffusion to Kaqchikel and K'iche'

Berendt or the anonymous author of the 1685 Kaqchikel calendar derives Izcal and Tacaxepual from the Nahuatl of the Aztecs, whose first and third months were Izcalli and Tlacaxipehualiztli (Edmonson 1988: 221, 222). Miles (1957: 744–45) likewise assigns a "Mexican" origin to Izcal as well as to the Kaqchikel and K'iche' forms of Tlacaxipehualiztli, but she also sees in the Mexican Pachtli months a source for the Kaqchikel and K'iche' Pach units. Edmonson (1988: 105, 145, 216–17, 220–26, 234, 237) attributes not only Izcal, Tacaxepual, and Tequexepual but also the Kaqchikel and K'iche' forms Pach and Mam to Late Postclassic Pipil Nahuatl, whose speakers invaded Guatemala, introducing their language and the Teotitlán calendar of Puebla and Veracruz. Ignoring Mam and Izcal, Campbell (1970: 5, item 44, 1977: 106, item 37, 107, item 53, 1978: 38–40, 47) relegates these names not to Pipil Nahuatl but to Gulf Coast Nahua.

It seems more likely that the K'iche' and Kaqchikel forebears, not the Pipils, entered the Highlands from coastal Lowland centers in the Tabasco-southern Veracruz area, the frontier between the Mexican and Maya cultural zones, about 1250 CE. The Pipil merchants left their central Mexico homelands perhaps as early as 800 CE, apparently due to the collapse of Teotihuacán (Henderson 1997: 235), and arrived on the shores of Guatemala and El Salvador between 800 and 925. The Pipils established enclaves across the coastal as well as piedmont zones of Guatemala, Honduras, Nicaragua, Costa Rica, and even Panama between 900 and 1200. Their variation of the Nahuat dialect and other lines of evidence support links with southern Veracruz (Carmack 1968: 71; Campbell 1970: 6–8, 1977: 103, 109; 1978: 38–39, 47; Coe 1993: 9, 28, 136–39; Edmonson 1988: 234).

The "Toltec" ancestors of the K'iche's and Kaqchikels probably constituted a new wave of Mexicanized Putun Maya, masters of both Chontal and Nahuat (Sharer 2006: 431), who are perhaps the eastern Nahua of Nonoalco, a land presumably in Veracruz and Tabasco (Campbell 1970: 7). They spoke an archaic Nahua dialect that not only used *t* and *u* instead of *tl* and *o*, as did the Nahuat of the Pipils, but also converted *iwi* to *i* and *kwaw* to *ko*. They used many Nahua terms, not just for military and ritual referents but for prestigious domestic and economic activities (Carmack 1968: 71, 1981: 44–50). "The fact that the Nahua loan words in Quichean languages show the results of obviously very late sound changes in the Gulf Coast dialects demonstrates that this contact was much later than the Pipil migrations, since Pipil shares none of these changes. . . . The Pipil apparently had nothing to do with the Toltec influence on Quichean languages" (Campbell 1978: 38–39, 47). The K'iche' and Kaqchikel forebears soon lost their languages and married local women. Their descendants went by the names of the native languages (Fox 1978: 2–3).

The K'iche'-Kaqchikel progenitors preserved several names from the Teotitlán calendar. This count commenced with the month known to the Mexicas or Aztecs as Tlacaxipehualiztli (Caso 1967: 38–40 T. 10 n. 15; Edmonson 1988: 222, 244). The Kaqchikels called it Tacaxepual and stressed its role in the colonial sources as their first month. The K'iche' people knew it as Tequexepual but assigned Nabe Mam (1722) to the premier slot instead, maybe after establishing K'umarcaaj or Utatlán, the capital of their kingdom, around 1400 (Carmack 1981: 122). The Kaqchikels revolted for independence in the late fifteenth century, most likely founding their capital Iximché in 1470 (Carmack 1981: 123; Fox 1978: 186) or 1480. The retention of the first month in the forebears' calendar, or possibly a return to it from the imposed K'iché first month, could have served the new power as a claim to "Toltec" or pedigreed legitimacy. In 1491 the Kaqchikels trounced the K'iche', capturing and sacrificing their two kings along with numerous nobles; they even imposed tribute on their former masters (Schele and Mathews 1998: 296–97).

The relative locations of the pairs and several single names in the 1662/1703 K'iche' roster most closely match those of pairs and synonyms in the Teotitlán calendar. The ancestral Tlacaxipehualiztli is cognate to the Kiché Tequexepual; both mean "Flaying of Men." In the fifth and fourth slots fall Tecuilhuitontli and Huei Tecuilhuitl "Little/Great Festival of Lords," counterparts perhaps in sense to Nabe Mam and V Cabmam "First/Second Grandfathers or Elders." Edmonson (1988: 237) interprets *mam* as "lord, elder," presumably his extension of the Nahua *mama* "to carry another on the back; to rule and govern others" (Kartunnen 1992: 134).

The Nahua Pachtontli/Huei Pachtli "Little/Great Moss" likely generated Nabe/V Cab Pach "First/Second Moss." Three more terms correspond to synonyms, specifically Quech'olli "Macaw, Flamingo, Spoonbill" and Tziquin Quih "Bird Days"; Panquetzaliztli "Rising-Up Flag" (Kartunnen 1992: 209) and 4,içi Lakam "Rising-Out Flag"; and finally, Cuahuitl Ehua "Tree Rises" and Che "Tree."

The distribution of numerous names on the Kaqchikel roster closely follows that of their counterparts in the Teotitlán list (Edmonson 1988: 221–24). The forebears' Tlacaxipehualiztli "Flaying of Men" is cognate to the Kaqchikel Tacaxepual. Tozoz-Tontli and Huei Tozoztli "Little/Great Vigil" parallel Nabei-Tumuzuz and Rucan Tumuzuz "First/Second Winged Ant"; Tozoz resembles Tumuzuz enough to merit suspicion as a source garbled in transmission. The ancestral Tecuilhuitontli and Huei Tecuilhuitl "Little/Great Festival of Lords" correlate to Nabeimam and Rucab-mam "First/Second Grandfathers or Elders." Pachtontli/Huei Pachtli "Little/Great Moss" must connect to Nabei/Rucan Pach "First/Second Moss." Quecholli "Macaw, Flamingo, Roseate Spoonbill" corresponds in meaning and location to Tziquin Σih "Bird Days"; Izcalli "Sprout" to Ytzcal "Sprout"; and Cuahuitl Ehua "Tree Rises" to Pariche "In the Trees."

If the forebears of the Kaqchikel rulers did not bring Izcal along from Gulf Coast Nahua about 1250 CE, they would have imported it from the Aztecs some 250 years later. In 1501 the Kaqchikel court at Iximché afforded a cordial welcome to a delegation of Pochteca or Aztec warrior-merchants and spies. The Kakchikels were eager for imperial support against their powerful, vindictive former K'iche' masters, from whom the Aztecs had wrested the rich provinces of Ayutla and Mazatlán only the year before during the conquest of cacao-rich Soconusco (Carmack 1981: 142). In 1510, nine years after the king had expelled them from his realm, Mexica emissaries returned to the K'iche' royal court at Utatlán to demand tribute. Weakened by endless wars with the Kaqchikels and others, the monarch paid in quetzal feathers, gold, precious stones, cacao, and cloth. From then on Mexica representatives, including women from the court of Tenochtitlán, maintained an almost continual presence there (Carmack 1981: 142, 143), and presumably among the more receptive Kaqchikels.

Perhaps under mounting Mexica pressure for tribute or trade, the Kaqchikels deemed it advisable to accept the name of the Aztec initial month Izcalli, dubbing their own seventeenth month Izcal; they kept Tacaxepual as their own first month. Though paying tribute, the K'iche' people did not incorporate Izcalli into their own calendar; instead they seem to have renumbered their months, advancing New Year's day fully eleven positions to Nabe Tzih in order to begin the year with their equivalent of the first month among the Texcocans, the Aztecs' partner in conquest. The calendrical machinations hint of a hierarchy founded on political prestige. Leaders of the Triple Alliance (a pact with their neighbor city-states Texcoco and Tacuba for conquest and tribute), the Aztecs dealt with the Kaqchikels, the junior but dominant power in the region, who had offered tribute or at least not refused to pay it. Second in rank in the pact, the Texcocans handled the K'iche' people, the senior but failed power, who had refused tribute but later relented, without offering battle. These relations intimate that calendrical manipulations were guided by a protocol linking the treatment of targeted peoples to a status based on power, seniority, cooperation, or resources. The same dynamic was repeated among the Tzotzils and Tzeltals.

Thus the K'iche' and Kaqchikel peoples apparently acquired their forms of the Tlacaxipehualiztli and Pachtli months not from the Pipil between 800 and 1500 CE or from the Aztecs after 1500, but from their own Chontal- and Nahuat-speaking immigrant forebears about 1300. This same calendar served as the source for two synonyms in K'iche' and one in Kaqchikel. Perhaps they devised the Mam months themselves, but the Kaqchikels did accept Izcalli from the Aztecs. It seems that the Aztec powers imposed either the name or the dates of their first months.

Diffusion to Poqom

As allies and kin of the expanding Central K'iche' people, both the Rabinal K'iche's and the Kaqchikels expelled the Poqomams from the Rabinal Valley about 1350 CE (Fox 1978: 255). After this point the name Pach "Moss" possibly entered Poqom from K'iche' or Kaqchikel as the synonym Makux "Lord Moss."

Diffusion to Tzeltal

Aztec interests and power loomed larger in the Highlands of Chiapas through the heightened activity of their merchant-spies as well as the subjugation of Comitán in 1498 and Zinacantán in 1511 or 1512. This new reality might have prompted the Tzeltals to shift the beginning of their year to Batzul, the equivalent of the first Aztec month, Izcalli. In 1548 1 Batzul corresponded to 17 January, as did 1 Izcalli (Edmonson 1988: 76, 105, 142–44, 221–22, 257–58). A unique first month, Ajelch'ak, also has a counterpart in a Nahua tradition that may have come into contact with the Tzeltals. That is Tlacaxipehualiztli, first in the Teotitlán calendar (Edmonson 1988: 221, 224, 257) of the K'iche' and Kaqchikel forebears who influenced the Tzeltal and Tzotzil counts between 1250 and 1350 CE.

Edmonson (1988: 218) suspects that the only other Tzeltal first month, Jok'en Ahaw "Water Hole Lord," was adopted from the Nahuatl Tecuilhuitontli "Little Festival of Lords" (Andrews 1975: 403), first month only in the Nahuatl count of Chalco (Edmonson 1988: 224). Geographically close to the Aztecs, Chalco might have served as an ally. If so, the shared initial months suggest that it too won the right to impose tribute on several Tzeltal towns.

Diffusion to Tzotzil

Nichikin as "Flower Days" recalls the Aztec Tlaxochimaco "All Give Flowers to Things"; when frozen they both occurred in July (Andrews 1975: 403; Caso 1967: 35, 73; Edmonson 1988: 217, 222). Beyond this any relationship remains speculative. The K'iche' Zih "White Flowers" is the closest source.

Two of the four Tzotzil first months are Batzul and Tzun. Apparently the expanding Aztecs, with their allies the Texcocans, imposed that position on them. Batzul mirrored Izcalli, the initial score at Tenochtitlán, while Tzun, the first month in other Tzotzil towns, corresponded to Tititl, the starting set of days at Texcoco (Edmonson 1988: 222, 258–59). The third year-starter, Muk'ta Sak, is the counterpart to the first month on the Teotitlán calendar, Tlacaxipehualiztli (Edmonson 1988: 224, 259). The warlike forebears of the K'iche'-Kaqchikel elite introduced this between 1250 and 1350 CE on their migration to Guatemala. Finally, Mok is an anomaly, a first month with no Nahuatl counterpart.

Innovation in the Lowlands

The Dresden Set

Written about 1450–1500 CE, the present manuscript of the Codex Dresden recorded several changes from the Classic ancestors. Though composed in Yucatán, the codex might record a form of Ch'olti'an.

Wo

On page 63 (C2) the scribe wrote a unique variant of the month name, following the sign for "black" with a double-looped knot instead of the usual crossed bands. Several contexts in the codex support the reading **TAAN** "center."

Tzek/Sek

The first instance of **SEC** "bat" (Lakandon) or **TZEC** "foundation; skull" appeared in this painted book (46b:C1; 62:A10). Spelled **tze/se-ka**, the name appeared only here and in the Yukatek tradition. If it is a reduced form of *tzek'* "skull," it recalls the rare death head forms of the Classic counterpart Caseu.

Xul

Two versions of this name are known, each distinguished by its animal head. The first presents a dog (*ok*, *tzul*, or *tz'i'* in Yukatek). Always suffixed with **-ni**, this first head could produce **o-ni** for **ÒON** "Avocado," or **OK-NI** "Dog-Nose" or "Enter-Nose," perhaps a term for the fire-hardened tip of the digging stick. The last suggestions are **tz'i-ni** for **TZ'ÍIN** "yucca" and **TZ'I'-NI** "dog-nose." Either allusion to the point of the stick serves as a virtual equivalent to Landa's **XUL** as *xúul* "planting stick." Assuming **ki** as an unwritten middle syllable, the compound spells **tz'i-ki-ni** or **TZ'IKIN** "bird," the Classic form of this name. The few examples in the Dresden each omit **ki**, as do all but one of the numerous instances in the monumental inscriptions.

The main sign of the second version resembles the head of a rat or mouse (*ch'o'*) so the compound could read **ch'o-ni** or **CH'ON**, a possible variant of *ch'oom* "vulture." The Classic counterpart *tz'ikin* means "bird." If this version also conceals its middle syllable and presents a synonym for "digging stick," it could record a counterpart to Tzeltalan *ch'oj tz'un* "piercing goad."

Ch'en

With its arcane superfix, the codex debuted a unique version of Ch'en "Well" (47c:D5). The intended name remains unknown.

Kankin

The occasional avocado form of the main sign suggests that the Yukatek scribe reinterpreted it as a similarly shaped variety of squash then inserted the vertebral column with ribs, read **tsu**, from *tsul* or *tsuy* "spine," as a cue to the Yukatek value **TSUH** "calabash, gourd" or more likely, with the ubiquitous -**wa** suffixed, to ***TSUW** as a possible variant of **TSUB'** "agouti."

Pax

One of the two rare versions of this name infixes **pa** and suffixes **xa** to the logograph to spell **PÁAX** "drum." The other appends just **xa** to indicate the same term. A full phonetic spelling appeared over 700 years earlier.

Kayab

With every instance of this name the scribe kept the parrot head main sign **ah** with its infix **K'AN** "yellow" but replaced the Classic **si-y** with **wa**. The sign group seems to represent **k'a-a-wa** for the Yukatek **K'Á'AW** "great-tailed grackle."

The Interim Set

About a century later Landa recorded four glyphic compounds with affixes and infixes indicative of new names. The inserted *X* and the attached -**ni** converted **YAX** to **YAX CUN** "green seat," via **cum** + **ni**. The same postfix turned **CHAC** to **CHAC TUN** "red stone." The replacement of **ka-** with **pa-** apparently transformed **TZEK** or **SEK** into **PAWA'** "net bag." The affixes to **CH'EN** spelled out its new variant as **i-ki-?-ma** or *ik ?-m* "black-?"; the unglottalized **k** suggests the rare inverted main sign begins with a **k'** or some other glottalized consonant.

Summary

Considerable evidence from archaeology, native chronicles, and linguistics bolsters the proposal that the ancestral K'iche'-Kaqchikel overlords subjugated the Highlands between 1250 and 1450, spoke Chontal Maya along with the Gulf Coast Nahua, and introduced their calendar with names that became the K'iche' Tequexepual, the Kaqchikel Tacaxepual, and the Pach of both systems. Perhaps after 1350 the Poqom, their neighbors or vassals to the east, accepted Pach "Moss" from these conquerors but only as their own synonym Makux "Lord Moss." About this time the Ixil to the west modified their own Mol "Group" or Mol Tche "Herd of Deer" with the Nahua term for "deer" to produce Mol Masat "Herd of Deer." The apparent final word from Nahua entered Kaqchikel as Izcal around 1510 from the Aztec Nahuatl Izcalli "Sprouts."

In the Lowlands during the Late Postclassic, many Yukatek month names arrived at their final forms in one to three stages. The Codex Dresden contains the first instance, when **TSEK** or **SEK**, **K'Á'AW**, and some undeciphered predecessor or form of **CH'EN** debuted. The second instance consists of the affixes and infixes with four glyphic compounds in the Landa manuscript, phonetic cues to **YAX CUN**, **CHAK TUN**, **PAWA'**, and **IC-?**, a variation of **CH'EN**. The same document preserves in alphabetic script and prefixed glyphic spellings the last step, the standard names. Of those rendered with letters, only **TZOZ** is unique, while **K'ANK'IN** is already known from one eighth-century text. Of those represented via prefixed logograms or syllables, **POP**, **ZIP**, **K'AYAB'**, and **KUMK'U** are new, while **WO** and **PAX** go back to the Late Classic.

The Period of the Conquest, 1520–1570 CE: Yukatek Seasonality?

In 1548 the Yukatek froze their calendars to the European Julian system and stopped counting leap-year days (Edmonson 1988: 76). Now **YAXK'IN**, **MOL**, **MAK**, and **MUAN** as well as the innovations **WO**, **SIP**, **XUL**, **KEH**, **K'ANK'IN**, and **KUMK'U** could relate to the seasons. The many fits between the names and the annual agricultural round suggest that the calendar priests invented new seasonals or reinterpreted older ones to do just that. **UO**, for example, was coined by 880 CE and **K'ANK'IN** by 743, yet could have been relevant in 1548.

THE LANDA SET: 1566

Landa prefixed one or two glyphs to eleven of his eighteen months that indicate Yukatek meanings that were rather different from those of the Classic and Dresden sets. For six he preposed syllabic spellings: **POP**, **WO**, **CHEN**, **PAX**, **K'AYAB**, and **KUMK'U**. To **SIP** he prefixed a phonetic complement, while to **TZEK**, **YAX**, and **KEH** he suffixed one. Finally, in a single case, **K'ANK'IN**, he significantly altered the main sign itself.

Tzek/Sek

Instead of the normal **tse/se-ka** for **TZÈEK** "skull," assuming *k'* can weaken word-finally to *k*, Landa substitutes **wa** for **ka**. Tsew and Sew produce no glosses in Yukatekan, Ch'olan, or Tzeltalan. An alternative understanding suggests that Landa's informant abbreviated the unique name innovated less than a hundred years earlier in the Codex Dresden on page 62. There the normal **tze-ka** for **TZÈEK** "skull" appears as **tze-ka-wa**, which can barely be parsed into **TZÈEK WÀAH** "skull tortilla," perhaps a kind of bread. Bricker (2000: 97) interprets this scrambled order as the error of a Yukatekan-speaking scribe copying a Ch'olan text. A third possibility views the **-wa** suffix as a cue to pronounce the compound as **KASEW**.

Ch'en

The inscriptional **IK' HA'** "black water" apparently changed in Yukatek first to **ÉEK' HA'* then to Landa's **CH'EN** or *ch'éen* "cave with water," the standard colonial form. Apparently **éek'* shifted from "black" to "pond," as in *kax ek'* "forest pond," where *éek'* refers to a body of water, likely as a "spot," also *éek'*. With the main sign **HA'** "water, pond" redundant, ***ÉEK'*** on its own produced *ch'éen* "well, grotto."

Yax

The written name and the affixes with this month reveal four stages of development. The main sign, when it did not show the *X* infix, represented the Classic **HAB'** "year"; **HA'** "year; rain; soapberry" and **SIHOM** "soapberry." With the *X* infix it yielded **KUM** "stone" in Ch'olan, first recorded in 874 CE. In the Landa manuscript it suffixes **-ni** and so reads **TÙUN** "stone." **YÁ'AX TÙUN** "green stone, jade" thus was the month name somewhere between about 1450 and 1550. By the time the bishop recorded the month names, scribes still wrote but no longer pronounced **TÙUN**, using just **YÁ'AX**.

Keh

Landa's **KEH** "deer" developed from the putative Ch'olan ***CHIK HAB'** "red season" after ***CHIK** shifted via homonymy from "red" to "deer" and then to *keej* "deer." Somewhere in the process redundancy rendered ***HAB'** mute, so "deer season" became just "deer."

K'ank'in

Landa's rendition of this compound preserves the **-wa** suffix present with the Dresden and Classic counterparts. His version also presents the two logograms that make up the name, **K'AN** "yellow" and **K'ÌIN** "sun, day." The prelate's scribe, however, mixed these three elements into a composite not at all like that of the other sets.

Pax

In front of the standard **pax-HAB'-ma** for **PÀAX-AM'** "tap instrument, drum" Landa's informant drew **pa** and an unidentified ovoid, **xV**, to yield the Yukatek **PÀAX** "drum."

K'ayab

Landa's scribe set down the normal Codex Dresden version of the name, read **k'a-a-wa** or **K'A'AW** "great-tailed grackle." To its left he supplied the apparent spelling **k'a-ya-b'a**, assuming that the brow or ridge in the upper left of the **k'a** sign reads **ya** (fig. 88 Kayab in appendix C). The resulting **K'AY-AB'** "sing-instrument, drum" bears no evident relationship to either its Dresden or Classic forerunner.

Kumk'u

Prior to Landa's time, the last month name transmuted its pronunciation and meaning. While retaining its written form **OL [WAH]** "ball [of food]," the glyphic compound acquired the reading **KÚUM K'UH** "thunder god" indicated through the prefixes **kum** and **k'u**. The semantic transfer resulted first from the reinterpretation of **OL WAH** or more likely just **OL** "ball" into the Ch'olan near-synonym **KU'M K'U'** "nest ball, egg," then the Yukatek transformation of this into the homophones **KÚUM** "thunder" and **K'UH** "god."

The Colonial Period, 1570–1800 CE

Variation and innovation confined to the tradition of origin characterize this period. Most of the changes qualify as substantial. Overall, the changes involve rather few names.

DIFFUSION TO K'ICHE'

The K'iche's, to all appearances, converted the K'atic "Burning the Underbrush" of their Kaqchikel neighbors to their own Tzij "To Set Afire." This they did by 1698, the earliest date of record for this rare name. It remained unrecorded elsewhere but was rediscovered in Santa Cruz del K'iche' during the 1930s. The same set of circumstances occurred with the term Hunobixquih, the unique 1698 expression mistakenly assigned to the Five Days.

DIFFUSION WITHIN KAQCHIKEL

Kaqchikel developed variants for just two names. The first, Hun Mam (1675), did not change the meaning of the standard form. The other name underwent significant modification, as the normal *pa-* of Pariche became *pay-*, changing "In the Granary" to "Joker Log" and then shortened to just *pay* "Joker." Finally, if not garbled, one name from the 1650 source, Moh "Crow, Hawk," represents either a loan of the Q'anjob'al Mo "Hawk" or an innovation.

DIFFUSION WITHIN POQOM

From early in the eighteenth century came the sole records of two Poqom names. One was the Poqomam Mol, from 1720, while Poqomchi' contributed the other, Pet Cat, in 1725.

DIFFUSION WITHIN Q'EQCHI'

In 1589, on the only known Ch'olan roster, the Q'eqchi' names Chichin and Ianguca appeared as loans. They never surfaced on a Ch'olan or Q'eqchi' list again.

DIFFUSION WITHIN TZELTAL

This tradition's first note on the months came from the early part of the period (1550–1800 CE). In a dictionary compiled about 1600 the phrase Chayquim Hohel Cacal described the span of five days as "lost days, five successive days." Toward the very end of the period the only other scraps from the roster surfaced, specifically the entries "*guincil* autumn" and "*jul-el* spring" in a vocabulary list from 1788. *Guincil* constitutes the root lexeme for the five consecutive months known as 7–3 Binkil. When the Tzeltals fixed their annual count to the Gregorian calendar immediately after 29 February 1584 (Edmonson 1988: 83), these five months might have constituted part of a 140-day countdown from 21 July to 8 December, with the winter solstice falling thirteen days later. The Tzeltals apparently instituted this series as a count to the month of the solstice and invented the names for its units in 1584. If they ever had 2–1 Binkil, the Tzeltals replaced them with Yaxkin and Pom at a date and for a reason still unknown.

Diffusion within Tzotzil

The sole name that changed entered the record in 1688 as Hok'in Ahval "Water Hole Lord." This version prevailed until 1723, when the natives reinterpreted it as Ok'in Ahwal "Crying Lord," a term that has remained standard to the present.

Two other names suffered outright deletion early on. Muy, the term for a certain tree as well as for ground wild tobacco, functioned as a queried alternate to Mux "Hominy" on the 1688 roster but as its replacement in the one of 1723. After this, however, Muy vanished from the record, and only Mux remains. The word denoting an enclosed area, Hoyoh, accompanied Ulol as an alternate on the 1723 roll, but this was its only appearance.

Finally, five names standard in all subsequent lists might represent products from the early years of the colonial era. After the Tzotzils froze their calendar to the Gregorian count in 1584 (Edmonson 1988: 83), seasonal names could function as commensurates. For many, though, the data are too sparse or the referents too broad to justify the assignment of dates relevant to the Tzotzil environment and the solar cycle. For others, the stemma reconstructions suggest that they entered Tzotzil long before 1584 and from different ecosystems, so their apparent seasonality can qualify as only coincidental. Running from 2 to 21 May, Julol "Arrive at the Middle" aptly describes the northern zenith passage. In the Tzotzil homeland this happens between 1 and 9 May, so perhaps this name was invented in 1584.

On the summer solstice of 21 or 22 June the sun rises at its extreme northerly point on the horizon then begins its trek back to the southern extremity. On the way it again crosses the zenith. In the Tzotzil-speaking region this occurs on 4–8 August. The series 1–4 Vinkil begins on 10 August in the frozen calendar, so it might constitute a count based on that event. Its end date of 28 October, though, does not correlate to any obvious celestial happening. The series might derive not from astronomy but from the four color months as altered by about 800 CE when the main sign first changed from Sihom to Winik, the Yukatek cognate to Vinkil.

Diffusion within Yukatek

An innovation of this tradition, **TZEK** debuted in the Codex Dresden, painted about 1450. Landa mentions the winal three times in his manuscript of 1566, spelling it in each instance as "Tzek." Afterward, this version appears in only two lists, rosters 1 and 9 of the Codex Pérez, a hybrid of Chilam Balam literature and archival sources dated to between 1550 and 1800. The otherwise-exclusive form throughout this period is "Sec," the reduced if not meaningless shape of Tzec; it first enters the corpus with the Sánchez de Águilar roster of 1613. Landa's unique rendition of the fourth name, Tzoz, might represent just a copyist's reversal of the usual Zotz (*sòotz'*) "Bat," but the enticing alternative casts it as a loan unaltered in form or meaning from K'iche' or Kaqchikel. Landa's creation is recorded only once in his manuscript and never again in any other source. The normal form "Zotz" debuts at last in the Chilam Balam of Chumayel, about 1750. The final Yukatek contribution adds a modifier to Yaxk'in, something that no other tradition does. The product, Tze Yaxk'in "Small Green-Days," also appears for the first time in the Chilam Balam of Chumayel.

The Modern Period, 1800–2000 CE: Minor Variations

In the last two centuries several traditions have provided a few modifications, reinterpretations, and innovations. Over the past ninety years Tzotzil has proven remarkably active.

Tzotzil Diffusion to Q'anjob'al

The unique name Cubihl "Whistle" finds a counterpart and apparent source in the Tzotzil Ok'in Ahwal "Whistle Lord." The Tzotzil invented this name between 1688 and 1723 by reinterpreting a loan from their neighbors the Tzeltals.

Tzotzil Diffusion to Tzeltal

In this period just one name underwent reinterpretation. Virtually all the Tzeltal sources presented Hok'in Ajaw "Water Hole Lord." The first recorded form, however, from 1887, was Hoquén-Hajab (Ok'-en Ahaw) "Crying Lord," a variant of the Tzotzil Ok'in Ahwal, itself originally set to paper in 1723. This alternate surfaced in Tzeltal just once more, as the Okinahaw of Tenejapa in 1944, a relic or a loan.

During the second half of the period one community experimented briefly with unique modifications on two names. For the first three decades Tenejapa alone used Bikituch "Little Opossum" for the usual Ch'in Uch of the same meaning. Through the next three decades Tenejapa abbreviated the normal Ahil Ch'ak "Marsh Flea" to Ch'ak "Flea."

Diffusion within Ixil

After probably centuries of existence the Ixil month names first found their way onto paper in 1940–42. Although only two of these six earliest lists presented the full complement of nineteen units, all of them displayed remarkable diversity, as each contained unique month names. Some thirty years later the three new rolls from Nebaj not only showed variation among themselves but also innovated yet more names, specifically Akmor, K'oolk'ooy, Keejepk'i, Leem, Mahab, Masnimchoo, Moochnimcho, Moochoo, Mooxnimchoo, Nee'choo, Onchiv, Pekxitx, Soojnoy, and Soonchooj.

Diffusion within Kaqchikel

During the period 1800–2000 Kaqchikel developed two variants, Çibix (1885) and Nabei ToΣ (1885). Each term really means the same as its original counterpart.

Diffusion within K'iche'

In his study and subsequent dictionary, Juan de León (1945, 1954) introduced several unique, perhaps archaic, esoteric month names from Santa Cruz del K'iche': Ajau Chap, Caam Kiij, Cak Che, Chacan, Chi Il Lacan, Cox Xe Puatl, Iquim Kij, Jop Bix, Jun Bix Kij, Li Quin Cab, Mamkuk and Zac Imuj. A few represent outright innovations, but many significantly modify their predecessors. For instance, 1/2 Mamkuk alters 1/2 Mam "1/2 Grandfather" to "1/2 Quetzal." Zac Imuj "Shadeless Days" as a reference to summer appears to represent the result of the progression from Sak "White" to *Sak Kij/*Sa-Ki "White Days, Summer" to Muj Kij "Summer," to the apparent hybrid.

Diffusion within Poqom

One unique name appeared early in the twentieth century. In 1914 Kaxik-Laj-Kij designated the unlucky span of five days.

Diffusion within Tzotzil

The earliest known version of one Tzotzil name was set down as Hok'in Ahval in 1688; this preserved the local translation of the standard Tzeltal form Hok'in Ajaw "Water Hole Lord." By 1723 the Tzotzil had recast it into their standard Ok'in Ahwal "Crying Lord," the only name to undergo noteworthy modification in this last period. By 1963 one community shortened this to Ok'in "Crying." In 1968 another added *k-* "our" to the second part, yielding Ok'en K-Ahwal "Our Lord Has Wept" or "Our Lord Is Crying." From the same town came the last variant, in 1981: H-Ok'en Ahwal "Wailer Lord" or "Drummer Lord," the sole version of this name in Tzotzil since 1688 to include the *h-* agentive prefix. In brief, Tzeltal has Hok'in Ajaw, with *h-* present but *-al* absent, while Tzotzil uses Ok'in Ahw-al, with *h* absent but *-al* present.

During the later years of the modern period several months developed additional alternate names. Okin Ahwal developed two alternates: by 1944 Chenalhó had coined Me' Okinahual "Mother Crying Lord" or "Mother Rain Days Lord," while by 1968 Chamula had minted Me'el Uch "Old Woman Opossum/Drizzle." In those same years, as an optional stand-in for Uch "Opossum" and in apparent opposition to their "Mother, Old Woman" counterparts, the same towns confected H Uch "Male

Opossum, Opossum Hunter," Mol Uch "Old Opossum, Old Man Drizzle," and Mol H Uch "Old Opossum Hunter, Old Man Drizzle Maker."

In one Tzotzil town two names briefly replaced their predecessors. Unenuch "Small Opossum/Drizzle" substituted for Uch from about 1910 through the 1930s at Magdalenas Tanhobel, while during the same decades Muk "Bury, Plant" was exchanged for Mok "Wall" at San Miguel Mitontic.

DIFFUSION WITHIN YUKATEK

The Itzaj priests of Tayasal on Lake Petén remained thoroughly conversant in the hieroglyphic script until 1697, when Spanish soldiers carried off their last codices. At this time some northern Yukateks still read the painted books. By 1750 literacy of the hieroglyphic script had likely vanished. The final manuscript form of the Chilam Balam of Chumayel dates to 1824–37. Its month list exhibits pictures beside the names, but only one, **MOL**, with its ring of dots, resembles its inscriptional counterpart. These scrawled recollections mark the demise of hieroglyphic literacy.

Frozen and unaltered since 1548, and apparently obsolete after about 1800, the Yukatek tradition briefly became the focus of the native revivalist shaman Hun Batz' Men (1983: 173–75). He amended the list of Landa, believing that he had been misled by the natives. He recast K'ayab "Drum" as K'ay Yab "Many Songs," a perhaps ungrammatical reference to spring. Uo "Bullfrog" and Zip "Err" he replaced with Uol and Zip', though without offering their respective definitions "Ball" and "Ripe."

Conclusions

In order to scrutinize the month names in their many forms and meanings, this study starts with the standard suite of the Classic inscriptions and follows through to the fourteen sets of the ethnographic present. The focus bears first on the names, then on the fixed dates. The results bring meaning to the words and dates and convey a high degree of relatedness among the traditions.

THE GLYPHIC CALENDARS

This analysis considers the innovations in terms of type, name, and set (table 1.9). Given the many years and numerous scribes, it comes as no surprise that the Classic set accounts for sixty-two of the ninety-two modifications (67.4 percent). The Codex Dresden follows, with twelve modifications (13.0 percent). The final span, combining the Interim Period with the work of Landa, accounts for the last eighteen alterations (19.6 percent). When the ninety-two modifications are classified and tallied by type, the category "complements" totals forty-four changes (47.8 percent). "Logographs" accounts for thirty-three (35.9 percent). Innovations due to the simultaneous use of complements and logographs are in the category "both" and total eight (8.7 percent). Seven of the ninety-two changes (7.6 percent) do not arise from the use of either complements or logographs and constitute the class "neither."

The last class is exclusive to Landa, whose names correspond partially or not at all to his glyphs for seven months, namely Tzoz, Tzec, Xul, Chen, Yax, Zac, and Ceh. Finally, scores for individual names deserve note. The three names that undergo the most innovations are K'ank'in with ten (10.9 percent), then Ch'en and Pax with eight each (8.7 percent). The name that changes the most via logographs is Sec: five times (5.4 percent). The names modified the most through complements are Mol and Pax, both also with five instances (5.4 percent). Only two names do not innovate across their long history: Yaxkin and Muan. They emerge as the most stable, while Kankin proves the least so. In any case innovations among the month names are sporadic in time and space, highly localized, and rather short-lived.

The Ethnographic Calendars

The linguistic-ethnic traditions with any calendrical data total fourteen. Ch'olti' preserves only three month names identical to their Ch'ol counterparts, while Poqomam and Poqomchi' very likely represent just one system, so the count comes to twelve. The traditions are Ch'ol, Ch'olti', Ch'orti', Chuj, Ixil, K'iche', Kaqchikel, Poqomam, Poqomchi', Q'anjob'al, Q'eqchi', Tzeltal, Tzotzil, and Yukatek.

With two sources Ch'ol has one calendar. The first dates to 1589, or perhaps 1548, while the second is recorded in 1631. No later example survives.

A Ch'olti' dictionary compiled in 1625 preserves four expressions for the last five days of the year. A census roll from 1712 offers a likely candidate for one month name. Baptismal records from about the same time yield the last traces of this tradition, two unimpeachable month names.

The Ch'orti' system includes two calendars. Both preserve the same winal name, though in different date sequences. One roster references 1588 or 1548, while the other is rooted in 1612.

The one Chuj calendar first enters the ethnographic record in 1930, with its dates evidently in order but many of the names in disarray. The only other mention of it is from 1975, correlating and naming the Five Days. Recession back to 1932 demonstrates that it is the equivalent of the Q'anjob'al calendar. The Chuj count was frozen in 1983.

Ixil proves the most prolific and the most unlike all the other systems as far as the names are concerned. In the six rosters from 1942, five innovate from three to six unique names each. The three lists from 1973 each preserve from one to four new designations. Altogether this system presents eight significant rolls. The sequence and dates for the 1942 calendars evince considerable disarray; sun counters in one town even tweaked their calendar into a nearly European format by switching months from twenty days to thirty. By the 1973 study the sequence had stabilized, and the dates were reliable to within two weeks because the count had become semi-frozen. Only New Year's Day varies each year, though constrained to between 1 and

15 March. This applies perhaps only to Nebaj, as the other major Ixil communities allegedly continue with the count nonfrozen.

The K'iche's over time devised three variant calendars, each starting with a different month. Two keep the same designation for the Five Days but insert it at different points. The third uses a unique name for the brief span but places it in the same spot as one of the other two. Except for this single difference, which might actually represent a Kaqchikel roster, the sequence of names is identical in all three rolls. The first full list complete with dates was recorded in 1698. This "is the only native calendar that has survived intact until the present time" (Edmonson 1988: 237). Kaqchikel and Ixil represent variations on the same calendar. But the Kaqchikel calendar cannot qualify as intact because no recent records indicate its present form. The Ixil calendar, though reported in 1973, has nearly frozen the count in one town, innovated several names in each roster, and grown uncertain of their sequence. Chuj also fails to qualify, first because it lacks one name, second because the sequence of names remains quite uncertain, and finally because the calendar has been frozen since 1983.

Seventeenth-century copies of a 1550 lexicon reproduce the earliest Kaqchikel names. The premier record is the calendar of 1685, replete with names, dates, and comments. The most recent vestige consists of two names in a dictionary from 1990.

The sole glimpse into the Poqomam tradition comes from a 1720 source. Here two names appear with dates. These suggest a sequence consonant with the Poqomchi' set.

The first documentation for the Poqomchi' tradition lies with an extensive dictionary of 1725. It yields two names with dates. Two more, without dates, slip as seasonal terms into a royal questionnaire filled out and remitted in 1788. The sole complete list of names with dates and comments appeared in its final form in 1914.

The seven rolls that represent the entire corpus for the Q'anjob'al calendar actually offer just one list. About half the names are not secure in their sequence and dates. The count was frozen in 1932.

The earliest information for the Q'eqchi' tradition consists of a complete calendar correlated with 1548 or 1589, but some regard it as Ch'ol instead. The first two indubitably Q'eqchi' names come from a vocabulary list of 1788. In 1979 the last roster was published. It presents a calendar composed of thirteen months of 28 days, plus 1 final day. This count corrects for leap years.

The Tzeltal system remained undocumented until 1788, when a local cleric included two month names on a vocabulary list. In 1887 a local scholar published the first full set of names along with their dates. Numerous reports have since entered the record, the latest in 1991. The variants stem from combinations of three factors: the name of the first month, its dates, and the position of the Five Days.

The Tzotzil calendar first entered the record in 1688 with a complete set of names and dates, likely a copy from some earlier document. The many sources since then, mostly from the nineteenth century, bring out the tradition's diversity. Its numerous variants result from the association of three features: the name of the first month, its dates, and the location of the Five Days. Only one version is unorthodox. The latest Tzotzil list appeared in a 1968 report.

In a treatise for his defense before the Inquisition, Bishop Landa in 1566 presented the first information on the Yukatek calendar, a complete roster of names and dates, together with detailed descriptions of related activities and rituals. The myriad sources after this all record the identical system. The name and dates of the first month and all the rest remain the same, as does the position of the Five Days. Yukatek stands out for its titulary of synonyms and descriptive phrases for the dreaded Five Days. The last primary native manuscript was finished between 1824 and 1837. The final compilation of calendrical data from Yukatek documents took place between 1843 and 1877.

The Names of the Months

Altogether the traditions generate over two hundred names. In a few of the sources the authors offer thirty-five apparent definitions, but only twenty-three of these are correct, about two out of three. Many comments accompanying the attempted definitions concern seasonality and agricultural activities.

In every tradition some month names prove unique. In eight systems a third or more qualify as such and in four over half. With 91 percent of its many names peculiar to itself, Ixil emerges as the most innovative tradition. Yukatek, Q'anjob'al, and Poqom fall in the middle, at about 50 percent. Exactly one third of the names qualify as sui generis in Ch'olan and Tzotzil. Chuj proves the least creative: just three of its nineteen names (15.8 percent) appear on no other list.

One or more of the names in every system allow variants. The traditions count 113 in all. By category these are cognates (45 percent), synonyms (33 percent), replacements (17 percent), and alternates (5 percent). Replacements are substitutes unrelated in form or meaning to the original names. Alternates function as additional optional companions to the established names. Ixil generates the most counterparts by far, with 35 percent of the total. The nearest contenders, Tzeltal and Tzotzil, place far behind, at 17 percent, while Chuj offers the least at 0.8 percent.

This study compares each name of every tradition to all the others. Initial results point up many possible equivalents. Apparent sense and need to make the fewest assumptions indicate the more likely counterparts.

The names group into thematic categories, as noted in chapter 3 and appendix F. About half the names focus on agriculture. Animals and religion constitute the other major concerns.

The similarity in the nineteen principal calendars of the fifteen traditions indicates five groups. The glyphs, Ch'olan, Yukatek, Poqom, and Q'eqchi' constitute the first group. The second consists of Tzeltal and Tzotzil. Chuj and Q'anjob'al constitute the third. Two K'iche' lists along with Kaqchikel make up the fourth. The fifth group consists of just the Ixil calendars.

THE DATES

Nearly all the fourteen postglyphic traditions record alternates in name and date for the First Month and the Five Days. The standard and variant calendars total thirty-six. Twenty of these are now frozen or fixed to the Gregorian count. The other sixteen run with no leap year adjustment and so recede one day every four years in regard to their Western equivalents. The number of calendars frozen and not frozen in each tradition is as follows: glyphic 0/1; Ch'ol 0/2, Ch'orti 1/0; Chuj 0/1; Ixil 0/3; K'iche' 0/3; Kaqchikel 0/1; Poqom 1/0; Q'anjob'al 0/1; Q'eqchi' 1/0; Tzeltal 2/4; Tzotzil 14/1; and Yukatek 1/0.

The calculations here relate three varieties of calendars and their values for the length of the year. The Vague Year of the Maya consisted of eighteen months of 20 days apiece, plus one period of 5 days, for a total of 365 days. This count did not correct for leap years, running continuously. In 45 BCE Julius Caesar instituted the Julian count of 365 days plus one more every fourth year (Bond 1889: 2–4), for a length of 365.25 days. This proved a bit too long, as solar events steadily occurred earlier. In 1582 Pope Gregory XIII corrected the accumulated ten-day error by decreeing the day after 4 October would be 15 October. He refined the leap year adjustment: every four years the count would add one day, in February. Beginning with 1600, each fourth century year would indeed add the day, but the other three century years would not. In 1600 and 2000, then, the bissextile was included but not in 1700, 1800, and 1900 (Bond 1889: 7, 9). These alterations improved the average to 365.2422 days, identical for present purposes to the solar or tropical round (Aveni 1980: 171). For a date to recede back to the same point requires 1507 years, one annual round divided by the discrepancy between the Vague Year and the tropical count or 365/0.2422 (Bricker 1982a: 103 n. 2). The reformed calendar debuted in Mexico on 15 October 1583 and in Guatemala on 29 January 1584. The Tzotzils and Tzeltals in turn fixed their systems to the new one in 1584 and 1585 (Edmonson 1988: 178); by accepting bissextiles, they kept each of their days aligned with its European counterpart.

New Year's

The Vague Year recedes one day every four years. If 10 Kän Jälib' fell on 24 July in 1588, for example, then it also fell on that day in 1589, 1590, and 1591, but in 1592 it came on 23 July. In calculations forward through time the Maya date drops behind. For equations back through time it draws ahead. The year reached always belongs to a set of four, by convention designated with either its first and last years or merely its first; in either case the opening member is always the leap year. Both 1589 and 1591, for instance, belong to the span 1588–91 or 1588 for short.

This relationship between the systems allows the recovery of unstated dates. For instance, in 1933 one scholar published the first correlated dates ever for Tzeltal but omitted the relevant year; in the town of Cancuc the annual round commenced on

1 Jokenajau or 7 May (Becerra 1933: 45–46 T. 8). About ten and twenty years later another investigator (Schulz 1942, 1953) shared complete dates from the same community. As 1 Jokenajhau corresponded to 30 April in 1941, the slippage of seven days confirmed that the calendar was nonfrozen, counting the bissextiles as ordinary days and so still receding; it also yielded the solution of 1941 minus 28 (7 times 4) or 1913 (that is, 1912–15), as the years when the Maya date corresponded to 7 May.

Recession also establishes the four years when dates were current: this offers useful historical knowledge because many original rosters survived only in much later copies. On 31 January 1698 Fray Basseta finished a K'iche' dictionary that featured a list of month names with dates but no referent year. In 1722 a scribe set down a calendar demonstrated contemporary with its composition date. The months of the 1698 and 1722 lists prove altogether comparable in their names, lengths, and sequence but not in their dates. For 1722 (1720–23) Nabe Mam began with 3 May, while in 1698 (1696–99) it started on 4 June. The difference of twenty-four years indicates six intercalary days; but as 1700 did not count a bissextile, the total really amounts to five. This is far less than the gap of thirty-two days between 4 June and 3 May, so the Basseta roster belongs to a different era. Recession carries the day from 1722 back 132 years to 1590, or 1588–91. This algorithm can establish when a name's denoted activity or sense corresponded to the seasons.

Finally, recession can ascertain when a calendar was frozen, unfrozen, or refrozen. The Tzeltal community of Oxchuc, for instance, equated its New Year's day 1 Batzul to the Western date in 1916–19. They still did in 1988–91 (Gómez Ramírez 1991: 215–16). If these Tzeltals froze their calendar in 1585, with 1 Batzul at 17 January (Edmonson 1988: 258), the sixteen-day retreat to 1 January, factoring in the nonbissextile 1900, indicates that sun priests unfroze the count in 1848–51. A faction in Oxchuc's dependency Dzajalchen stayed on this reckoning as late as 1941, when they equated New Year's to 26 December (Villa Rojas 1946: 573–75). As Villa Rojas (1946: 576) explicitly pointed out, the calendar ran unfrozen. Similarly, the Tzeltal Tenejapa recorded 1 Batzul as 26 December in 1943 then counted 29 February 1944 as one of the twenty days of a month (Barbachano 1945b: 116), so here too the calendar ran unfrozen. It continued at least through 1965, when the Gregorian equivalent was recorded as 20 December (Berlin et al. 1974: 119 TT. 5.16 and 5.17, 120–24). The same date appears thirteen years later in San Cristóbal de las Casas in a calendar published for 1979 (Edmonson 1988: 257–58). This indicates that in 1964–67 the town both adopted the count from Tenejapa and refroze it.

The last example comes from the Tzotzil municipality San Miguel Mitontic. In 1916 its 1 Tzun still fell on 28 December (Schulz 1953: 115), as it had ever since the daykeepers froze the count to the Gregorian calendar in 1584 (Edmonson 1988: 259–60). In 1943, however, the equivalent had slipped to 26 December (Barbachano 1945a: 16), while Muktasak began right on 29 February 1944. Clearly the sun priests had unfrozen the annual round. The two-day discrepancy indicates the span 1932–35 for the change.

The Five Days

The number of names and positions for the Five Days varies. Four systems (Kaqchikel, Q'anjob'al, Q'eqchi', and Yukatek) have just one standard term and one position. The Tzeltals and Tzotzils employ an identical name, but subtraditions insert it in two different places. Ixil has just one name but finds it in three distinct places. Ch'ol, Ch'orti', and Poqom use one or two names, each in a different slot. K'iche' shares one designation for the period with its close relative Kaqchikel but innovates a second term and puts the pair at two or three places in the sequence. The glyphic calendar presents two distinct forms; both always appear at the end of the round. Finally, Chuj distributes two or three names between a pair of sites.

Among the calendars the tally of months that mark the onset of the year ranges from one to four. The systems with just one initial month are glyphic, Ch'ol, Ch'orti', Chuj, Kaqchikel, Poqom, Q'anjob'al, Q'eqchi', and Yukatek. K'iche' lists two or three initial months, but three function for the Ixil and Tzeltal rolls. The Tzotzil rosters use four: in each case all the months retain their names and dates, while only the position numerals shift. Distinct first months can operate consecutively through time or simultaneously in various community subtraditions. For the K'iche' and Kaqchikel they seem to reflect immigrant origins and later political adjustments. Among the Tzeltal and Tzotzil the varied starts might result first from tributary relations to Nahua-speakers between 1250 and 1550 CE and since then from migration or marriage between two communities that had each been subject earlier to one city of the Aztec Triple Alliance.

Except for the feature of freezing, most of the thirty-seven individual calendars have retained their native structure of eighteen months plus five days. Three, however, have altered even this. The least revised version comes in the fixed 1901 Tzotzil roster. This commences on 1 January. Its shortest month, 18 days long, corresponds to the month of February. Its longest months, seven in all and 21 days long, overlap substantially with the European calender's seven months of 31 days. Its remaining ten months each run the normal 20 days. The Five Days have been absorbed into the longer months.

More assimilated is the Ixil roster from Nebaj (table 2.8), which has an initial month of 34 days, ten more consecutive months of 30 days, a final month of 26 days (a clear counterpart to the concurrent February), and then the final five days.

The most deviance surfaces in the Q'eqchi' town Cahabón. This frozen count consists of thirteen months of 28 days, plus 1 day. It clearly adjusts every four years for leap year by adding an extra day right after 31 December and equating it to 1 January.

Of the thirty-seven distinct calendars from the fourteen traditions, twenty are frozen. Fifteen of these start up between late December and the end of March. The five left begin in mid- or late July. Most of the Five Days fall in February and March. Up to twenty-five of the thirty-three fixed initial dates for First Months and the Five Days might find explanations in annual solar events or the church's season of Lent. Table 4.13 lists all the dates, by tradition, and matches them with possible referents.

The frozen initial dates of late December seem related with possible days of the winter solstice, 21, 22, or 23 December. The Tzotzil First Months beginning on 26, 27, and 28 December start the year five days later, but arguably they also distinguish New Year's as the day of the sun's first detectable movement. In late December this initial motion is perceived and real, but after that it must be fictive or ritual, because the natives assign it to impossibly late dates. The range that the Tzotzil and Ch'orti' assign it, from 26 December to 8 February, includes other start dates, specifically 31 December; 9, 15, 16, 17 and 25 January; and 3 February.

The March initial dates, all Tzotzil, might refer to the vernal equinox. As this falls on 21, 22, or 23 March, 2 and 3 March alert to it twenty days ahead of time, 11 March ten days, 16 and 18 March five days. The event itself is marked by the month start of 23 March.

1 May could presage the northern zenith passage among the Tzeltals of Oxchuc by five days. This occurs across Tzeltal country between 6 and 8 May.

One initial date appears to focus on the summer solstice. The Poqoms of Amatitlán begin their Five Days on 30 May, a countdown to 4 June, first day in the month of the summer solstice of 21 or 22 June.

Instead of the winter solstice or the spring equinox, 3 February and 2 and 3 March might indicate instead a church event. Falling anytime from 3 February to 9 March, Carnival or Shrove Tuesday can embrace all the initial dates from 3 February through 3 March. There are twelve of these, eleven Tzotzil and one Ch'orti'.

Only the Ch'orti' community Jocotán begins its year on the feast of its patron saint. The day of Santiago falls on 24 July. This precedes by exactly one native month the local southern zenith passage on 13 August.

History might explain up to eleven first month dates among the Tzeltals and Tzotzils. Some of the Tzeltals in Oxchuc start the year on 10 February; this also marked New Year's for the Gulf Coast Nahua ancestors of the K'ichean rulers, who passed through in 1250–1350 CE on their way to Guatemala. Some clans of these interlopers perhaps remained behind to establish hegemony and tribute rights over the Tzotzil towns of Chalchihuitán, Chamula, Chenalhó, Guitiupa, Istacostoc, Mitontic, Pantelhó, and Tanhobel, because each begins the year with the equivalent of the invaders' first month, be it 27 or 28 February or 2 or 3 March. Some in Chamula, Chenalhó, Istacostoc, and Mitontic start with 26, 27, or 28 December, the local counterparts to New Year's among the Aztecs' partner in battle, the Texcocans. Finally, some in Chenalhó, Comitán, Istacostoc, and Utrilla commence on 9, 15, or 17 January, their version of the Aztec New Year.

In sum, historical encounters could account for maybe eleven frozen initial dates, ecclesiastical foci for up to thirteen, and solar events for as many as twenty. This leaves three conspicuously without correlates, probably tied to the same event: 11, 14, and 16 July.

A month name is seasonal because it denotes an activity, event, or period significant in the annual round of nature or human concerns. Altogether the fifteen traditions present fifty-five such terms. These represent forty-two innovated forms along with thirteen derived from glyphic antecedents. Of all the seasonal names thirty-five are "commensurates": in each case the activity or event denoted occurs on the same day as the month's initial date or includes it and the rest of the month within its own span. The "noncommensurates" or seasonals not consonant with appropriate dates total twenty. More than any other topic the seasonals focus on maize agriculture. Of the twenty-four themes presented, eighteen clearly concern cultivation.

The frequency and status data on the seasonal names can be briefly summarized. Out of fifty-five terms, thirty-six occur in just one tradition each. Of the twenty known from more than one roster, a dozen appear in two, four in four, one in six, two in eight, and one in nine. The most frequent are Tz'ikin (nine), Mol (eight), Yaxkin (eight), Mac (six), Caseu (four), Olh (four), Sutz (four), and Uniu (four).

With twelve apparent commensurates, Landa's Yukatek system fits the seasons well, first because it is frozen and second because seven of the twelve are innovations, crafted perhaps for a fixed round. Some of these and the other five succeed through reinterpretation or homophony and represent fresh associations. So many fits suggest that the calendar priests invented commensurates specifically for the freezing of the count. The same might apply to Tzeltal and Tzotzil, because each has nine commensurates, two glyphic and seven new.

Except for interpretations for a few terms tied to specific events or general periods, seasonals are of limited value to a reconstruction of the history of the names. They present problems as soon as they produce two or more referents for a single definition or admit to more than one interpretation. With either too many data or too few, the correct meaning or referent often eludes confirmation. Even frozen calendars have some seasonals that are not commensurate. The very existence of names that are neither seasonal nor commensurate suggests that for them and for the seasonals, if not perhaps for the commensurates as well, symbolism and ritual might determine the sense, form, and order of names on the lists more than agriculture and the weather do. Perhaps the month names do not so much track the year as incorporate it within the cosmology.

Stemmas or descent lines order the development of the names in time and space and trace the transfers of sound or meaning. Four out of five names relate as cognates, homonyms, or synonyms, while seven of ten ultimately derive from the glyphs. The stemmas suggest a sketch of their history. The Late Classic (600–900 CE) witnessed first the flow of glyphic names out from their homelands to the Highlands and north to Yucatán and second the spread of the Yukatek invention **SIHOOM** to the glyphs and most other traditions. The Early Postclassic (900–1250 CE) saw the results of the deletion of **SIHOOM** from nearly every list, a remarkable drop in reinterpretations as well as innovations, and the rise of a brisk internal exchange of names among the Highland traditions. During the Late Postclassic (1250–1520 CE) Nahua arrived in the Highlands of Chiapas and Guatemala, while more innovations developed in the northern Lowlands. The K'iche's and Kaqchikels acquired rulers and names from the Gulf Coast Nahuas about 1300. After 1500, along with the Tzotzils and the Tzeltals, they altered their calendars under Aztec impact. After the Conquest (1520–1550 CE) the Yukateks, Tzeltals, Tzotzils, and Poqoms all froze their counts in 1548. Then or during the colonial period (1570–1800 CE) they fashioned new names or reinterpreted old ones to mark annual seasonal phenomena, agricultural activities, or solar events. In the modern period (1800–2000 CE) a few of the fourteen traditions have frozen, unfrozen, or revived their counts, but most have changed slightly or not at all.

Two facts about the Maya calendars stand out. The first is the virtually monolithic uniformity of the hieroglyphic tradition throughout the 700 years of the Classic and across the vast region defined by the monuments with dated texts. The second is the prolific diversity of the nonglyphic traditions along the southern borders of this sphere, systems first recorded in the colonial or modern period. The invariant inscriptional count of the months co-occurs in a principally Ch'olan-Yukatekan area, where numerous concentrations of ruins across the hilly or flat terrain indicate state-level polities often with sizable territory and lesser satellites, societies ruled by divine kings and linked by a common script. The tortured, mountainous marches to the south nurtured many distinct languages and smaller towns run by human chiefs or lineage heads who apparently communicated, if at all, not through written but oral means. The uniform calendar could represent just a by-product of a standard script or it might instead serve as a shared solution to the needs of urban elites and bureaucrats for a mutually intelligible scheduling of commerce, tribute, warfare, ritual, dynastic marriages, and state visits.

Such a standard tool would seem out of place among a congeries of rival polities because it more often results instead from the uniformity imposed by a conqueror. Why should the southern Lowland Maya of the Classic resemble the China of Qin more than they do the Greek city states before Philip and Alexander? "There was no general Greek calendar: every state had its own; and each of the four possible points for beginning a new year [solstices, equinoxes] was adopted somewhere in Greece; even the months changed their names across frontiers" (Durant 1966: 341). Why does this apply more to the Maya on the rocky fringes to the south? Geography seems to suggest one answer. The rivers and easier-to-traverse terrain of the Lowlands would facilitate contact and thus the retention of the common language and culture. Its water, rain, and soil apparently sustained the larger populations that tend to coalesce into myriad interactive states with bureaucracies that prefer convenient uniformity. Maybe a basically similar environment, a shared ideology, and a cultural inertia also promoted the standard set of month names.

Closing Thoughts: Key Findings

The standard hieroglyphic month names remained virtually unchanged across the vast geographic and temporal span of the Classic. The many innovations proved remarkably scattered, localized, and ephemeral; only five lasted long enough to be recorded by the Spanish. With the end of the Classic the names vanished altogether from the ceramics and almost entirely from the stone monuments.

The Classic compounds form distinct classes based on the mandatory or optional presence or absence of logographs, purely phonetic syllables, or phonetic complements. The question of why each is not written with every possible combination eludes resolution.

Toward the end of the Late Postclassic the names surfaced again in the glyphs of the Codex Dresden. Three underwent minor alterations, but in five the changes were major. By the time of Bishop Landa most of the compounds had passed through one or two further modifications. By 1750 the hieroglyphic names and the writing system had entered oblivion.

Across the long history of the names only two never really changed. The others all did, in form or meaning or both, to develop into fourteen traditions. These consist of some 250 names, with about 575 plausible senses. Just over 80 percent of the names seem related to one another, while almost 75 percent appear to descend directly or circuitously from the glyphs.

With its innovation of the name **SIHOOM** Yukatekan made the single most important contribution to the glyphic set as well as the highland traditions emerging during the Classic. Gulf Coast Nahua inserted several names into the Kaqchikel and K'iche' rosters in the Late Postclassic. After 1500 the Aztecs foisted names on the Kaqchikels.

Various events appear tied to the numerous initial dates of the ethnographic counts. The winter solstice, vernal equinox, and northern solar zenith passage lead the list. Not far behind come Shrove Tuesday or Carnival and the start of Lent for the variable dates in February and March. Losses to the Gulf Coast Nahuas then later to the Aztecs and their allies might explain some K'iche', Tzeltal, and Tzotzil New Year's dates, shifts likely imposed for scheduling tribute.

This study demonstrates both the conservatism and the creativity of the Maya calendar. Throughout the Classic numerous new names suddenly appeared at various sites but quickly vanished. The prestige culture of the southern Lowlands retained only five innovations. As cognates, synonyms, homonyms, and even equivalents to homonyms, these and the standard forms appeared to the north and south during the Late Classic or survived there later. The Classic tradition impacted the others earlier and more intensely than suspected. Up to 75 percent of their names might descend from its hieroglyphic designations.

The few non-native forms seem to be impositions by invaders. As the first months in counts with these loans and in other rosters might also result from foreign pressure, the suspicion arises that the calendar served a dominant power as a tool for defining the status of tributaries or allies.

The political nature of the calendar proves difficult to demonstrate before the Late Postclassic, but its ritual nature is evident from the beginning. The names concern mythic fundamentals, not practical phenomena tied to specific dates. Only after the freezing of several calendars do some names designate annual solar and meteorological events or agricultural activities. Seasonal, ritual, or political, the month names as a whole outline Maya cosmology and history.

Appendix A

*Phonemes of Languages Relevant to
Maya Hieroglyphic Writing*

Abbreviations and sources: pc: proto-Ch'olan (Kaufman and Norman 1984: 112); py: proto-Yukatek (Fisher 1973; Fox 1978); y: Yukatek (Bricker et al. 1998: xi–xii); cc: Classic Ch'olti'an (Houston et al. 2000). The phoneme *x* is like English *sh*. Noteworthy differences are the sixth or schwa vowel, *ä*, in proto-Ch'olan and proto-Yukatek; the four vowel tones in both forms of Yukatek, on all vowels but schwa (*a* neutral, *áa* high, *àa* low, *á'a* glottalized); the *h:j* contrast in all but modern Yukatek; and the five long vowels in Classic Ch'olti'an.

TABLE A.1

Phonemes of languages relevant to Maya hieroglyphic writing

proto-Ch'olan	proto-Yukatek	Yukatek	Classic Ch'olti'an
'	'	'	'
b'	b'	b'	b'
tz	tz	tz	tz
tz'	tz'	tz'	tz'
ch	ch	ch	ch
ch'	ch'	ch'	ch'
h	h	h	h
j	j	—	j
k	k	k	k
k'	k'	k'	k'
l	l	l	l
m	m	m	m
n	n	n	n
p	p	p	p
p'	p'	p'	p'
s	s	s	s
t	t	t	t
t'	t'	t'	t'
w	w	w	w
x	x	x	x
y	y	y	y
a	a, áa, àa, a'a	a, áa, àa, a'a	a, aa
ä	ä	—	—
e	e, ée, èe, e'e	e, ée, èe, e'e	e, ee
i	i, íi, ìi, i'i	i, íi, ìi, i'i	i, ii
o	o, óo, òo, o'o	o, óo, òo, o'o	o, oo
u	u, úu, ùu, u'u	u, úu, ùu, u'u	u, uu

Appendix B

Sites with Affiliated Languages

TABLE B.1

Sites with affiliated languages

Ch'olan	Source	Yukatek	Source	Bilingual	Source
Bonampak	SF	Bonampak	JF2		
Dos Pilas	JF1-SF	Dos Pilas	FJ		
Machaquilá	SF	Machaquilá	J1		
Piedras Negras	SF	Piedras Negras	JF2		
Xcalumkin	SM	Xcalumkin	J1		
Yaxchilán	H-SF	Yaxchilán	J1-JF2		
Quiriguá	SF	Uxmal	SF		
Naranjo	H	Naranjo	JF2	Naranjo	SF
Tikal	H	Tikal	J1	Tikal	SF
				Río Azul	SF
Aguateca	JF1-SF	Chichén Itzá	J1-SF	Calakmul	SF
Copán	H-JF1-SF	Dz'ibilchaltun	SF	Caracol (?)	SF
Palenque	H-JF1-SF	Nah Tunich	SF	El Perú	SF

Source key: FJ: Fox and Justeson, in Ringle 1985: 158; H: Hammond 1982: 94; J1: Justeson et al. 1985: 14, 16; J2: Justeson et al. 1985: 67; JF1: Justeson and Fox, in Sharer 1994: 589; JF2: Justeson and Fox, in Sharer 1994: 590; S: Sharer 1994: 589; SF: Schele and Freidel 1990: 22–25, 51: maps; SM: Schele and Mathews 1998: 363 n. 29.

For much more detail on the diagnostic traits, sites, dates, and regions for Classic Yukatek, Western Ch'olan, and Eastern Ch'olan, see Lacadena and Wichmann 2002; and Wichmann 2006, especially table 1 and fig. 1.

Appendix C

Innovative Hieroglyphic Month Names

Figure 1. Xul, Hauberg Stela: A2. Drawing by Linda Schele © David Schele. Courtesy of Los Angeles County Museum of Art.

Figure 2. Xul, Stela 6: C1, Itzimté. Drawing by Eric von Euw, Corpus of Maya Hieroglyphic Inscriptions. © President and Fellows of Harvard College, Peabody Museum of Archaeology and Ethnology, PM# 2004.15.6.8.5 (digital file# 99280023).

Figure 3. Pop, as drawn in the extant copy of Landa's *Relación de las cosas de Yucatán*. Courtesy of George Stuart and the Biblioteca Real, Madrid.

Figure 4. Pop, Palace Tablet: N8, Palenque. Drawing by Linda Schele © David Schele. Courtesy of Los Angeles County Museum of Art.

Figure 5. Uo, Unprovenienced Lintel: B1, Yaxchilán. Drawing by Peter L. Mathews, in Mayer (1987: plate 74).

Figure 6. Uo, Stela E: east, 27B, Quiriguá. Drawing by Matthew G. Looper. Courtesy of Matthew G. Looper.

Figure 7. Uo, Vessel K6751: M4, Kan Dynasty. Modified by author after photograph by Justin Kerr © Justin Kerr. Courtesy of Justin Kerr.

Figure 8. Uo, Monjas Lintel 4: A2-4, Chichén Itzá. Drawing by John S. Bolles. © 1977 University of Oklahoma Press. Used with permission. All rights reserved.

Figure 9. Uo, as drawn in the extant copy of Landa's *Relación de las cosas de Yucatán*. Courtesy of George Stuart and the Biblioteca Real, Madrid.

Figure 10. Uo, Vessel K6751: J6, Kan Dynasty. Modified by author after photograph by Justin Kerr © Justin Kerr. Courtesy of Justin Kerr.

Figure 11. Uo, Tablet of the 96 Glyphs: G2, Palenque. Drawing by Linda Schele (1985: 98).

Figure 12. Uo, Codex Dresden, 24:C25. Drawing by Carlos A. Villacorta, in Villacorta and Villacorta (1977: 58).

Figure 13. Uo, Codex Dresden, 63:C2. Drawing by Carlos A. Villacorta, in Villacorta and Villacorta (1977: 136).

Figure 14. Sip, Vessel K6751: H4, Kan Dynasty. Modified by author after photograph by Justin Kerr © Justin Kerr. Courtesy of Justin Kerr.

Figure 15. Sip, Monument 126:B6, Toniná. Drawing by Kees Grootenboer.

Figure 16. Sip, Temple of Insciptions, West: D7, Palenque. Drawing by Linda Schele © David Schele. Courtesy of Los Angeles County Museum of Art.

Figure 17. Sip, Monument Fragment, Copán. Drawing by Linda Schele, in Schele (1987b: fig. 2).

Figure 18. Sip, Hieroglyphic Stairway 5: 158, Yaxchilán. Drawing by Ian Graham, Corpus of Maya Hieroglyphic Inscriptions. © President and Fellows of Harvard College, Peabody Museum of Archaeology and Ethnology, PM# 2004.15.6.7.42 (digital file# 992000033).

Figure 19. Sip, Capstone 6: A3–4, Ek Balam. Drawing by Alfonso Lacadena García-Gallo, in Grube et al. (2003: 16).

Figure 20. Sip, as drawn in the extant copy of Landa's *Relación de las cosas de Yucatán*. Courtesy of George Stuart and the Biblioteca Real, Madrid.

Figure 21. Sotz, Unprovenienced Vessel K0955: G1, Museum of Fine Arts Boston. Modified by author after photograph by Justin Kerr © Justin Kerr. Courtesy of Justin Kerr.

Figure 22. Sotz, Hieroglyphic Stairway 3: AX23, El Resbalón. Preliminary drawing by Eric von Euw. Courtesy of the Corpus of Maya Hieroglyphic Insciptions.

Figure 23. Sotz, Monument 16, Panel 7: B1, Quiriguá. Drawing by Ann Hunter, in Maudslay (1974: 2:plate 63).

Figure 24. Sec, Ballcourt Ring: H, Oxkintok. Drawing by Miguel Rivera Dorado. Courtesy of Acta Mesoamericana.

Figure 25. Sec, Stela 12: D1, Yaxchilán. Drawing by Linda Schele © David Schele. Courtesy of the Foundation for the Advancement of Mesoamerican Studies, Inc., www.famsi.org.

Figure 26. Sec, Lintel 41: B1, Yaxchilán. Drawing by Ian Graham, Corpus of Maya Hieroglyphic Inscriptions. © President and Fellows of Harvard College, Peabody Museum of Archaeology and Ethnology, PM# 2004.15.6.6.12 (digital file# 992000032).

Figure 27. Sec, Monument 24: W2b, Quiriguá. Drawing courtesy of Matthew G. Looper.

Figure 28. Sec, Randall Stela: C1, Usumacinta region. Drawing by Berthold Riese. Courtesy of Karl H. Mayer.

Figure 29. Sec, Codex Dresden, 62: A10. Drawing by Carlos A. Villacorta, in Villacorta and Villacorta (1977: 134).

Figure 30. Tzec, as drawn in the extant copy of Landa's *Relación de las cosas de Yucatán*. Courtesy of George Stuart and the Biblioteca Real, Madrid.

Figure 31. Xul, Altar 1: C1, Naranjo. Drawing by Ian Graham, Corpus of Maya Hieroglyphic Inscriptions. © President and Fellows of Harvard College, Peabody Museum of Archaeology and Ethnology, PM# 2004.15.6.3.22 (digital file# 992000028).

Figure 32. Xul, Altar 1: L4, Yaxchilán. Unattributed drawing in Fox and Justeson (1984: fig. 24c).

Figure 33. Xul, Codex Dresden, 63: B2. Drawing by Carlos A. Villacorta, in Villacorta and Villacorta (1977: 136).

Figure 34. Xul, Codex Dresden, 49c: D1. Drawing by Carlos A. Villacorta, in Villacorta and Villacorta (1977: 108).

Figure 35. Mol, Temple XIV Tablet: B1, Palenque. Drawing by Linda Schele © David Schele. Courtesy of Los Angeles County Museum of Art.

Figure 36. Mol, Unprovenienced Vessel K1344: G2, Museum of Fine Arts Boston. Modified by author after photograph by Justin Kerr © Justin Kerr. Courtesy of Justin Kerr.

Figure 37. Mol, Stela 18: A3, Yaxchilán. Drawing by Tatiana Proskouriakoff, in *Maya Hieroglyphic Writing*, 3rd edition, by J. Eric S. Thompson. © 1971 University of Oklahoma Press. Used with permission. All rights reserved.

Figure 38. Mol, Lintel 14: E1, Yaxchilán. Drawing by Ian Graham, Corpus of Maya Hieroglyphic Inscriptions. © President and Fellows of Harvard College, Peabody Museum of Archaeology and Ethnology, PM# 2004.15.15.1.144 (digital file# 99280023).

Figure 39. Mol, Del Río Throne: B1, Palenque. Unattributed drawing in Mathews and Schele (1974: fig. 5).

Figure 40. Chen, Unprovenienced Altar: B2, Chicago Art Institute. Preliminary field drawing by Ian Graham. Courtesy of the Corpus of Maya Hieroglyphic Inscriptions.

Figure 41. Chen, Temple of the Inscriptions, East: L6, Palenue. Drawing by Linda Schele © David Schele. Courtesy of Los Angeles County Museum of Art.

Figure 42. Chen, Lintel 30: F2, Yaxchilán. Drawing by Ian Graham, Corpus of Maya Hieroglyphic Inscriptions. © President and Fellows of Harvard College, Peabody Museum of Archaeology and Ethnology, PM# 2004.15.6.6.2 (digital file# 99160088).

Figure 43. Chen, Casa Colorada: 29, Chichén Itzá. Drawing after Hermann Beyer, in *Maya Hieroglyphic Writing*, 3rd edition, by J. Eric S. Thompson. © 1971 University of Oklahoma Press. Used with permission. All rights reserved.

Figure 44. Chen, as drawn in the extant copy of Landa's *Relación de las cosas de Yucatán*. Courtesy of George Stuart and the Biblioteca Real, Madrid.

Figure 45. Chen, Codex Dresden, 47c: D5. Modified by author after Förstemann (1880). Courtesy of Ancient Americas at the Los Angeles County Museum of Art.

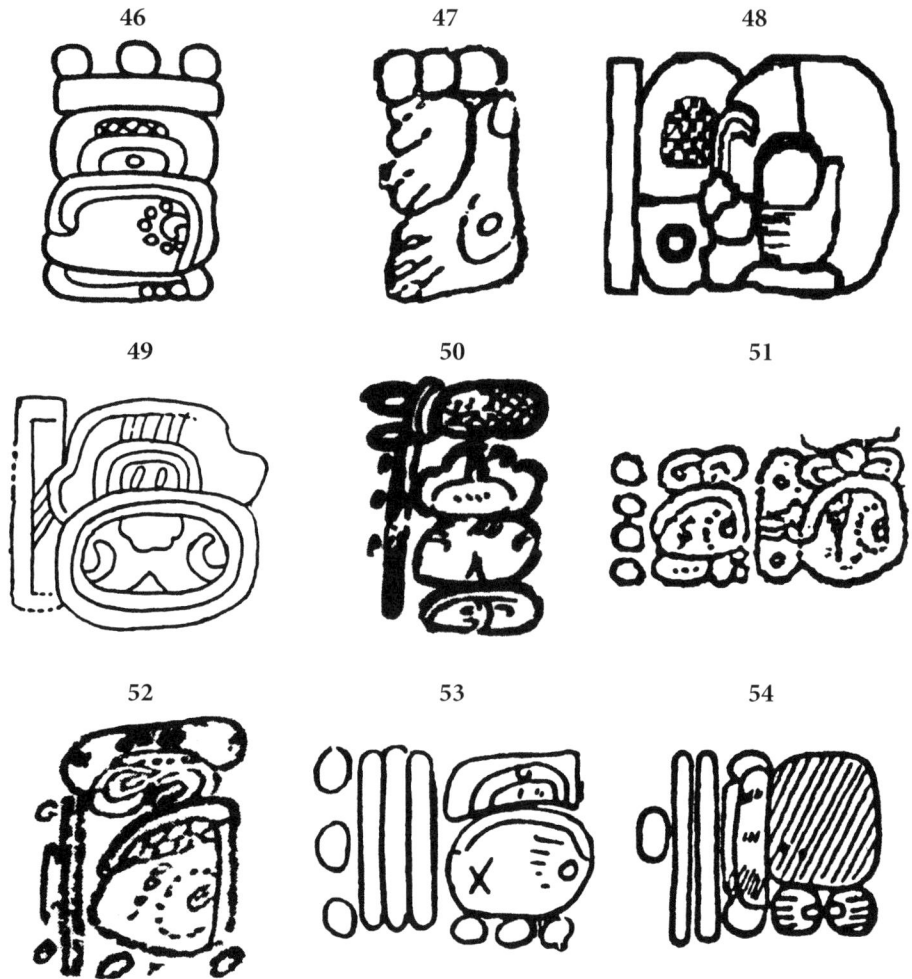

Figure 46. Chen, Monument 9 (Stela I): D2a, Quiriguá. Drawing by Matthew G. Looper in Looper (2003: fig. 3.7). © 2003. By permission of the University of Texas Press.

Figure 47. Chen, Temple of the Inscriptions, West: S3b, Palenque. Drawing by Linda Schele © David Schele. Courtesy of Los Angeles County Museum of Art.

Figure 48. Chen, Hieroglyphic Stairway 2, Central Section, Step 6: E1, Dos Pilas. Unattributed drawing in Fahsen (2002: fig. 6).

Figure 49. Yax, Stela 53: pC2, Copán. Drawn by Linda Schele (Schele and Freidel 1990: fig. 16b).

Figure 50. Yax, Capstone 18: A2, Ek Balam. Drawing by Alfonso Lacadena García-Gallo (2003: fig. 16).

Figure 51. Yax, Unprovenienced Vessel K1606: B1-C1, Princeton Art Museum. Modified by author after photograph by Justin Kerr © Justin Kerr. Courtesy of Justin Kerr.

Figure 52. Yax, Unprovenienced Vessel K0955: H4, Museum of Fine Arts, Boston. Modified by author after photograph by Justin Kerr © Justin Kerr. Courtesy of Justin Kerr.

Figure 53. Yax, Stela 1: B2, Comitán. Unattributed drawing in Blom and LaFarge (1927: fig. 352). Courtesy of the Middle American Research Institute.

Figure 54. Yax, Temple of the Four Lintels, Lintel 1: E1, Chichén Itzá. Drawing by Hermann Beyer. Courtesy of the Carnegie Institute for Science.

55 **56** **57**

58 **59** **60**

61 **62** **63**

Figure 55. Yax, Chichén Itzá "23, Block 18." Drawing after Hermann Beyer, in *Maya Hieroglyphic Writing*, 3rd edition, by J. Eric S. Thompson. © 1971 University of Oklahoma Press. Used with permission. All rights reserved.

Figure 56. Yax, as drawn in the extant copy of Landa's *Relación de las cosas de Yucatán*. Courtesy of George Stuart and the Biblioteca Real, Madrid.

Figure 57. Sac, Stela E: A10, Copán. Drawing by Linda Schele © David Schele. Courtesy of Los Angeles County Museum of Art.

Figure 58. Sac, Stela E, east: 22b, Quiriguá. Drawing by Lambert and Ann Hunter in Maudslay (1974: 2:plate 32).

Figure 59. Sac, Monument 69: C, Toniná. Drawing by Ian Graham, Corpus of Maya Hieroglyphic Inscriptions. © President and Fellows of Harvard College, Peabody Museum of Archaeology and Ethnology, PM# 2004.15.6.15.60 (digital file# 125870001).

Figure 60. Sac, Stela 25: D2, Piedras Negras. Drawing by John Montgomery. Courtesy of and © the Foundation for the Advancement of Mesoamerican Studies, Inc., www.famsi.org.

Figure 61. Sac, Drawing 82: D3, Naj Tunich. Drawing by Andrea J. Stone, in Stone (1995: fig. 7.29). © 1995. By permission of the University of Texas Press.

Figure 62. Ceh, as drawn in the extant copy of Landa's *Relación de las cosas de Yucatán*. Courtesy of George Stuart and the Biblioteca Real, Madrid.

Figure 63. Mac, Lintel 43: B1, Yaxchilán. Drawing by Ian Graham, Corpus of Maya Hieroglyphic Inscriptions. © President and Fellows of Harvard College, Peabody Museum of Archaeology and Ethnology, PM# 2004.15.6.6.14 (digital file# 99160089).

64 **65** **66**

67 **68** **69**

70 **71** **72**

Figure 64. Mac, Stela 8: A2, Aguateca. Drawing by Ian Graham. Courtesy of the Middle American Research Institute.

Figure 65. Mac, Lintel: D1, Hecelchakan Museum. Drawing by David Stuart. Courtesy of David Stuart.

Figure 66. Mac, Temple XIX, platform: J5b, Palenque. Drawing by David Stuart (2005: fig. 54).

Figure 67. Kankin, Altar 1: A3, Naranjo. Drawing by Ian Graham, Corpus of Maya Hieroglyphic Inscriptions. © President and Fellows of Harvard College, Peabody Museum of Archaeology and Ethnology, PM# 2004.15.6.3.22 (digital file# 99200028).

Figure 68. Kankin, Altar 1: H8, Naranjo. Drawing by Ian Graham, Corpus of Maya Hieroglyphic Inscriptions. © President and Fellows of Harvard College, Peabody Museum of Archaeology and Ethnology, PM# 2004.15.6.3.22 (digital file# 99200028).

Figure 69. Kankin, Temple of the Sun: P15, Palenque. Drawing by Linda Schele © David Schele. Courtesy of the Foundation for the Advancement of Mesoamerican Studies, Inc., www.famsi.org.

Figure 70. Kankin, Panel 2: A14, Xcalumkin. Drawing by Eric von Euw, Corpus of Maya Hieroglyphic Inscriptions. © President and Fellows of Harvard College, Peabody Museum of Archaeology and Ethnology, PM# 2004.15.6.10.36 (digital file# 99200034).

Figure 71. Kankin, Stela 52 (CPN 2722): B1, Copán. Drawing by B. W. Fash. Courtesy of Instituto Hondureño de Antropología e Historia (IHAH). Reproduced with permission.

Figure 72. Kankin, Temple XIV Tablet: D5, Palenque. Drawing by Linda Schele © David Schele. Courtesy of Los Angeles County Museum of Art.

Figure 73. Kankin, Temple XVIII, fragment, Palenque. Drawing by Linda Schele and Peter L. Mathews (1979: number 467).

Figure 74. Kankin, Codex Dresden, 24: C25. Drawing by Carlos A. Villacorta, in Villacorta and Villacorta (1977: 134).

Figure 75. Kankin, as drawn in the extant copy of Landa's *Relación de las cosas de Yucatán*. Courtesy of George Stuart and the Biblioteca Real, Madrid.

Figure 76. Pax, Stela 25: D1, Naranjo. Drawing by Ian Graham, Corpus of Maya Hieroglyphic Inscriptions. © President and Fellows of Harvard College, Peabody Museum of Archaeology and Ethnology, PM# 2004.15.6.3.2 (digital file# 99200035).

Figure 77. Pax, Altar 1: D1, Mountain Cow. Drawing by Ian Graham. Courtesy of the Corpus of Maya Hieroglyphic Inscriptions.

Figure 78. Pax, Stela 4: B1, Ixtutz. Drawing by Ian Graham, Corpus of Maya Hieroglyphic Inscriptions. © President and Fellows of Harvard College, Peabody Museum of Archaeology and Ethnology, PM# 2004.15.6.4.17 (digital file# 99200030).

Figure 79. Pax, Altar 12: A2, Caracol. Drawing by Nikolai Grube.

Figure 80. Pax, Codex Dresden, 47b: C1. Drawing by David Stuart. Courtesy of the Center for Maya Research.

Figure 81. Pax, Codex Dresden, 61c: B2. Drawing by David Stuart. Courtesy of the Center for Maya Research.

Figure 82. Pax, Drawing 65: G4, Naj Tunich. Drawing by Andrea J. Stone in Stone (1995: fig. 7–9a). © 1995. By permission of the University of Texas Press.

Figure 83. Pax, as drawn in the extant copy of Landa's *Relación de las cosas de Yucatán*. Courtesy of George Stuart and the Biblioteca Real, Madrid.

Figure 84. Pax, Drawing 52: B1, Naj Tunich. Drawing by David Stuart. Courtesy of National Geographic Creative.

Figure 85. Pax, Stela 2: E1, Aguateca. Preliminary field drawing by Ian Graham. Courtesy of the Middle American Research Institute.

Figure 86. Kayab, Temple of the Cross: P9, Palenque. Drawing by Linda Schele © David Schele. Courtesy of Los Angeles County Museum of Art.

Figure 87. Kayab, Temple of the Cross: P9, Palenque. Drawing by Linda Schele © David Schele. Courtesy of Los Angeles County Museum of Art.

Figure 88. Kayab, as drawn in the extant copy of Landa's *Relación de las cosas de Yucatán*. Courtesy of George Stuart and the Biblioteca Real, Madrid.

Figure 89. Kayab, Unprovenienced Vessel K1003: B1, Princeton Art Museum. Modified by author after photograph by Justin Kerr © Justin Kerr. Courtesy of Justin Kerr.

Figure 90. Kayab, Codex Dresden, 62: B10. Drawing by Carlos A. Villacorta, in Villacorta and Villacorta (1977: 134).

91 **92** **93**

94 **95** **96**

97 **98** **99**

Figure 91. Kayab, Structure 10L-26, Upper Shrine Temple Inscription: B3, Copán. Drawing by David Stuart (2012: fig. 2).

Figure 92. Cumku, Codex Dresden, 63: B18. Drawing by Huberta Robinson, in *Maya Hieroglyphic Writing*, 3rd edition, by J. Eric S. Thompson. © 1971 University of Oklahoma Press. Used with permission. All rights reserved.

Figure 93. Cumku, Stela 3, East, Block 7, bottom left, Copán. Drawing by B. W. Fash. Courtesy of Instituto Hondureño de Antropología e Historia (IHAH). Reproduced with permission.

Figure 94. Cumku, Stela 23: C8. Drawing by Sylvanus G. Morley (1920: fig. 26a). Courtesy of the Carnegie Institute for Science.

Figure 95. Cumku, Robey Panel: E. Drawing courtesy of Peter L. Mathews.

Figure 96. Cumku, Codex Dresden, 51a: A3. Drawing by Carlos A. Villacorta, in Villacorta and Villacorta (1977: 112).

Figure 97. Cumku, Panel A: D2, La Corona (formerly Site Q). Preliminary field drawing by Ian Graham. Courtesy of the Corpus of Maya Hieroglyphic Inscriptions.

Figure 98. Cumku, unknown provenience. Drawing courtesy of Linda Schele.

Figure 99. Cumku, Hieroglyphic Stairway: J1, Copán. Adapted from George Byron Gordon (1902: plate 12). Reproduced courtesy of the Peabody Museum of Archaeology and Ethnology, Harvard University.

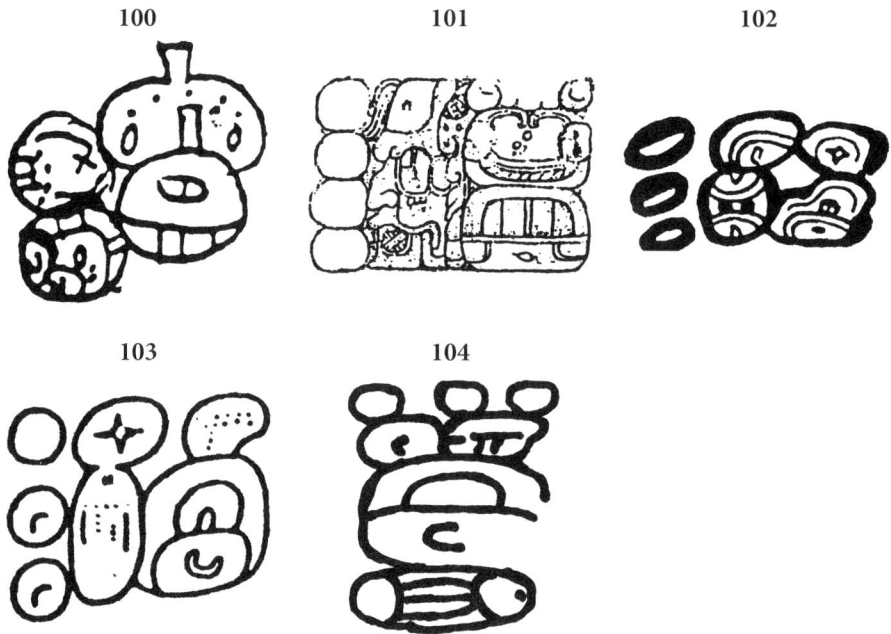

Figure 100. Cumku, as drawn in the extant copy of Landa's *Relación de las cosas de Yucatán*. Courtesy of George Stuart and the Biblioteca Real, Madrid.

Figure 101. Uayeb, Hieroglyphic Stairway, Step 41:1 Copán. Unattributed drawing in Stuart and Schele (1986: fig. 13).

Figure 102. Uayeb, Burial Plate, Tikal. Drawing by Simon Martin. Courtesy of Simon Martin.

Figure 103. Uayeb, Stela 14: A7, Caracol. Drawing by Virginia Greene, in Beetz and Satterthwaite (1981: fig. 14a). Courtesy of Penn Museum.

Figure 104. Uayeb, Stela 6: E7a, Caracol. Drawing by David Stuart (2005: fig. 41c). Courtesy of David Stuart.

Appendix D

Stemmas

Each stemma delineates the changes in form, meaning, and language. The arrow (>) means "corresponds to or gives rise to." Parentheses enclose optional words or steps. Asterisks denote hypothetical forms. If no gloss follows a name, the one for the immediately preceding term still applies. If no name follows a language abbreviation, the last preceding name from the same language is to be understood. If no name and no definition follow an abbreviation, supply the last instance of each from the same language. For example, in g3.3 and g3.4 below the sequence "g chäk red > c chac" means the glyphic name *chäk* "red" corresponds to *chac* in the Ch'olan tradition, which also denotes "red." The section "> c chic red > c deer > qq tzaack game animal;" indicates that the Ch'olan variant *chic* "red" corresponds to a Ch'olan homophone *chic* meaning "deer," source for the Q'eqchi' synonym *tzaack* "game animal"; the semicolon marks the end of one branch in a stemma. The final segment, "s > y céeh deer," encodes the proposition that Ch'olan *chic* "deer" generates the Yukatek synonym *céeh* "deer."

The stemmas assume a model of economy or "least moves" for the form, sense, geographical proximity, and number of shifts. Steps inside parentheses are optional only as transitions to the immediately following point. There are many examples of the Ch'olan names with glyphic predecessors: any name attributed to such a Ch'olan source may just as well spring from the glyphic counterpart. In g7.1–2, for instance, the Poqom *kasew* might derive from the Ch'olan or the glyphs; the Ch'olan name, however, descends from the glyphs. After dates "+" means "or later."

Abbreviations: an: Aztec Nahuatl; c: Ch'olan (Ch'ol, Ch'olti', Ch'orti', proto-Ch'olan); cj: Chuj; g: Glyphs; gn: Gulf Coast Nahuat; gy; Glyphic Yukatek; ix: Ixil; k: K'iche'; kq: Kaqchikel; m: Mam; mz: Mixe-Zoque; n: Nahua; p: Poqom; q: Q'anjob'al; qq: Q'eqchi'; te: Tzeltal; to: Tzotzil; y: Yukatek, proto-Yukatekan. Boldface marks the earliest forms. These appear alphabetically, each as the head of a stemma. The letter after the stemma numeral relates the terms as follows: c: cognate; h: homonym; s: synonym.

g1.1   h mz **hip** scrape, sharpen, polish > g **hi**

g1.2   s mz **hip** > y **ha'** scrape, sharpen, polish

g1.3   h y **ha'** scrape, sharpen, polish > y **ha'** water, rain > g **ha'/ha-ab'** water/rain time

g1.4   c y **éek'/yá'ax/säk/chäk ha'** black/green/white/red water > g **ik'/yäx/säk/chäk ha'** black/green/white/red water

g1.5   c g **ha'/*ha'-ab'** water, rain/rain time > ix *a ki* water days

g1.6   s g **ha'/ *ha'-ab'** water, rain/rain time > q *nabitc* rain time > c cj *nabich* rain time

g2.1   h y **ha'** water > y **ha'** soapberry > s y **sihom** soapberry

g2.2   s y *éek'/yá'ax/säk/chäk ha'* black/green/white/red soapberry > y COLORS *sihoom* COLORS soapberry (593/727 CE)

g2.3   > g *ik'/yäx/säk/chäk sihoom* COLORS soapberry

g2.4   > c q *k'eq/yax/zaq/k'aq sihom* COLORS soapberry

g2.5   s q *yax sihom* green soapberry > q *q'an sihom* yellow soapberry

g2.6   s y COLORS *sihoom* human > y COLORS *wíinik* man

g2.7   h > g COLORS *winik* month (300–500, 794 CE) > y COLORS [month] (744 CE)

g2.8   h g *sihoom* soapberry > ix *sijom zenzontle* mockingbird > s ix *pactzi* kind of bird

g2.9   c q *k'eq/zaq/k'aq sihom* black/white/red soapberry > m *q'aq/saq/kaq siijan* COLORS soapberry > h k 1/2/3 *si'jan* 1/2/3 white flower

g2.10   c > k 1/2/3 *si'j* 1/2/3 white flower > s p *utzumckigh* flower days and s to *nichikin* flower days

g3.1   h g ***ha':** *chäk ha'* red water > g **chäk ha-am'* red season

g3.2   h > kq, k *cak am* red spider (500–700 CE?)

g3.3   c c *chac* red > c *chic* red > h c *chic* deer > s y *céeh* deer

g3.4   > s qq *tzaack* game animal

g4.1   h y *éek'* black > y *éek'* spot > c y *éek'* waterhole > y s *ch'een* well

g4.2   h > qq *cheen* mosquito

g4.3   s y *ch'een* > s y, c **zihora* wellspring > s te *hok'in 'ahaw* waterhole lord

g4.4   c > to *hok'in 'ahval* waterhole lord > h to *ok'in ahwal* crying lord (1723 CE)

g4.5   c > te *ok'in 'ahaw* crying lord (1887, 1944 CE) > c to *ok'in* crying

g4.6   c to *ok'in ahwal* crying lord > to *me'okinahual* mother-rain-days lord

g4.7   s > to *me'el 'uch* old woman drizzle

g4.8   s to *ok'in ahwal* crying lord > q *cubihl* whistle (1723+ CE)

g4.9   h te *hok'en 'ahaw* waterhole lord > te *h-ok'-en 'ajaw* drummer lord

g4.10   s > y *kayab* horizontal drum > cs y *k'ay yab* song of many

g5.1   c g ***ha':** **säk ha'* white water > te *sakil ha* white water, atole > cj *savul* hominy

g5.2   s te *sakil ha* white water, atole > te *mush* hominy > c to *mush* hominy g5.3 s g ***ha'**: *säk ha' white water > te *sakil uch white drizzle and white/small opossum

g5.4 > s te *alaj/yaj/bikit/chin/uch* little opossum > s ix *ne' choo* baby-sized rat

g5.5 c te *bikit/yaj uch* little/baby-sized opossum > to *bikit/unen uch* little/baby-sized opossum

g5.6 s te *bikit/chin uch* little opossum > ix *soonchoj* smallest rat and ix tal choo small rat

g5.7 s to *bikit/unen uch* little/baby-sized opossum > to *h-'uch* male drizzle maker

g5.8 > c to *mol h-'uch* old man drizzle maker c > to *mol 'uch* old man drizzle

g5.9 h g *****ha'-ab**': *säk hab' white season > te *sakilab* clear seer/white stalk

g5.10 c > to *mucta sac/si sac* great white corn/cold white corn

g5.11 h g *****sac sihom** white soapberry > p *saq siij johm* offering gourd-bowl

g5.12 > s p *saq johm* white gourd s > p *sac-gojk* white gourd-squash

g5.13 c y *sac* white > c *sac* c > p *sac* > c k *sak* white c > k *sac imuj* splendid canopy

g6.1 c g *****ha'**: *****yäx ha'** green water > y *yá'ax ha'* green water/chocolate

g6.2 c > p *rax kikow* green/elote chocolate > p *aj kikow* elote/royal chocolate

g6.3 h > p *aj kikow* cacao merchant/beggar > c *ah qui cou* cacao merchant/beggar

g6.4 s > qq *rakol* merchant

g6.5 s g *****ha'**: *****yäx ha'** green rain > te *yax uch* green/large drizzle

g6.6 h > te *yax uch* green/large opossum > s te *muc(-ul) 'uch* large opossum

g6.7 c > to *mucta uch* large opossum

g6.8 s te *muc(-ul) 'uch* large opossum > ix *nimcho* big rat and *masnimcho* ? big rat

g6.9 s and ix *mochnimcho* large-headed big rat and ix *tchoochcho* useless rat

g6.10 s ix LARGE *cho* LARGE rat/baby > k 1 *mam* = older *mam* = grandfather

g6.11 s ix SMALL *cho* SMALL rat/baby > k 2 *mam* = younger *mam* = grandchild

g6.12 c k 1/2 *mam* grandfather/grandchild > kq 1/2 *mam* grandfather/grandchild

g6.13 c k 1/2 *mam* grandfather/grandchild > k 1/2 *mamkuk* 1/2 male quetzal

g6.14 c k 1 *mam* grandfather > ix *mam a ki* grandfather prayermaker

g6.15 c y *yaax* green > p *yax* green > h qq *yax* crab pincers

g6.16 h c *yax* green > q *yox* crab > s q *tap* crab > cj *tap* crab (900+ CE) and qq *tap* crab

g7.1 c g **kaseew** *pacaya* palm > c *cazeu* *pacaya* palm

g7.2 c *cazeu* *pacaya* palm > p *kasew* *pacaya* palm

g8.1 c g **k'än ah si'** yellow woodchopper > c *ccanazi* and p *canazi* yellow woodchopper g8.2   c g **k'än ah si'** > ix *zil'ki* firewood grove days and ix *tzil'ki* split-the-fire-wood days

g8.3   s q *sivil* woodchopper > c cj *sivil* woodchopper

g9.1   c g **k'än häl-aab'** yellow loom > c **k'an halab'* yellow loom

g9.2   c > p *kanjalam* yellow loom (1000+ CE) and s ix *tzu'ki* loom bar days

g9.3   s g **k'än häl-aab'** yellow loom > y *poop* mat (900 CE) and s c **pop* mat

g9.4   c > qq *pop* mat > c k *tz'iba pop* painted mat > s ix *mu* throne

g10.1   c g **chäk k'ät** red wicker (761 CE) > c *chacc^c at* red wicker

g10.2   h g **chäk k'ät** red wicker > g **chik kät* deer clay vessel > s y **keeh k'ät* deer idol

g10.3   h > y *siip* deer god "miss-shot" > y *zip'* ripe

g10.4   s g **chäk k'ät** red/great wicker > g *yäx k'ät* great/first wicker (761+ CE)

g10.5   s h > p *pet cat* first burn

g11.1   c g **ik' k'ät** black wicker (727 CE) > c *icat* black wicker and h qq *gkatoc* set-on-fire

g11.2   c > kq *k'atic* burn brush > s k *tziij* set on fire > ix *xukul* (Σih) piece of fire-wood (days)

g12.1   c g **mähk** enclosure > q *mak* enclosure and > te *mac* wall, enclosure > to *moc* wall (600+ CE)

g12.2   c g **mähk** shell; enclosure > y *maac* turtle shell; pit; granary

g12.3   c g **mähk** enclosure > c *chantemac* granary on tall posts and ix *chanteemak*

g12.4   c c *chantemac* > qq *chantemac* and p *chantemak*

g12.5   c p *chantemak* granary on tall posts > k *chee* granary > c kq *pariche* in the granary

g12.6   s k *chee* log > ix *lajab* deadfall trap and s ix *kucham* tree trunk and ix *kahabtsé* log beehive

g12.7   c ix *kahabtsé* log beehive > ix *kahab ('ki)* honey(comb) (days)

g12.8   s > to *pom* honey; incense > c te *pom* incense

g12.9   h kq *pariche* in the granary > kq *payriche* joker log > kq *pay* joker

g13.1   c g **mol** harvest; pile > c *mol* harvest; pile and y *mool* harvest; pile

g13.2   c and qq *moloc* collect and p *mol* group and q *mol* group

g13.3   c q *mol* group > cj *mol* group

g13.4   c g **mol** harvest; pile > ix *mol* group > c ix *molch'u* flock of grackles

g13.5   c and ix *mol tche* herd of deer > s *mol masat* herd of deer (1300+ CE)

g14.1   c g **muwaan** hawk > c *muhan* hawk and p *muan* hawk and qq *muan* hawk

g14.2   c and y *muan* hawk, owl and s q *mo* hawk > cj *mo* hawk/owl and kq *moh* hawk, crow

g14.3   s g **muwaan** > ix *moocho* brown hawk > s ix *pekxiitx* "cacao wing" hawk

g15.1   c g **ool** bud, sprout > c *olh* sprout and p *ojl* sprout > s k *tz'izil lacam* sprout flags

g15.2   s g **ool** > ix *muen (chin)* seedling (bag)

g15.3   s g **ool** sprout > ix *hui'ki* tip days > s te *batsul* tip of amaranth > to *batsul*

g15.4   c g **ool** sprout; pierce > c *olh* sprout; pierce and p *ojl* sprout; pierce

g15.5   s > kq 1/2 *to∑ic* 1/2 bloodletting > s k *tz'izil lacam* sewing/bloodletting banner

g15.6   s > k *chi il lacan* in-the-zenith/flaying banner

g15.7   c kq 1/2 *to∑ic* 1/2 bloodletting > kq 1/2 *to∑* 1/2 bleed with lancet and h 1/2 loincloth

g15.8   s > q *wex* loincloth> cj *bex* loincloth

g15.9   h kq 1/2 *to∑ic* 1/2 bloodletting > p **tokgüik* piercer; bite; dog > s p *tsi* dog

g15.10   s g **ool** sprout; pierce > q *(k'aq) xuhem* (red) threader, bloodletter > cj *xujim* bloodletter

g15.11   s g **ool**/c *olh* sprout; pierce > te *hul-ol* piercing > c to *jul-ol* piercing

g15.12   h g **ool** bud, sprout > te *hul-ol* arrival (at middle) > to *jul-ol* arrival (at middle)

g15.13   c g **ool**/c *olh* center, heart > te *olalti* heart mouth > to *olalti* heart mouth

g15.14   h te *olalti* heart mouth > te *alalti* small mouth

g15.15   s g **ool**/c *olh* ball > c **ku'm k'u'* nest ball, egg > y *cumku* thunder god

g15.16   s > p *kchip* last child (thunder-lightning god)

g15.17   s c **ku'm k'u'* nest ball, egg > ix *soj ki* nest days

g16.1   c g **PAS-HAB'** sprout season > g y ***PA'X-HAB'** break open season

g16.2   h > g y **PAAX-AB'** tap instrument > s yg **PAAX** upright drum

g16.3   > g **PAAX-xa** (600–650 CE)/ **-xi** (615 CE) drum

g17.1   c g **suutz'** bat > y *sòotz'* bat (1750 CE or earlier) > y *tzoz* bat (1566)

g18.1   c g **tan** ashes (615/662 CE) > q *baktan* bone ashes > c cj *tam* ashes

g18.2   c p *chab* arrow > k *ch'ab* arrow > k *ajau chap* archer

g18.3   c g **tan: ik' tan** black ashes (633 CE) > y **éek' ta'an* black ashes

g18.4   s > y * *éek' säbäk* soot; black ink > y **säbäk-t-ah* use black ink, write

g18.5   s > y * *éek' tz'ib'* write; writing > s y **woj* letter, glyph > h y *wo'* toad

g18.6   h > y *uol* ball

g19.1   c g **tz'ikin** birds > q **tsikin* birds > cj **tziquin* birds

g19.2   c g **tz'ikin** bird (603 CE) > c **tziquin* bird > p *tzikin kij* bird days

g19.3   c > kq *4,iquin ∑ih* hummingbird days > s ix *tzunun ki* hummingbird days (1000+ CE)

g19.4   s > q *wa tsikin* squeak bird/hummingbird > cj *huatziquin* squeak bird/hummingbird

g19.5   c kq *4,iquin ∑ih* bird days > k *tz'iquin q'ih* bird days

g19.6   c k *tz'iquin q'ih* bird days/g **tz'ikin** bird > ix *tz'ikin ki* bird days > ix *soj ki* nest days

g19.7   s > ix *yowal* head-ring

g19.8   s g **tz'ikin**/c **tziquin* bird > qq **xul* bird > c q *yacul* blue jay > cj *yaxul* blue jay

g19.9   s qq **xul* bird > qq *chichin* bird and h y *xuul* digging stick g19.10   s y *xuul* digging stick > p *tik-che-ik* plant-with-stick and s te *tsun* plant with digging stick

g19.11   c te *tsun* > to *tsun* plant with digging stick and s ix *awax ki* corn-sowing days

g19.12   c > qq *auuck* corn-sowing

g20.1   c g **uniiw** avocado (378?; 681 CE) > q *oneu* avocado > cj *oneu* avocado

g20.2   c g **uniiw** avocado > c *uniu* avocado > p *uniw* avocado

g20.3   c g **uniiw** avocado > s **un-ts'uts'ub'* avocado of coati > **un-ts'uts'uw* > g ts'uts'uw coati; > s **tsub'* agouti > *tsuw* agouti (?)

g20.4   s g **un** avocado > p **ooj* avocado> p **oo* chilacayote > s p *sac-gock* white chilacayote

g20.5   s g **un** > y **un k'ìin* avocado days > h **uh k'ìin* bead days > s *k'an k'ìin* bead days

g20.6   h > y *kankin* yellow (corn) days (690?; 743 CE) > p *kanajal* yellow corn ear

g20.7   c and qq *gkan* yellow > c q *kanal* yellow corn ear > cj *kanal* yellow corn ear

g21.1   h g **way-hab'** sleep/vision/nagual of the year > g **way-ab'** bed > y *uayab* bed

g21.2   h > y *uuayayab* vision of the year > y u *uayab haab* poison of the year

g21.3   s > c *mahi* poison > h qq *mayejick* offerings time

g22.1   c g **yäx k'in** green days > ix *yax'ki* green days

g22.2   c and cognate te *yashkin* green days > c to *yaxkin* green days

g22.3   c and y *yaxkin* green days > y *tzeyaxkin* small green days

g22.4   c g **yäx k'in** green/thatch-grass days > c *yazquin* green/ thatch-grass days

g22.5   s and q *yaxakil* green grass times > c cj *yaxaquil* green grass times

g22.6   s g **yäx k'in** green/thatch-grass days > qq *raxgkim* green thatch-grass

ix1.1   s ix **koj ki** clear-the-land days > qq *gkalec* clear and weed the field

k1.1   c k **botam** rolled up > k *ibota* large roll of the year > c k *obota* five rolls of the year

kq1.1   c kq *botam* doubled over > s ix *mek'aj* bent over stalks > s ix *paksi* bend firewood

k2.1   c k **1/2 lik'in kaaj** stretched out sky > kq *liΣinΣá* stretched out sky

k3.1   c k **tz'api q'ih** shut door/disaster days > kq *tz'api q'ih* shut door/disaster days

k4.1   c k **huno bix gih** five song days > k *jun bix kij* one song days/trill song days

kq2.1   s kq **tumuzuz** blue bean beetle; maize bug > ix *leem* bean beetle; maize worm

kq2.2   s > q *bak* corncob bug/worm > c cj *bak* corncob bug/worm

kq2.3   s kq **tumuzuz** blue bean beetle > p *mox kij* dung beetle

kq2.4   s kq **tumuzuz** winged ants > ix *akmór* wasp swarm

n1.1   c gn **pach** moss > k *pach* moss (1250+ CE)> kq *pach* moss (1250+ CE)

n1.2   s p *makux* lord moss (1350+ CE)

n2.1   c gn **tequexepual** flaying of men/slaves > k *tequexepual* flaying of men/slaves (1250+ CE)

n2.2   c k *tequexepual* flaying of men/slaves > k *cox xe puatl* yellow-flay-skin

n2.3   c kq *tacaxepegual* flaying of men/slaves (1250+ CE)

n3.1   c an **izcal (Σih)** sprout (days) > kq *izcal (Σih)* sprout (days) (1510)

q1.1   c q *****uchum** opossum > kq *uchum* opossum

qq1.1   c qq *****ianguca** pour divine honey/jugs of water > c *ianguca* pour divine honey/jugs of water

te1.1   c te *ch'ay k'in* lost days/days of destruction > to *ch'ay k'in* lost days

to1.1   c to **muy** ground wild tobacco > cj *sacmay* white tobacco > q *saqmay* white danger

Appendix E

Index to Counterparts to Stemmas

Boldface marks the earliest, often simplest forms. *Italic* indicates names derived from a source other than the glyphs. An asterisk, *, designates reconstructed or hypothetical forms.

Glyphs *ha' g1; *ha'-ab' g1; **ha':** *chäk ha' g2, g3; **ha':** *chäk ha-am' g3; **ha':** *säk ha' g2, g5; **ha':** *yäx ha' g2, g6; k'ät: *chik k'ät g10; k'ät: yäx k'ät g10; *ik' ha' g2; **ik' tan** g18; **k'än ah si'** g8; **k'än häl-aab'** g9; **k'ät: chäk k'ät** g10; **k'ät: ik' k'ät** g11; **kaseew** g7; **mähk** g12; **mol** g13; **muwaan** g14; **ool** g15; **PAX/pa-?-wab'/pa'-tum** g16; *säk hab' g5; *sihoom: ik'/yäx/säk/chäk g2; **suutz'** g17; **tan/tam** g18; **tz'ikin** g19; **uniiw** g20; **way-hab'** g21; wayab' g21; **yäx k'in** g22

Ch'olan ah qui cou g6; cazeu g7; ccanazi g8; chac g3; chacc^cat g10; chantemac g12; *chäk g3; chic g3; chichin g19; hitz quin g21; hitz yum g21; icat g11; *ku'm k'u' g15; *k'an halab' g9; mahi g21; mol g13; muhan g14; olh g15; *pok g18; *pop g9; sac g5; sihom/zihora g4; *tzic-es g3; *tziquin g19; uniu g20; *un k'in g20; *yajir k'in g21; yax g6; yazquin g22; *ianguca* qq1

Chuj bak kq2; bex g15; huatziquin g19; kanal g20; mo g14; mol g13; nabich g1; oneu g20; savul g5; sivil g8; tam g18; tap g6; *tziquin g19; xujim g15; yaxaquil g22; yaxul g19; *sacmay* to1

Ixil a ki g1; akmór kq2; awax ki g19; chanteemak g12; hui'ki g15; kahab ('ki) g12; kahabtsé g12; kucham g12; lajab g12; LARGE cho g6; mam a ki g6; masnimcho g6; mochnimcho g6; mol g13; molch'u g13; mol masat g13; mol tche g13; moocho g14; mu g9; muen (chin) g15; ne' choo g5; nimcho g6; pactzi g4; pekxiitx g14; *sijom g4; SMALL cho g6; soj ki g15, g19, soonchoj g5; tal choo g5; tchoochcho g6; tzil'ki g8; tzu'ki g9; tzunun ki g19; tz'ikin ki g19; xukul (Σih) g11; yax'ki g22; yowal g19; zil'ki g8; **koj ki** ix1; *leem* kq2; *mekíaj* kq1; *paksi* kq1

K'iche' ajau chap g18; cak am g3; ch'ab g18; chee g12; chi il lacan g15; 1/2 mam g6; 1/2 mamkuk g6; sac imuj g6; sak g5; 1/2/3 si'j g2; *1/2/3 si'jan g2; 1/2/3 tziij g11;

tz'iba pop g9; tz'iquin q'ih g19; tz'izil lacam g15; **botam** k1; **1/2 lik'in kaaj** k2; **huno bix gih** k4; **tz'api q'ih** k3; *ibota* k1; *obota* k1; *jun bix kij* k4; *cox xe puatl* n2; *1/2 pach* n1; *tequexepual* n2

Kaqchikel cak am g3; k'atic g11; 1/2 mam g6; moh g14; pariche g12; payriche g12; pay g12; 1/2 toΣ g15; 1/2 toΣic g15; 4,iquin Σih g19; **1/2 tumuzuz** kq2; *botam* kq1; *liΣinΣá* k2; *tz'api q'ih* k3; *uchum* q1; *izcal (Σih)* n3; *1/2 pach* n1; *tacaxepegual* n2

Mam *q'aq/saq/kaq siijan g2

Nahua an **izcal (Σih)** n3; gn **pach** n1; gn **tequexepual** n2

Poqom *aj kikow g6; canazi g8; chab g18; chantemak g12; kanajal g20; kanjalam g9; kasew g7; kaxik-laj-kij g21; kchip g15; mol g13; muan g14; ojl g15; *ooj g20; *oo g20; pet cat g10; *rax kikow g6; sac g5; sac-gock g5, 20; *saq johm g5; *saq siij johm g5; tam g18; tik-che-ik g19; *tokgüik g15; tsi g15; tzikin kij g19; uniw g20; *utzumckigh g2; yax g6; *makux* n1; *mox kij* kq2

Q'anjob'al baktan g18; cubihl g4; kanal g20; mak g12; mo g14; mol g13; nabitc g1; oneu g20; k'eq/yax/q'an/zaq/k'aq sihom g2; sivil g8; tap g6; *tsikin g19; wa tsikin g19; wex g15; (k'aq) xuhem g15; yacul g19; yaxakil g22; *bak* kq2; *saqmay* to1

Q'eqchi' auuck g19; chantemac g12; cheen g4; *chichin g19; gkan g20; gkatoc g11; mayejick g21; moloc g13; muan g14; pop g9; rail cutan g21; rakol g6; raxgkim g22; tap g6; tzaack g3; *xul g19; yax g6; ***ianguca** qq1; *gkalec* ix1

Tzeltal alalti g15; batsul g15; h-ok'-en 'ajaw g4; hok'en 'ahaw g4; hok'in 'ahaw g4; hul-ol g15; mac g12; mush g5; ok'in 'ahaw g4; olalti g15; pom g12; sakilab g5; sakil ha g5; tsun g19; uch: alaj/yaj uch g5; uch: bikit/chin/alaj/yaj uch g5; uch: muc(-ul) 'uch g6; uch: *sakil uch g5; uch: *yax uch g6; yashkin g22; **ch'ay k'in** te1

Tzotzil batsul g15; hok'in 'ahval g4; jul-ol g15; *mac g12; me'el 'uch g4; me'okinahual g4; moc g12; mucta sac g5; mush g5; nichikin g2; ok'in g4; ok'in ahval g4; olalti g15; pom g12; si sac g5; tsun g19; uch: bikit/unen uch g5; h-'uch g5; uch: mol 'uch g5; uch: mol h-'uch g5; uch: muctauch g6; yaxkin g22; **muy** to1; *ch'ay k'in* te1

Yukatek ceeh g3; *chäk g3; *chäk sihom g3; cheen g4; cumku g15; *éek' g4; *éek' tá'an g18; *éek' sihom g4; *ha': *éek'/yá'ax/säk/chäk ha' g2; kankin g20; kayab g4; *keeh k'ät g10; k'ay yab g4; *k'än k'iin g20; maac g12; mool g13; muan g14; paax g16; *pok g18; poop g9; sac g5; *säb'äk g18; *säk sihom g5; *sihoom: *éek'/yá'ax/säk/chäk sihoom g2;*sihom/*zihora g4; siip g10; sòotz' g17; tzeyaxkin g22; tzoz g17; uayab g21; *uh k'iin g20; uuayayab g21; u uayab haab g21; uyail kin g21; uo g18; uol g18; wayeb g21; xuul g19; yá'ax g6; *yá'ax ha' g6; *yá'ax sihom g6; yaxkin g22; zip' g10

Appendix F
Themes with Traditions

For definitions see chapter 3. A name counts once for every tradition it occurs in. Optional or required, each member of a series, combined with the root, counts as a distinct name. Abbreviations: cj: Chuj; cn: Ch'olan; g: Glyphic; ix: Ixil; k: K'iché; kq: Kaqchikel; p: Poqom (= Poqomchi', Poqomam); q: Q'anjob'al; qq: Q'eqchi'; te: Tzeltal; to: Tzotzil; y: Yukatek. Names appear in all CAPITAL letters in their first or most typical form.

Agriculture: Activities AKMÓR ix; AUUCK qq; AWAX KI ix; BAKTAN q; BOTAM kq; BOTAM k; CAAM KIJ k; CANAZI p; CCANAZI cn; CHANTEMAC cn qq; CHANTEMAK p, ix; CH'AY K'IN to; ÇIBIX kq; GKALEC qq; GKATOC qq; HOYOH to; IANGUCA cn; JULAL te; KAHABTSEE ix; KANAL q; KOJ KI ix; KÄN AJ SI' g; K'ANK'IN y; K'ATIC kq; LAJAB'KI ix; MAAC y; MAK g, q; MEK'AJ ix; MOL cn, cj, ix, p, q; MOLOC qq; MUEN ix; MUEN CHIN ix; MUK to; PAKSI k; PET CAT p; RAKOL qq; SAVUL cj; SIBIXIC kq; SIVIL, SIWIL cj, q; TAM cj; TIK CHEIK p; TSUN te; TSUN to; TZIL KI ix, TZ'API Q'IH k; 4,API ΣIH kq, UOL y; XUKUL (KI) ix; ZIH: NABEY/U KAB/R OX ZIH k; ZIL'KI ix

TOTAL: 60

Agriculture: Plants BAK' q; CAΣAN kq; MUCTACAC to; MUEN ix; MUY to; ONEU cj, q; SAC-GOJK p; SACMAY cj; SAQMAY q; SISAC to; TOΣIC: NABEY/RU CAB kq; UN g; UNIU cn, p; UNIW g

TOTAL: 17

Agriculture: Seasons and Weather A KI ix; BATAM k; BATSUL te; BATS'UL to; BIKIT UCH to; CAJAB KI ix; CHAM p; CUMKU y; GKAN qq; HA': CHÄK/ IK'/ SÄK/YÄX g; HAB': CHÄK/IK'/SÄK/YÄX g; HNICHK'IN to; H'OK'EN 'AHWAL to; HUI KI ix; IANGUCA cn; IZCAL (ΣIH) kq; JULAL te; JUN BIX KIJ k; KANAJAL p; K'ANK'IN y; ME'EL 'UCH to; ME'OKINAHUAL to; MOL cn, g; MOL H'UCH to; MU ix; MUC UCH to; NABITC q; NOL (KI) ix; OCH KI ix; OHL cn; OJL p; OK'IN to; PACH: NABEY/ RU CAB kq; PACH: NABEY/ U KAB k; PACH: KAM'/QAM

PACH kq; PETZETZ KI ix; RAXGKIM qq; SIIP y; ZIP' y; TUMUZUZ: NABEY/ RUCAB TUMUZUZ kq; TSUN te; TZU KI ix; TZ'E YAXKIN y; TZ'IQUIN Q'IH k, p, ix; TZ'IZIL LACAM k; 4,IQUIN ΣIH kq; UTZUMCKIGH p; YASHKÍN te; YAX-AKIL q; YAXAQUIL cj; YAXKIN y; YASHKÍN te; YAX KI ix; YÄX KIN g; YAXK'IN to; YAZQUIN cn; ZIH: NABEY/ U KAB/ R OX ZIH k

TOTAL: 72

Animals: Amphibians MAAC y; WOO y

TOTAL: 2

Animals: Birds CHICHIN cn; *CUMKU cn; HUATZIQUIN cj; K'ÄN ASIJ g; MO cj, q; MOCH'Ú, MOLCHU, MOLCH'Ú, ix; MOH kq; MOOCHOO ix; MUAN p, y; MUHAN cn; MUWAN g; MAMKUK: NABÉ/U CAB k; OLALTI to; PACTZI ix; PEKXÍTX ix; TSIKIN cn; TZ'IKIN g; TZ'IKIN KI ix; *TZ'IQUIN KIN cn; TZ'IQUIN Q'IH k; TZUNUN KI ix; 4,IQUIN ΣIH kq; WA TSIKIN q; YACUL q; YAXUL cj; *XUL qq

TOTAL: 31

Animals: Crustaceans TAP cj, q, qq; YASHKÍN te; YAX qq; YAX KI ix

TOTAL: 6

Animals: Fish SAC y; XUUL y

TOTAL: 2

Animals: Insects AGELCHAC te; CAJAB KI ix; CAKAM, CAKAN kq; CCANAZI cn; CHEEN qq; HNICHK'IN to; KAHÁB ix; KAQAM k; LEEM ix; MO cj, q; MOX-KIJ p; SAKIL HA' te; TUMUZUZ: NABEY/ RUCAB kq; TZ'E YAXKIN y

TOTAL: 17

Animals: Mammals CEEH y; CHAC cn; CHO: MASNIM-, MOCHNIM-, NE'-, NIM-, TCHOOCH- ix; 'ELECH to; H'ELECH to; H'UCH to; KOJ KI ix; K'OOLK'OOY ix; MOL H'UCH to; MOL MASAT ix; MOL TCHE ix; MUC(-UL) UCH te; MUKTAUCH to; SEK y; SOONCHOJ ix; *SOO y; SUTZ' g; TALCHOO ix; TSI p; TZOZ y; TZ'TUZ'UW g; UCH: ALA/UNEN to; UCH: ALAJ-/BIKIT-/ CHIN-/YAHAL- te; 'UCH: (MOL) (H) to; UCHUM kq

TOTAL: 34

Animals: Worms BAC cj BAK q; LEEM ix; NOL (KI) ix; SOJNÓY ix; 1/2 TUMUZUZ kq

TOTAL: 7

Architecture CHEE k; CUMKU y; MAC te; MAHÁB ix; MOC to; PARICHE' kq; TZEEC y

TOTAL: 7

Artifacts BEX cj; BOTA kq; BOTAM k, kq; CAAM KIJ k; *CANHALAB cn; CAN-HALIB cn; CHEE k; CHI IL LACAN k; HOK'EN AHAW te; HOQUÉN-HAJAB te; (K'AQ) CUHEM q; ICAT cn; KANJALAM p; KAQAM k; KAYAB y; KÄN JÄL-AB' g; K'ANK'IN y; MAAC y; MAC te; MAK g, q; MU ix; MAMKUK: NABÉ/ U CAB k; OBOTA k, kq; OLH cn; PAAX g, y; PAAX-IL g; PA'AX-AB' g; PA'-TUN g; POOP y; POP qq; SAC IMUJ k; SAKIL HA' te; TSUN to; TZ'IBA POP k; UOL y; WAY-AB' g; WEX q; XUJIM cj, XUUL y

TOTAL: 45

Astronomy CAAM KIJ k; CAΣAN kq; CHI IL LACAN k; HUNO BIX GIH k; IQUIM KIJ k; JULOL to; KANAL cj, q; KAQAM k; LIK'IN KA: NABEY/U KAB k; TZ'E YAXKIN y; YAZQUIN cn

TOTAL: 13

Body Parts ALALTI te; BAC cj, BAK q; COX XE PUATL k; OLALTI te; to; TZEEC y

TOTAL: 7

Colors CAKAN kq, k; CHAC cn, qq; GKAN (K'AN) qq; SAC cn, k, p, y; YAX cn, p, y

TOTAL: 12

Drink and Food AH QUI COU cn; CAJAB KI ix; KAHÁB ix; MUSH te; to; OL g; OL WAJ g; POM to; SAVUL cj; TSANAKVÁY ix

TOTAL: 10

Geography CH'EEN y; *ÉEK' y; HOKIN AHAW te; ZIHORA cn

TOTAL: 4

Gods CUMKU y; HOKIN AHAW te; HOQUÉN-HAJAB te; H'OK'EN 'AHWAL to; (MOL) H'UCH to; KCHIP p; MAKUX p; MAM A KI ix; MAMKUK: NABÉ/U CAB k; 'OK'EN KAHVAL to; 'OK'IN 'AHWAL to; PAYRICHÉ kq; UAYAB: y

TOTAL: 15

Hunting AJAU CHAP k; CHAB p; CUBIHL q; H'ELECH to; HOQUÉN-HAJAB te; MAC y; MAK q; MOL MASAT ix; MOL TCHE ix, (MOL) H'UCH to; SIIP: y; TZAACK (TZAAK) qq

TOTAL: 13

Illnesses CAKAM, CAKAN kq, k; CHACC^{C}AT cn; ICAT cn; KAXIK-LAJ-KIJ p; OJA KWAL cj; ONCHIL, ONCHÍV ix; XUUL y; ZIP' y; ZOJ KI ix

TOTAL: 13

Kinship MAM: NABEY/RU CAB kq; MAM: NABEY/U KAB k; MAM: HUN/ QAM kq

TOTAL: 6

Plants: General IBOTA k; MAKUX p; MEAK cj; MOOL y; PACH: NABEY/U KAB k; PACH: NABEY/RU CAB kq; PAY kq; SAC y; SIHOM: CHÄK/IK'/SÄK/YÄX g; SIHOM: K'EQ/YAX/K'AQ/SAX/K'AN q; TZ'E YAXKIN y; WOO y; YAAX y

TOTAL: 22

Plants: Trees CAK CHE k; CAZEU cn; CHEE k; KASEW g, p; KUCHAM ix; MAKUX p; MUY to; XUUL y

TOTAL: 9

Ritual and Religion AJAU CHAP k; A KI ix; ALALTI te; BATAM k, BOTAM k, kq, BOTAN k; CAK CHE k; CAKAM k; CAKAN k; CAKAN, CAΣAN kq; ÇAPYUM cn; CHACAT cn; CHACCᶜAT cn; CHAIKÍN te; CHAM p; CHÄK K'ÄT g; CHI IL LACAN k; CH'AB k; COX XE PUATL k; (K'AQ) CUHEM q; CUMKU y; HULOL te; HUNO BIX GIH k; IBOTA k, kq; IK' K'ÄT g; JOP BIX k; JULAL te; JUL EL te; JULEL te; JULOL to; JUN BIX KIJ k; KANAL q; KAQAM k; KCHAM p; K'ATIC kq; K'AY YAB y; LIK'IN KA: NABEY/U CAB k; LIΣINΣÁ kq; MAHI cn; MAMKUK: NABÉ/U CAB k; METCH'KI ix; OL g; OL WAJ g; OLALTI te, to; OJL p; OLH cn; POM te, to; SAC IMUJ k; SAKILAB te; TACAXEPEGUAL kq, TEQUEXEPUAL k; TOΣIC: NABEY/RU CAB kq; TZIJEP ix; TZ'API Q'IH k; TZ'IZIL LACAM k; 4,API ΣIH kq; UAYAB y; XUJIM cj; YASHKÍN te; ZIH: NABEY/U KAB/R OX k

TOTAL: 72

Society AH QUI COU cn, p; AKMÓR ix; MOL ix

TOTAL: 4

Time BOTA kq; CH'AY K'IN to; IBOTA k; K'AY YAB y; OBOTA kq, k; OYEBIN cj; SORE 'AHTAL te; WAY-HAB' g; XET KI ix; XUUL y

TOTAL: 11

Trade AH QUI COU cn, p; IZCAL kq

TOTAL: 3

Appendix G

Zenith Passage Dates by Culture, Site, and Latitude

Matching site latitudes with apparent declinations in the *Astronomical Almanac for the Year 1997* generates the dates for the solar zenith passages (zn, zs) (Aveni 1980: 118, table 11). **Boldface** marks explicitly stated data and their sources.

TABLE G.1
Zenith passage dates by culture, site, and latitude

Culture	Site	Latitude N	zn	zs	days N/S	Source
au: nm	Takalik Abaj, El Baúl	14°20'	28-IV	15-VIII	109/256	E
au: nm	Izapa	14°42'	1-V	14-VIII	105/260	E
au	Copán	14°50'	30-IV	13-VIII	105/260	AH, M
au	Quiriguá	15°18'	1-V	12-VIII	103/262	E
au	Bonampak	16°44'	7-V	6-VIII	91/274	L
au	Yaxchilán	16°54'	8-V	5-VIII	89/276	M
au	Naranjo	17°08'	8-V	5-VIII	89/276	M
au	Tikal	17°13'	9-V	4-VIII	87/278	M
au	Palenque	17°31'	10-V	3-VIII	85/280	M
au: nm	Cerro de las Mesas; Tuxtla	18°30'	13-V	31-VII	79/266	E
Aztec	Tenochtitlán	19°24'	17-V	26-VII	70/295	R; N 3, 14, 18
au: nm	Teotihuacán	19°41'	17-V	26-VII	70/295	A80
Chòl	Lanquín	15°34'	3-V	10-VIII	99/266	ANG
Chòrti'	southern limit	14°25'	30-IV	13-VIII	105/260	G
Chòrti'	Esquipulas	14°34'	29-IV	13-VIII	106/259	ANG
Chòrti'	Quezaltepeque	14°38'	30-IV	13-VIII	105/260	ANG
Chòrti'	Santiago Jocotán	14°49'	30-IV	13-VIII	105/260	ANG
Chòrti'	northern limit	15°10'	1-V	12-VIII	103/262	G
Chuj	Solomá	15°39'	3-V	10-VIII	99/266	ANG
Chuj	Santa Eulalia	15°44'	3-V	10-VIII	99/266	ANG

(continues)

TABLE G.1 Zenith passage dates by culture, site, and latitude (*continued*)

Culture	Site	Latitude N	zn	zs	days N/S	Source
Chuj	San Mateo Ixtatán	15°50'	4-V	9-VIII	97/268	ANG
Ixil	Nebaj	15°24'	2-V	11-VIII	101/264	ANG
Ixil	San Juan Cotzal	15°26'	2-V	11-VIII	101/264	ANG
Ixil	Chajul	15°29'	3-V	10-VIII	99/266	ANG
K'iche'	Quezaltenango	14°50'	30-IV	13-VIII	105/260	ANG
K'iche'	Chichicaste-nango	14°57'	1-V	12-VIII	103/262	ANG
K'iche'	Santa Cruz del Quiché	15°02'	1-V	12-VIII	103/262	ANG
K'iche'	Utatlán	**15°04'**	**30-IV**	**13-VIII**	**104/261**	**E**
K'iche'	Momostenango	**15°05'**	**1-V**	**12-VIII**	103/262	T
K'iche'	Momostenango	**15°05'**	**2-V**	**11-VIII**	101/264	T
Kaq	Iximché	14°46'	30-IV	13-VIII	105/260	ANG
Pc	Cobán	15°28'	2-V	11-VIII	101/264	ANG
Pm	San Juan Amatitlán	14°29'	29-IV	14-VIII	107/258	ANG
Pm	San Luis Jilotepeque	14°39'	30-IV	13-VIII	105/260	ANG
Pm	Chinautla	14°42'	30-IV	13-VIII	105/260	ANG
Q'anjb'l	Santa Eulalia	15°44'	3-V	10-VIII	99/266	ANG
Q'eq	Lanquín	15°34'	3-V	10-VIII	99/266	ANG
Tzeltal	S Cristóbal de las Casas	16°45'	**6-V**	7-VIII	93/272	R; **N 17**
Tzeltal	S Ildefonso Tenejapa	16°49'	7-V	6-VIII	91/274	GNS

Tzeltal	Oxchuc	16°51'	7-V	6-VIII	91/274	GNS
Tzeltal	Cancuc	16°55'	8-V	5-VIII	89/276	GNS
Tzotzil	Comitán, Chiapas	15°01'	**1-V**	12-VIII	103/262	**N 13, 17**
Tzotzil	Santo Domingo Comitán = Comitán de Domínguez	16°15'	5-V	8-VIII	95/270	GNS, R
Tzotzil	S Lorenzo Zinacantán	**16°45'**	**5-V**	**7-VIII**	94/271	**V**
Tzotzil	S Juan Chamula	16°47'	7-V	6-VIII	91/274	GNS
Tzotzil	San Miguel Mitontic	16°51'	7-V	6-VIII	91/274	GNS
Tzotzil	San Andrés Larráinzar-Istacostoc	16°53'	7-V	6-VIII	91/274	GNS
Tzotzil	S Pedro Chenalhó	16°53'	7-V	6-VIII	91/274	GNS
Tzotzil	María Magdalena Aldamas-Tanhobel	16°56'	8-V	5-VIII	89/276	GNS
Tzotzil	Santa Catalina Pantelhó	16°56'	8-V	5-VIII	89/276	GNS
Tzotzil	Santa Marta Manuel Utrilla-Yolotepec	16°58'	8-V	5-VIII	89/276	GNS

(*continues*)

TABLE G.1 Zenith passage dates by culture, site, and latitude (*continued*)

Culture	Site	Latitude N	zn	zs	days N/S	Source
Tzotzil	S Pablo Chalchihuitán	17°00'	8-V	5-VIII	89/276	GNS
Tzotzil	N Sra de la Asunción de Guitiupa: = Huitiupán	17°13'	9-V	4-VIII	87/278	R
Yukatek	Río Bec	18°21'	13-V	31-VII	79/286	M
Yukatek	Becan	18°31'	14-V	30-VII	77/288	M
Yukatek	Puuc south extreme	**19°40'**	**19-V**	**26-VII**	69/296	**AH 58**
Yukatek	Edzna	**19°37'**	19-V	**26-VII**	69/296	**MS 64**
Yukatek	no site named	19°54'	**20-V**	24-VII	65/300	**CB**
Yukatek	Xcalumkin	20°09'	21-V	23-VII	64/301	P
Yukatek	Labna	20°10'	21-V	23-VII	64/301	M
Yukatek	Sayil	20°11'	**22-V**	**22-VII**	62/303	**AH 39**
Yukatek	Tekax (CB)	20°12'	21-V	23-VII	63/302	DB 46
Yukatek	Kabah	20°15'	22-V	22-VII	62/303	M
Yukatek	Oxkutzcab	20°18'	22-V	22-VII	62/303	DB 49
Yukatek	Mérida [1579: error]	★**20°20'**	**20-V**	22-VII	63/302	**MG**
Yukatek	Maní (Pérez) (CB)	20°21'	22-V	22-VII	61/304	NF, Y
Yukatek	Calkiní	20°22'	22-V	22-VII	62/303	DB 50
Yukatek	Uxmal	**20°22'**	**22-V**	**22-VII**	62/303	**AH T. 2**
Yukatek	Teabo (CB Nah)	20°24'	22-V	22-VII	61/304	NF, Y

Yukatek	Chumayel (CB)	20°26′	22-V	22-VII	61/304	NF; Y
Yukatek	Coba	20°30′	23-V	21-VII	59/306	AH
Yukatek	Chan Kom	20°34′	23-V	21-VII	60/305	DB 47
Yukatek	Oxkintok	20°34′	**24-V**	**21-VII**	59/306	**AH 47**
Yukatek	Kaua (CB)	20°37′	23-V	20-VII	58/307	NF; Y
Yukatek	Mayapán	20°37′	23-V	21-VII	60/305	AH
Yukatek	Chichén Itzá	20°40′	**26-V**	**20-VII**	55/310	**AA 25**
Yukatek	Valladolid	20°41′	**23-V**	21-VII	60/305	DB 46; **N 20**
Yukatek	Mérida	20°54′	**25-V**	18-VII	**55/310**	DB 50; **N 20**
Yukatek	no site named	21°03′	26-V	**18-VII**	53/312	**PP**
Yukatek	Dziibilchaltun	**21°06′**	**25-V**	**17-VII**	54/311	**M 63**
Yukatek	Motul	21°06′	26-V	18-VII	53/312	DB 49
Yukatek	Tizimín (CB)	21°08′	26-V	18-VII	53/312	AH; DB 46
Yukatek	Ixil (CB)	21°09′	26-V	17-VII	52/313	DB 49

Abbreviations: A: *Astronomical Almanac for the Year 1997*; AA: Aveni 1997; AH: Aveni and Hartung 1986: table 1; ANG: *Atlas Nacional de Guatemala 1972*; au: archaeological, uncertain (probably Cholan or Yukatekan); A80: Aveni 1980: 63, fig. 25; CACM: *Chiapas: Atlas Cultural de México* 1987; CB: Chilam Balam books; DB: Díaz Babio 1977; E: Edmonson 1988: 120; G: Girard 1966: 1, 21, 206; GNS: *Global Net Survey* at http://geonames.nga.mil/ gns/html/ (data kindly supplied by Molly Malloy); Kaq: Kaqchikel; L: Lounsbury 1982: 149; M: Morley 1938: 35: appendix 3; MG: Garza et al. 1983: 73; MS: Milbrath 1999; N: Nuttall 1928; n: north; NF: *Mexico Air Navigation Map: Merida: N-F16-South*; nm: non-Maya; P: Pollock 1980: 418; Pc: Poqomchi; Pm: Poqomam; PP: Pío Pérez 1843: 280; Qanjbl: Qanjobal; Qēq: Qēqchi'; R: *Rand McNally: The New International Atlas*; T: Tedlock 1992b: 27–28; V: Vogt 1997: 111; VM: Malmström 1997: 3, 61; Y: *Yucatán: Atlas cultural de México*; zn: zenith passage north; zs: zenith passage south. Latitudes are rounded to the minute. Dates are Gregorian, accurate to plus or minus one day.

References

Academia de las Lenguas Mayas de Guatemala
1988   *Lenguas mayas de Guatemala*. CD. Guatemala City: Instituto Indigenista Nacional, Ministerio de Cultura y Deportes.

Ajpacaja Tum, Pedro Florentino, Manuel Isidro Chox Tum, Francisco Lucas Tepaz Raxuleu, and Diego Adrian Guarchaj Ajtzalam
1996   *Diccionario del idioma k'iché*. Antigua, Guatemala: Proyecto Lingüístico Francisco Marroquín.

Alvarado López, Miguel
1975   *Léxico médico quiché-español*. Guatemala City: Instituto Indigenista Nacional.

Álvarez, Juan
n.d.   Vocabulario en la lengua 4,iche Otlatecas. MS. Berlin, Ibero-Amerikanisches Institut, Lehmann Bibliothek.

Andrade, Manuel J.
1946   Materials on the Mam, Jacaltec, Aguacatec, Chuj, Bachahom, Palencano, and Lacandon Languages. Microfilm Collection of Manuscripts on American Cultural Anthropology No. 10. Chicago: University of Chicago Library.

Andrews, J. Richard
1975   *Introduction to Classical Nahuatl*. Austin: University of Texas Press.

Ángel, Fray
18th century a   Vocabulario de la lengua cakchikquel. No. 497. American Philosophical Society, Philadelphia.
18th century b   Vocabulario de la lengua kaqchikel. Fonde Américaine 41 (R7466). Bibliothèque nationale, Paris.

Anonymous
1675   Vocabulario de la lengua kaqchikel. Fonde Américain No. 43 (R. 7493). Bibliothèque Nationale, Paris. Photographic copy in Latin American Library, Tulane University, New Orleans.

Anonymous
1685   Calendario de los indios de Guatemala 1685. Photostatic copy of C. H. Berendt's MS copy of 1878, Brigham Young University Library, Provo, Utah. Identified as pages 21–25 from

the "Crónica Franciscana," an anonymous 283-folio MS in possession of the Sociedad Económica de Guatemala in 1878, now lost.

Anonymous
17th century   Bocabulario en lengua 4iche y castellana. Box 59, folders 1, 2. Brigham Young University Library, Provo, Utah.

Anonymous
1787   Diccionario de Quiché. MS. Copied by Fermín Joseph Tirado in 1787 at Santo Domingo Zacapula [Sacapulas]. Gates Collection. Brigham Young University Library, Provo, Utah.

Anonymous
1813   Vocabulario de la lengua kaqchikel y española, con un arte de la misma lengua. Fonde Américain No. 47 (R. 7508). Bibliothèque nationale, Paris.

Anonymous
1837   Vocabulario en lengua castellana y guatemalteca, que se llama cakchiquel chi. MS. Fond Américaine No. 7. Bibliothèque nationale, Paris.

Anonymous
n.d.   Vocabulario copioso de las lenguas cakchiquel y 4iche, also titled Vocabulario en lengua vakchi4el y 4iche otlatecas. Microfilm copy of MS in Gates Collection, Brigham Young University Library, Provo, Utah.

Ara, Domingo de
1986   *Vocabulario de lengua tzeldal según el Órden de Copanabastla*. Edición de Mario Humberto Ruz. Fuentes para el Estudio de la Cultura Maya, 4. Mexico City: Universidad Nacional Autónoma de México: Instituto de Investigaciones Filológicas, Centro de Estudios Mayas.

Armas, Isaías
1897   Vocabulario breve de la lengua maya recogido en el pueblo de San José y San Luís. MS 53 11. Latin American Library, Tulane University, New Orleans.

Arzapalo Marín, Ramón
1987   *El ritual de los Bacabes*. Fuentes para el Estudio de lla Cultura Maya, 5. Mexico City: Universidad Autónoma de México.

Asicona Ramírez, Lucas, Domingo Méndez Rivera, and Rodrigo Domingo Xinic Bop
1998   *Diccionario ixil de San Gáspar Chajul*. Antigua, Guatemala: Proyecto Lingüístico Francisco Marroquín and Asociación Chajulense Va'l aq Quyol.

Atlas Nacional de Guatemala
1972   Guatemala City: Instituto Geográfico Nacional, Ministerio de Comunicaciones y Obras Públicas.

Attinasi, John J.
1973   Lak T'an: A Grammar of the Chol (Mayan) Verb. Ph.D. dissertation, University of Chicago, Chicago.
1975   Phonology and Style in Chol Maya Ritual. *Columbia University Working Press in Linguistics* 1:1–28.

Aulie, Evelyn W.

1948   *Chol Dictionary*. Microfilm Collection of Manuscripts on American Cultural Anthropology 26. Chicago: University of Chicago Library.

Aulie, H. Wilbur, and Evelyn W. de Aulie

1978   *Diccionario ch'ol-español español-ch'ol*. Serie de Vocabularios y Diccionarios Indígenas Mariano Silva y Aceves, No. 21. Mexico City: Instituto Lingüístico de Verano.

Avendaño y Loyola, Fray Andrés de

1996   *Relación de las dos entradas que hice a la conversión de los gentiles ytzáex, y cehaches*. Mexicon Occasional Publications 3. Möckmuehl, Germany: Verlag Anton Sauerwein.

Aveni, Anthony F.

1980   *Empires of Time: Calendars, Clocks and Cultures*. New York: Basic Books.

1989   *World Archaeoastronomy*. Cambridge: Cambridge University Press.

1997   *Stairways to the Stars*. New York: John Wiley and Sons.

Aveni, Anthony F., and Horst Hartung

1986   *Maya City Planning and the Calendar*. Transactions, No. 76, Part 7. Philadelphia: American Philosophical Society.

Ayala Falcón, Marciela

1997   Who Were the People of Tonina? In *The Language of Maya Hieroglyphs*, ed. Martha J. Macri and Anabel Ford, pp. 69–75. San Francisco: Pre-Columbian Art Research Institute.

Baaijens, Thijs

1995   The Typical "Landa Year" as the First Step in the Correlation of the Maya and the Christian Calendar. *Mexicon* 17(3): 50–51.

Baer, Phillip, and William R. Merrifield

1971   *Two Studies on the Lacandones of Mexico*. Summer Institute of Linguistics, Pub. 33. Norman: University of Oklahoma Press.

Barbachano, Fernando Cámara

1945a   *Monografía de los tzotziles de San Miguel Mitontik, Chiapas, México*. Microfilm Collection of Manuscripts on American Cultural Anthropology Series 1, No. 6. Chicago: University of Chicago Library.

1945b   *Monografía sobre los tzeltales de Tenejapa, Chiapas, México*. Microfilm Collection of Manuscripts on American Cultural Anthropology Series 1, No. 5. Chicago: University of Chicago Library.

Barela, Francisco

n.d.   Vocabulario kakchiquel. MS. Museo Nacional de México, Mexico City.

Barrera Marín, Alfredo, Alfredo Barrera Vásquez, and R. M. López-Franco

1976   *Nomenclatura etnobotánica maya*. Colección Científica No. 36. Mexico City: Instituto Nacional de Antropología e Historia.

Barrera Vásquez, Alfredo, Juan Ramón Bastarrachea Manzano, and William Brito Sansores

1980   *Diccionario maya Cordemex*. Mérida, Yucatán, Mexico City: Ediciones Cordemex.

Barrera Vásquez, Alfredo, and Silvia Rendón

1948   *El libro de los libros de Chilam Balam*. 3rd reprint. Mexico City: Fondo de Cultura Económica.

Barthel, Thomas

1952   Der Morgensternkult in den Darstellungen der Dresdener Mayahandschrift. *Ethnos* 17:73–112.

Basseta, Domingo de

2005   *Vocabulario de la lengua quiché* (1698). René Acuña, ed. Instituto de Investigaciones Filológicas, Centro de Estudios Mayas, Fuentes para el Estudio de la Cultura Maya, 18. Mexico City: Universidad Nacional Autónoma de México

Bassie Sweet, Karen

1991   *From the Mouth of the Dark Cave: Commemorative Sculpture of the Late Classic Maya*. Norman: University of Oklahoma Press.

Bastarrachea Manzano, Juan Ramón, Ermilo Yah Pech, and Fidencio Briceño Chel

1992   *Diccionario básico español maya español*. Biblioteca Básica del Mayab. Mérida, Yucatán, Mexico: Maldonado Editores.

Baudez, Claude-François

1994   *Maya Sculpture of Copan*. Norman: University of Oklahoma Press.

Becerra, Marcos

1933   *El antiguo calendario chiapaneco: Estudio comparativo entre éste i los calendarios precoloniales maya, Quiché i Nahoa*. Mexico City: Imprenta Mundial.

1934   Vocabulario de la lengua chol, que se habla en el distrito de Palenque del Estado de Chiapas, de la República Mexicana, acopiado por el profesor Marcos E. Becerra, en noviembre y diciembre de 1934. *Boletín del Museo Nacional de Arqueología, Historia y Etnografía*, series 6, 1.

1937   Vocabulario de la lengua chol. *Anales del Museo Nacional*, 5a series, 2: 249–78.

Beetz, Carl P., and Linton Satterthwaite

1981   *The Monuments and Inscriptions of Caracol, Belize*. University Museum Monograph No. 45. Philadelphia: University Museum, University of Pennsylvania.

Berendt, C. H.

1877   Calendario de los indios de Guatemala, 1722 Kiché. Copy of lost original MS. See Chol Poal Σih Maceval Σih and Ahilabal Σih below.

Berlin, Brent

1968   *Tzeltal Numeral Classifiers: A Study in Ethnographic Semantics*. Janua Linguarum. Series Practica, 70. The Hague: Mouton.

Berlin, Brent, Dennis E. Breedlove, and Peter H. Raven

1974   *Principles of Tzeltal Plant Classification*. New York: Academic Press.

Berlin, Brent, and Terrence Kaufman

1962   Diccionario del Tzeltal de Tenejapa. MS. Stanford University, Stanford, Calif.

Berlin, Heinrich

1951　The Calendar of the Tzotzil Indians. In *The Civilization of Ancient America*, ed. Sol Tax, pp. 155–61. Selected Papers of the 29th International Congress of Americanists. Chicago: University of Chicago Press.

1968　Estudios epigráficos II. *Antropología e Historia de Guatemala* 20(1): 13–24.

1987　Vericuetos mayas. In *Homenaje a José Mata Gavidia*, ed. Rigoberto Juárez-Paz, pp. 9–24. Guatemala City: Facultad de Humanidades, Universidad de San Carlos de Guatemala.

Beyer, Hermann

1937　Studies on the Inscriptions of Chichén Itzá. Contributions to American Archaeology, Vol. 4, No. 21. *Carnegie Institution of Washington Publication* 483:29–175.

Blair, Robert

1964　Yucatec Maya Noun and Verb Morphosyntax. Ph.D. dissertation, Indiana University, Indianapolis.

Blair, Robert, John S. Robertson, Larry Richman, Greg Sansom, Julio Salazar, Juan Yool, and Alejandro Choc

1981　*Diccionario español-cakchiquel-inglés.* Language and Intercultural Research Center, New World Languages Research Division, Brigham Young University, Provo, Utah. New York: Garland Publishing.

Blom, Frans, and Oliver LaFarge

1927　*Tribes and Temples.* 2 vols. Middle American Research Institute, Publication 1. New Orleans: Tulane University.

Bolles, David

1990　The Mayan Calendar: The Solar-Agricultural Year, and correlation Questions. *Mexicon* 12(5): 85–89.

Bolles, John S.

1977　*Las Monjas: A Major Pre-Mexican Architectural Complex at Chichen Itza.* Norman: University of Oklahoma Press.

Bond, John J.

1889　*Handy-Book of Rules and Tables for Verifying Dates with the Christian Era.* 4th ed. London: George Bell and Sons.

Book of Chilam Balam of Tizimín

n.d.　Photostatic copy in Latin American Library, Tulane University, New Orleans. Original in Museo Nacional de Antropología, Mexico City.

Brasseur de Bourbourg, Charles Étienne

1864　*Relation des choses de Yucatan de Diego de Landa.* Collection de Documents dans les Langes Indigènes, pour Servir à l'Étude de l'Histoire et de la Philologie de l'Amérique Ancienne 3. Paris: A. Durand.

1872　*Dictionnaire, grammaire et chrestomathie de la langue maya.* Paris: Maisonneuve et Cie, Libraires-Éditeurs.

1961　*Gramática de la lengua quiché.* Guatemala City: Editorial del Ministerio de Educación Pública "José de Pineda Ibarra."

Breedlove, Dennis E., and Robert M. Laughlin

1993   *The Flowering of Man: A Tzotzil Botany of Zinacantan*. Vol. 1. Smithsonian Contributions to Anthropology, No. 35. Washington D.C.: Smithsonian Institution Press.

Bricker, Harvey M., and Victoria R. Bricker

2007   When Was the Dresden Codex Venus Table Efficacious? In *Skywatching in the Ancient World: New Perspectives in Cultural Astronomy*, ed. Clive Ruggles and Gary Urton, pp. 95–119. Boulder: University Press of Colorado.

2011   *Astronomy in the Maya Codices*. Philadelphia: American Philosophical Society.

Bricker, Victoria R.

1978   Antipassive Construction in Yucatec Maya. In England 1978:3–24.

1982a   The Origin of the Mayan Solar Calendar. *Current Anthropology* 23(1): 101–3.

1982b   Reply. *Current Anthropology* 23(3): 354–55.

1986   *A Grammar of Mayan Hieroglyphs*. Middle American Research Institute, Publication 56. New Orleans: Tulane University.

1990a   *A Morpheme Concordance of the Book of Chilam Balam of Chumayel*. Middle American Research Institute, Publication 59. New Orleans: Tulane University.

1990b   *A Morpheme Concordance of the Book of Chilam Balam of Tizimín*. Middle American Research Institute, Publication 58. New Orleans: Tulane University.

1997   The Structure of Almanacs in the Madrid Codex. In *Papers on the Madrid Codex*, ed. V. R. Bricker and Gabrielle Vail, pp. 1–25. Middle American Research Institute, Publication 64. New Orleans: Tulane University.

2000   Bilingualism in the Maya Codices and in the Books of Chilam Balam. *Written Language and Literacy* 3(1): 77–115.

Bricker, Victoria R., and Helga-Marie Miram

2002   *An Encounter of Two Worlds: The Book of Chilam Balam of Kaua*. Middle American Research Institute Publication 68. New Orleans: Tulane University.

Bricker, Victoria, Eleuterio Poʼot Yah, and Ofelia Dzul de Poʼot

1998   *A Dictionary of the Maya Language as Spoken in Hocabá, Yucatán*. Salt Lake City: University of Utah Press.

Brinton, Daniel Garrison

1882   *The Maya Chronicles*. Brinton's Library of Aboriginal American Literature, No. 1. Philadelphia. Reprint: New York: AMS Press, 1969.

1885   *The Annals of the Cakchiqueles*. Philadelphia. Reprint: New York: AMS Press, 1969.

1890   *Essays of an Americanist*. Philadelphia: Porter and Coates.

1893   The Native Calendar of Central America and Mexico. A Study in Linguistics and Symbolism. *Proceedings of the American Philosophical Society* 31(1):258–314.

1895   *A Primer of Mayan Hieroglyphics*. University of Pennsylvania Series in Philology, Literature and Archaeology, Vol. 3, No. 2. Boston: Ginn and Company.

Brown, Cecil H., and Staley R. Witkowski

1982   Growth and Development of Folk Zoological Life Forms in the Mayan Language Family. *American Ethnologist* 9:97–112.

Bruce, Robert D.

1974   *El libro de Chan K'in*. Mexico City: Instituto Nacional de Antropología e Historia.

1979   *Lacandon Dream Symbolism*. Mexico City: Ediciones Euroamericanas Klaus Thiele.

Cadena, Carlos

1787   Untitled Vocabulary List for "Castellano Pocomán." In Fernández 1892:25–30.

1788   Untitled Vocabulary Lists for "Castellano, Quiché, Cacchii and Poconchí." In Fernández 1949:107–15.

Calderón, Hector

1957   "Calendario cakchiquel de los indios de Guatemala, 1685." *Antropología e Historia de Guatemala* 9:17–29.

Calendario Sna Holobil

1979   (Tzeltal) San Cristóbal de las Casas, Mexico: Sna Holobil.

Campbell, Lyle

1970   Nahua Loan Words in Quichean Languages. *Chicago Linguistics Society* 6:3–11.

1977   *Quichean Linguistic Prehistory*. University of California Publications in Linguistics, 81. Berkeley: University of California Press.

1978   Quichean Prehistory: Linguistic Contributions. In England 1978:25–54.

1984a   El pasado lingüístico del sureste de Chiapas. *Investigaciones recientes en el área maya* 1:165–84. Mexico City: Sociedad Mexicana de Antropología.

1984b   The Implications of Mayan Historical Linguistics for Glyphic research. In Justeson and Campbell 1984:1–16.

1988   *Linguistics of Southeast Chiapas, Mexico*. Papers, 50. New World Archaeological Foundation. Provo: Brigham Young University.

Carlson, John B.

1990   America's Ancient Skywatchers. *National Geographic* 177(3): 76–107.

Carmack, Robert

1968   *Toltec Influence on the Postclassic Culture History of Highland Guatemala*. Middle American Research Institute, Publication 26:49–92 (preprint). New Orleans: Tulane University.

1973   *Quichean Civilization*. Berkeley: University of California Press.

1979   *Evolución del reino quiché*. Guatemala City: Piedra Santa.

1981   *The Quiché Mayas of Utatlan*. Norman: University of Oklahoma Press.

Carmack, Robert M., and James L. Mondloch, trans.

1983   *El título de Totonicapán*. Fuentes para el Estudio de la Cultura Maya, 3. Mexico City: Instituto de Investigaciones Filológicas, Centro de Estudios Mayas, Universdidad Nacional Autónoma de México.

Caso, Alfonso

1967   *Los calendarios prehispánicos*. Mexico City: Universidad Nacional Autónoma de México.

1971   Calendrical Systems of Central Mexico. In *Handbook of Middle American Indians*, 10:333–48. Austin: University of Texas Press.

Cassell, Jonathon
1974   *Lacandon Adventure*. San Antonio: Naylor Company.

Charencey, Hyacinthe de
1883   *Des suffixes en langue quichée*. Louvain, Belgium: Typografie de Ch. Peeters.
1885   *Vocabulaire de la langue tzotzil. Extrait des Mémoires de l'Académie Nationale des Sciences, Arts et Belles Lettres de Caen*, 251–89. Caen, France: Imp. F. Le Blanc-Handel.

Chel, Antonio Cedillo, and Juan Ramírez
1999   *Diccionario del idioma ixil de Santa María Nebaj*. Antigua, Guatemala: Proyecto Lingüístico Francisco Marroquín.

Chiapas: Atlas Cultural de México, Cartográfico II
1987   Mexico City: Secretaría de Educación Pública, Instituto Nacional de Antropología e Historia, Grupo Editorial Planeta.

Chol Poal Σih Maceval Σih and Ahilabal Σih
1722   MS. Photostatic copy in Latin American Library, Tulane University, New Orleans.

Ciudad Real, Antonio de
1984   *Calepino maya de Motul, I, II*. René Acuña, ed. Mexico City: Universidad Nacional Autónoma de México, Instituto de Investigaciones Filológicas.
n.d.   Vocabulario de la lengua de maya [Spanish-Maya]. Xerox copy of positive prints from microfilm of manuscript in John Carter Brown Library, Brown University, Providence, Rhode Island.

Clancy, Flora Simmons
2009   *The Monuments of Piedras Negras, an Ancient Maya City*. Albuquerque: University of New Mexico Press.

Coe, Michael D.
1973   *The Maya Scribe and His World*. New York: Grolier Club.
1975   Native Astronomy in Mesoamerica. In *Archaeoastronomy in Pre-Columbian America*, ed. A. F. Aveni, pp. 3–31. Austin: University of Texas Press.
1977   Supernatural Patrons of Maya Scribes and Artists. In *Social Process in Maya Prehistory*, ed. Norman Hammond, pp. 327–47. London: Academic Press.
1982   *Old Gods and Young Heroes*. Jerusalem: American Friends of the Israel Museum.
1993   *The Maya*. London: Thames and Hudson.

Coe, Michael D., and Justin Kerr
1998   *The Art of the Maya Scribe*. New York: Harry N. Abrams.

Cogolludo, Diego López de
1971   *Los tres siglos de la dominación española en Yucatán o sea historia de esta provincia*. 2 vols. Graz: Akademische Druck- und Verlagsanstalt.

Cojti Macario, Narciso, Martín Chacach Cutzal, and Marcos Armando Cali
1998   *Diccionario del idioma kaqchikel, kaqchikel-español*. Antigua, Guatemala: Proyecto Lingüístico Francisco Marroquín.

Colby, Benjamin
1976   Anomalous Ixil—Bypassed by the Postclassic? *American Antiquity* (Washington, D.C.)
41(1): 71–84.

Colby, Benjamin N., and Lore M. Colby
1981   *The Daykeeper: The Life and Discourse of an Ixil Diviner*. Cambridge, Mass.: Harvard
University Press.

Colby, Benjamin, and Pierre L. van den Berghe
1969   *Ixil Country: A Plural Society in Highland Guatemala*. Berkeley: University of California
Press.

Colby, Lore
1966   Esquema de la morfología tzotzil. In Vogt 1966:373–95.

Coto, Fray Tomás de
1983   *Vocabulario de la lengua cakchiquel v(el) Guatemalteca, nuevamente hecho y recopilado
con summo estudio, trabajo y erudición (Thesaurus Verbrv.)*. René Acuña, ed. Mexico City:
Universidad Nacional Autónoma de México.

Cowan, Marion M.
1968   *Tzotzil Grammar*. Norman, Okla.: Summer Institute of Linguistics.

Craine, Eugene R., and Reginald C. Reindorp
1979   *The Codex Pérez and the Book of Chilam Balam of Maní*. Norman: University of Oklahoma Press.

Curley García, Francisco
1967   *Vocabulario del dialecto o lengua gkecchí*. Guatemala City: Tipografía Nacional.

Davies, Nigel
1987   *The Aztec Empire*. Norman: University of Oklahoma Press.

Delgaty, Colin C., and Agustín Ruiz Sánchez
1978   *Diccionario tzotzil de San Andrés con variaciones dialectales: Tzotzil-español, español-
tzotzil*. Serie de Vocabularios Indígenas Mariano Silva y Aceves, No. 22. Mexico City:
Instituto Lingüístico de Verano.

Díaz Babio, F.
1977   Cartografía, coordenadas y geografía física del estado de Yucatán. In *Enciclopedia Yucatanense*, 1:43–66. 2nd ed. 12 vols. Mexico City: Edición Oficial del Gobierno de Yucatán,

Díaz Olivares, Jorge
1980   *Manual del tzeltal*. Mexico City: SEP and UNICEF, Universidad Autónoma de Chiapas.

Diego, Mateo Felipe
1998   *Diccionario del idioma chuj, chuj-español*. Antigua, Guatemala: Proyecto Lingüístico
Francisco Marroquín.

Diego Antonio, Diego de, Francisco Pascual, Nicolas de Nicolas Pedro, and Carmelino Fernando
Gonzales
1996   *Diccionario del idioma q'anjob'al*. Antigua, Guatemala: Proyecto Lingüístico Francisco
Marroquín.

Dienhart, John M.

1989   *The Mayan Languages: A Comparative Vocabulary*. Vols. 2 and 3. Odense, Denmark: Odense University Press.

Durant, Will

1966   *The Story of Civilization: Vol. 2, The Life of Greece* (1939). New York: Simon and Schuster.

Edmonson, Munro Stirling

1965   *Quiché-English Dictionary*. Middle American Research Institute, Publication 30. New Orleans: Tulane University.

1967   Classical Quiché. In *Handbook of Middle American Indians*, vol. 5:249–67. Austin: University of Texas Press.

1971   *The Book of Counsel: The Popol Vuh of the Quiche Maya of Guatemala*. Middle American Research Institute, Publication 35. New Orleans: Tulane University.

1973   *Meaning in Mayan Languages: Ethnolinguistic Studies*. American Anthropological Association. The Hague: Mouton.

1976   The Mayan Calendar Reform of 11.16.0.0.0 *Current Anthropology* 17(4): 713–17.

1982   *The Ancient Future of the Itza: The Book of Chilam Balam of Tizimin*. Austin: University of Texas Press.

1986   *Heaven Born Meridaa and Its Destiny: The Chilam Balam of Chumayel*. Austin: University of Texas Press.

1988   *The Book of the Year: Middle American Calendrical Systems*. Salt Lake City: University of Utah Press.

1997   The Count of the Cycle and the Number of the Days. In *Quiché Dramas and Divinatory Calendars*, pp. 113–50. Middle American Research Institute, Publication 66. New Orleans: Tulane University,

England, Nora, ed.

1978   *Papers in Mayan Linguistics*. University of Missouri Miscellaneous Publications in Anthropology, No. 6. Studies in Mayan Linguistics, No. 2. Columbia: University of Missouri.

Ernst, Carl H., and Roger W. Barbour

1989   *Turtles of the World*. Washington, D.C.: Smithsonian Institution Press.

Euw, Eric von

1977   *Corpus of Maya Hieroglyphic Inscriptions: Volume 4, Part 1. Itzimte, Pixoy, Tzum*. Cambridge, Mass.: Peabody Museum of Archaeology and Ethnology, Harvard University.

Fahsen, Frederico

2002   Rescuing the Origins of Dos Pilas Dynasty: A Salvage of Hieroglyphic Stairway #2, Structure L5–49. Grantee Report, Foundation for the Advancement of Mesoamerican Studies, Inc. http://www.famsi.org/reports/01098/01098Fahsen01.pdf.

Fernández, Jesús

1937   Diccionario poconchí. *Anales de la Sociedad de Geografía e Historia de Guatemala* 14:47–70, 184–200.

Fernández, León

1892   *Lenguas indígenas de Centro América en el siglo XVIII según copia del archivo de Indias.*
        San José, Costa Rica: Tipografía Nacional.

1949   Lenguas de Guatemala en el siglo XVIII. *Anales de la Sociedad de Geografía e Historia de
        Guatemala* 24(1–2): 107–58.

Fisher, William Morrison

1973   Towards the Reconstruction of Proto-Yucatec. Ph.D. dissertation, University of Chicago.

1976   On Tonal Features in the Yucatecan dialects. In *Mayan Linguistics*, ed. Marlys McClaran,
        1:29–43. Los Angeles: UCLA American Indian Studies Center.

Flores, Ildefonso Joseph

1753   *Arte de la lengua metropolitana del reyno cakchiquel o guatemálico.* Guatemala City: n.p.

Forrest, Morelos, and Jean Brewer

1962   *Vocabulario mexicano de Tetelcingo.* Vocabularios Indígenas "Mario Silva y Aceves" No. 8.
        Mexico City: Summer Institute of Linguistics.

Förstemann, Ernst Wilhelm

1880   *Die Mayahandschrift der Königlichen Öffentlichen Bibliothek zu Dresden.* Leipzig: Verlag
        der A. Naumann'schen Lichtdruckerei. Digital copy formerly in the possession of Linda
        Schele, retrievable from the Foundation for the Advancement of Mesoamerican Studies
        (FAMSI) at http://www.famsi.org/mayawriting/codices/pdf/dresden_fors_schele_all.pdf.

Fought, John Guy

1967   Chorti (Mayan) Phonology, Morpophonemics, and Morphology. Ph.D. dissertation. Yale
        University, New Haven.

1972   *Chorti (Mayan) Texts, 1.* Sarah S. Fought, ed. Philadelphia: University of Pennsylvania
        Press,

1984   Cholti Maya: A Sketch. In *Supplement to the Handbook of Middle American Indians*,
        vol. 2, *Linguistics*, ed. M. S. Edmonson pp. 43–55. Austin: University of Texas Press.

Fox, James Allen

1978   Proto-Mayan Accent, Morpheme Structure Conditions, and Velar Innovations. Ph.D.
        dissertation. University of Chicago.

Fox, James Allen, and John S. Justeson

1980   Mayan Hieroglyphs as Linguistic Evidence. In *Third Palenque Round Table, 1978 Part 2*,
        ed. Merle Greene Robertson, pp. 204–16. Austin: University of Texas Press.

1984   Polyvalence in Mayan Hieroglyphic Writing. In Justeson and Campbell 1984:17–76.

Freeze, Ray

1975   *A Fragment of Early K'ekchi' Vocabulary with Comments on the Content.* Columbia:
        Museum of Anthropology, University of Missouri.

Freidel, David, Linda Schele, and Joy Parker

1993   *Maya Cosmos.* New York: William Morrow.

García Hernández, Abraham, and Santiago Yac Sam

1989   *Diccionario quiché-español.* Guatemala City: Instituto Lingüístico de Verano.

Garza, Mercedes de la

1995   *Aves sagradas de los mayas.* Mexico City: Facultad de Filosofía y Letras, Centro de Estudios Mayas del Instituto de Investigaciones Filológicas, Universidad Nacional Autónoma de México.

Garza, Mercedes de la, Ana Luisa Izquierdo, Ma del Carmen León, and Tolita Figueroa, eds.

1983   *Relaciones histórico-geográficas de la gobernación de Yucatán (Mérida, Valladolid y Tabasco), I.* Mexico City: Instituto de Investigaciones Filológicas, Centro de Estudios Mayas, Universidad Nacional Autónoma de México.

Gates, William E.

n.d. [1922]   Documents from the files of William Gates. MS. Microfilm. Millwood, N.Y.: Kraus Microform. J. P. Harrington Papers, Vol. 7, Section 1. Smithsonian Institution, Washington, D.C.

1931   The Thirteen Ahaus in the Kaua Manuscript. *Maya Society Quarterly* (Baltimore) 1(1):2–20.

1932a   A Lanquin Kekchi Calendar. *Maya Society Quarterly* (Baltimore) 1:29–32.

1932b   Pokomchi Calendar. *Maya Society Quarterly* (Baltimore) 1:1:75–77.

Gavarrete, Juan

1868   *Geografía de Guatemala.* 2nd ed. Guatemala City: n.p.

Girard, Rafael

1940   *El chortí.* Revista del Archivo y Biblioteca Nacionales 19. Tegucigalpa, Honduras: n.p.

1949   *Los chortís ante el problema maya.* 5 vols. Mexico City: Editorial Cultura, T.G., S.A.

1962   *Los mayas eternos.* Mexico City: Antigua Librería Robredo.

1966   *Los mayas: Su civilización, su historia, sus vinculaciones continentales.* Mexico City: Libro México.

Gómez Ramírez, Martín

1991   *Ofrenda de los ancestros en Oxchuc.* 2nd ed. Consejo Estatal de Fomento a la Investigación y Difusión de la Cultura, DIF/Chiapas—Instituto Chiapaneco de Cultura. Tuxtla Gutiérrez, Mexico: Gobierno del Estado de Chiapas.

Gordon, George Byron

1902   *The Hieroglyphic Stairway, Ruins of Copan.* Memoirs of the Peabody Museum of American Archaeology and Ethnology, Harvard University, Vol. 1, No. 6. Cambridge, Mass. Reprint, New York: Kraus Reprint, 1970.

Gossen, Gary H.

1974a   *Chamulas in the World of the Sun: Time and Space in a Maya Oral Tradition.* Cambridge, Mass.: Harvard University Press.

1974b   A Chamula Solar Calendar Board from Chiapas, Mexico. In *Mesoamerican Archaeology New Approaches,* ed. Norman Hammond, pp. 218–53. Austin: University of Texas Press.

Graham, Ian

1967   *Archaeological Explorations in El Peten, Guatemala.* Middle American Research Institute, Publication 33. New Orleans: Tulane University.

1978   *Corpus of Maya Hieroglyphic Inscriptions: Vol. 2, Part 2 Naranjo Chunhuitz Xunantunich.* Cambridge, Mass.: Peabody Museum of Archaeology and Ethnology, Harvard University.

1979   *Corpus of Maya Hieroglyphic Inscriptions: Vol. 3, Part 2 Yaxchilan*. Cambridge, Mass.: Peabody Museum of Archaeology and Ethnology, Harvard University.

1980   *Corpus of Maya Hieroglyphic Inscriptions: Vol. 2, Part 3 Ikxun Ucanal Ixtutz Naranjo*. Cambridge, Mass.: Peabody Museum of Archaeology and Ethnology, Harvard University.

1982   *Corpus of Maya Hieroglyphic Inscriptions: Vol. 3, Part 3 Yaxchilan*. Cambridge, Mass.: Peabody Museum of Archaeology and Ethnology, Harvard University.

Graham, Ian, and Eric von Euw

1992   *Corpus of Maya Hieroglyphic Inscriptions: Vol. 4, Part 3 Uxmal Xcalumkin*. Cambridge, Mass.: Peabody Museum of Archaeology and Ethnology, Harvard University.

Graham, Ian, and Peter Mathews

1996   *Corpus of Maya Hieroglyphic Inscriptions: Vol. 6, Part 2 Tonina*. Cambridge, Mass.: Peabody Museum of Archaeology and Ethnology, Harvard University.

Graulich, Michel

1981   The Metaphor of the Day in Ancient Mexican Myth and Ritual. *Current Anthropology* 22(1): 45–60.

Greene, Merle, Robert Rands, and John A. Graham

1972   *Maya Sculpture*. Berkeley: Lederer, Street and Zeus.

Grube, Nikolai

1990   *Die Entwicklung der Mayaschrift*. Acta Mesoamericana Vol. 3. Berlin: Verlag von Flemming.

1994   Epigraphic Research at Caracol, Belize. In *Studies in the Archaeology of Caracol, Belize*, ed. Diane Chase and Arlen Chase, pp. 83–122. San Francisco: Pre-Columbian Art Research Institute.

Grube, Nikolai, Alfonso Lacadena, and Simon Martin

2003   Chichen Itza and Ek Balam: Terminal Classic Inscriptions From Yucatan. Part II of *Notebook for the XXVIIth Maya Hieroglyphic Forum at Texas March 2003*. Austin: Maya Workshop Foundation.

Grube, Nikolai, and Werner Nahm

1994   A Census of Xibalba: A Complete Inventory of *way* Characters on Maya Ceramics. In *The Maya Vase Book, Volume 4*, ed. Justin Kerr, pp. 686–715. New York: Kerr Associates.

Guiteras-Holmes, Calixta

1946   *Informe de San Pedro Chenalhó*. Microfilm Collection of Manuscripts on American Cultural Anthropology No. 14. Chicago: University of Chicago Library.

1961   *Perils of the Soul: The World View of a Tzotzil Indian*. New York: Free Press of Glencoe.

Guzmán, Pantaleón de

1984   *Compendio de nombres en lengua cakchiquel*. René Acuña, ed. Filología: Gramáticas y Diccionarios 1. Mexico City: Instituto de Investigaciones Filológicas, Universidad Nacional Autónoma de México,

Haeserijn V., Estéban

1979   *Diccionario k'ekchi'-español*. Guatemala City: Piedra Santa.

Hammond, Norman
1982   *Ancient Maya Civilization.* New Brunswick, N.J.: Rutgers University Press.

Hanks, William F.
1990   *Referential Practice: Language and Lived Space among the Maya.* Chicago: University of Chicago Press.

Harrington, John B.
1957   Valladolid Maya Enumeration. In *Smithsonian Institution, Bureau of American Ethnology, Bulletin 164: Anthropological Papers, No. 54,* pp. 241–78. Washington, D.C.: U.S. Government Printing Office.

Harris, William H., and Judith S. Levey, eds.
1975   *The New Columbia Encyclopedia.* New York: Columbia University Press.

Hassig, Ross
1988   *Aztec Warfare.* Norman: University of Oklahoma Press.

Haviland, John B.
1981   *Sk'op sotz'leb: El tzotzil de San Lorenzo Zinacantan.* Mexico City: Universidad Nacional Autónoma de Mexico.
1988   It's My Own Invention: A Comparative Grammatical Sketch of Colonial Tzotzil. In *The Great Tzotzil Dictionary of Santo Domingo Zinacantan, Volume 1: Tzotzil-English,* by Robert M. Laughlin, pp. 79–123. Smithsonian contributions to Anthropology, No. 31. Washington, D.C.: Smithsonian Institution Press.

Haviland, William A.
1992   From Double-Bird to Ah Cacao: Dynastic Troubles and the Cycle of Katuns at Tikal, Guatemala. In *New Theories on the Ancient Maya,* ed. Elin C. Danien and Robert Sharer, pp. 71–80. Monograph 77, University Museum Symposium Series, Vol. 3. Philadelphia: University Museum, University of Pennsylvania.

Helfrich, Klaus
1973   *Menschenopfer und Tötungsrituale im Kult der Maya.* Monumenta Americana, 9. Berlin: Gebr. Mann Verlag.

Henderson, John S.
1997   *The World of the Ancient Maya.* 2nd ed. Ithaca: Cornell University Press.

Herbruger, Alfredo, Jr., and Eduardo Díaz Barrios
1956   *Método para aprender a hablar, leer y escribir la lengua cakchiquel.* Vol. 1. Guatemala City: Tipografía Nacional de Guatemala.

Herrera y Tordesillas, Antonio de
1952   *Historia general de los hechos de los castellanos en las Islas y Tierra Firme del mar Océano que llaman Indias Occidentales* (1615). Madrid: Editorial Maestre.

Hofling, Charles Andrew
1997   *Itzaj Maya-Spanish-English Dictionary.* Salt Lake City: University of Utah Press.
2011   *Mopan Maya-Spanish-English Dictionary.* Salt Lake City: University of Utah Press.

Holland, William R.

1963   *Medicina maya en los altos de Chiapas: Un estudio del cambio socio-cultural.* Colección de Antropología Social. Mexico City: Instituto Nacional Indigenista.

Hopkins, Nicholas

1967a   The Chuj Language. Ph.D. dissertation, University of Chicago.

1967b   A Short Sketch of Chalchihuitan Tzotzil. *Anthropological Linguistics* 9(4): 9–25.

1968   Chuj Dictionary Outline. Unpublished MS in possession of the author.

1973   Compound Place Names in Chuj and Other Mayan Languages. In Edmonson 1973:165–81.

1980   Chuj Animal Names and Their Classification. *Journal of Mayan Linguistics* 2(1): 13–39.

2012   *A Dictionary of the Chuj (Mayan) Language as Spoken in San Mateo Ixtatán, Huehuetenango, Guatemala.* Tallahassee, Fla.: Jaguar Tours.

Hopkins, Nicholas, and J. Kathryn Josserand

1988   *Chol (Mayan) Dictionary Database, Part III, Fascicle 11*: Chol Monosyllabic Dictionary Database. Final Performance Report, National Endowment for the Humanities Grant RT-20643–86, 30 September 1988. N.p.

Hopkins, Nicholas, J. Kathryn Josserand, and Ausencio Cruz Guzmán

2011   *A Historical Dictionary of Chol (Mayan): The Lexical Sources from 1789 to 1935.* Tallahassee, Fla.: Jaguar Tours.

Houston, Stephen D.

1988   The Phonetic Decipherment of Mayan Glyphs. *Antiquity* 62:126–35.

1992   Classic Maya Politics. In *New Theories on the Ancient Maya*, ed. Elin C. Danien and Robert Sharer, pp. 65–69. Philadelphia: University Museum, University of Pennsylvania.

1993   *Hieroglyphs and History at Dos Pilas: Dynastic Politics of the Classic Maya.* Austin: University of Texas Press.

Houston, Stephen, John Robertson, and David Stuart

1998   Disharmony in Maya Hieroglyphic Writing: Linguistic Change and Continuity in Classic Society. In *Anatomía de una civilización: Aproximaciones interdisciplinarias a la cultura maya*, ed. Andrés Ciudad Ruíz et al., pp. 275–96. Madrid: Sociedad Española de Estudios Mayas.

2000   The Language of Classic Maya Inscriptions. *Current Anthropology* 41(3): 321–56.

2001   *Quality and Quantity in Glyphic Nouns and Adjectives.* Research Reports on Ancient Maya Writing *47*. Washington, D.C.: Center for Maya Research.

Houston, Stephen, David Stuart, and Karl A. Taube

1989   Folk Classification of Classic Maya Pottery. *American Anthropologist* (Washington, D.C.) 91(3): 720–26.

Hruby, Zachary X., and Mark B. Child

2004   Chontal Linguistic Influence in Ancient Maya Writing. In Wichmann 2004:13–26.

Hull, Kerry Michael

2003   Verbal Art and Performance in Ch'orti' and Maya Hieroglyphic Writing. Ph.D. dissertation, University of Texas at Austin.

Hun Batz' Men
1983   *Los calendarios mayas y hunab k'u.* Mexico City: Ediciones Horizonte.

Hunn, Eugene S.
1977   *Tzeltal Folk Zoology.* New York: Academic Press.

Hunt, Eva
1977   *The Transformation of the Hummingbird: Cultural Roots of a Zinacantecan Mythical Poem.* Ithaca, N.Y.: Cornell University Press.

Hurley Delgaty, Alfa, and Agustín Ruiz Sánchez
1978   *Diccionario tzotzil de San Andrés, con variaciones dialectales.* Serie de Vocabularios y Diccionarios Indígenas Mariano Silva y Aceves, No. 22. Mexico City: Instituto Lingüístico de Verano.

Johnsgard, Paul A.
1983   *The Hummingbirds of North America.* Washington, D.C.: Smithsonian Institution Press.

Justeson, John S.
1975   The Identification of the Emblem Glyph of Yaxha, El Petén. In *Studies in Ancient Meso-america, II*, ed. John A. Graham, pp. 123–29. Contribution 27. Berkeley: University of California Research Facility.
1978   Mayan Scribal Practice in the Classic Period: A Test-Case of an Explanatory Approach to the Study of Writing Systems. Ph.D. dissertation, Department of Anthropology, Stanford University. Ann Arbor: University Microfilms.
1984   Interpretations of Maya Hieroglyphs. In Justeson and Campbell 1984:315–62.
1988   The Non-Maya Calendars of Tabasco and Veracruz and the Antiquity of the Long Count and the Month Count. *Journal of Mayan Linguistics* 6(1): 1–21.
1989a   Ancient Maya Ethnoastronomy: An Overview of Hieroglyphic Sources. In *World Archae-oastronomy*, ed. Anthony Aveni, pp. 76–129. Cambridge: Cambridge University Press.
1989b   The Representational Conventions of Mayan Hieroglyphic Writing. In *Word and Image in Maya Culture: Explorations in Language, Writing and Representation*, ed. William F. Hanks and Donald S. Rice, pp. 25–38. Salt Lake City: University of Utah Press.

Justeson, John S., and Lyle Campbell, eds.
1984   *Phoneticism in Mayan Hieroglyphic Writing.* Institute for Mesoamerican Studies, Publication No. 9. Albany: State University of New York.
1997   The Linguistic Background of Mayan Hieroglyphic Writing: Arguments against a "Highland Mayan" Role. In *The Language of Maya Hieroglyphs*, ed. M. J. Macri and A. Ford, pp. 41–67. San Francisco: Pre-Columbian Art Research Institute.

Justeson, John S., William M. Norman, Lyle Campbell, and Terrence Kaufman
1985   *The Foreign Impact on Lowland Mayan Language and Script.* Middle American Research Institute, Publication 53. New Orleans: Tulane University.

Kartunnen, Frances
1992   *An Analytical Dictionary of Nahautl.* Norman: University of Oklahoma Press.

Kaua, Book of Chilam Balam of
n.d.   Photostatic copy in Gates Collection, Brigham Young University, Provo, Utah (see also Bricker and Miram 2002).

Kaufman, Terrence
1971 *Tzeltal Phonology and Morphology*. University of California Publications in Linguistics, No. 61. Berkeley: University of California Press.
1972 *El proto-tzeltal-tzotzil*. Centro de Estudios Mayas 5. Mexico City: Universidad Nacional Autónoma de México.
1974a *Ixil Dictionary*. Technical Report No. 1, Laboratory of Anthropology. Irvine: University of California.
1974b Meso-American Indian Languages. In *Encyclopaedia Britannica*, pp. 956–63. 15th ed. Chicago: Helen Hemingway Benton.
1978 Archaeological and Linguistic Correlations in Mayaland and Associated Areas of Meso-America. *World Archaeology* 8(1): 101–18.

Kaufman, Terrence, and William Norman
1984 An Outline of Proto-Cholan Phonology, Morphology and Vocabulary. In Justeson and Campbell 1984:6677.

Keller, Kathryn C.
1955 The Chontal (Mayan) Numeral System. *International Journal of American Linguistics* 21(3): 258–75.

Keller, Kathryn C., and Plácido Luciano G.
1997 *Diccionario chontal de Tabasco*. Serie de Vocabularios y Diccionarios Indígenas "Mariano Silva y Aceves," No. 36. Tucson: Summer Institute of Linguistics.

Kluepfel, Charles
1986 Planets. Astronomical software package marketed by its author.

Knorozov, Yuri
1955 *Diego de Landa: Soobshchenie o delakh v Yukatane 1566 g.* Moscow/Leningrad: Izdatel'stvo Akademii Nauk SSR.
1967 *Selected Chapters from the Writings of the Maya Indians*. Russian Translation Series, Vol. 4. Cambridge, Mass.: Peabody Museum of Archaeology and Ethnology, Harvard University.
1982 *Maya Hieroglyphic Codices*. Sophie D. Coe, trans. Institute for Mesoamerican Studies, Publication No. 8. Albany: State University of New York.

Knowles, Susan Marie
1984 A Descriptive Grammar of Chontal Maya (San Carlos Dialect). Ph.D. dissertation. Tulane University, New Orleans.

Krupp, Edwin C.
1991 *Beyond the Blue Horizon*. New York: HarperCollins.

Kurbjuhn, Kornelia
1989 *Maya: The Complete Catalogue of Glyph Readings*. Kassel: Schneider and Weber.

Lacadena García-Gallo, Alfonso
2003 El corpus glífico de Ek' Balam, Yucatán, México. Grantee Report 01057. Los Angeles: Los Angeles County Museum of Art Foundation for the Advancement of Mesoamerican Studies.

2011   Mayan Hieroglyphic Texts as Linguistic Sources. In *New Perspectives in Mayan Linguistics*, ed. Heriberto Avelino, pp. 343–73. Newcastle upon Tyne, UK: Cambridge Scholars Publishing.

Lacadena García-Gallo, Alfonso, and Søren Wichmann

2002   The Distribution of Lowland Maya Languages in the Classic Period. In *La organización social de los mayas: Memorial de la Tercera Mesa Redonda de Palenque*, edited by Vera Tiesler, René Cobos, and Merle Greene Robertson, 2:275–314. Mexico City: Instituto Nacional de Antropología e Historia and Universidad Nacional Autónoma de Yucatán.

LaFarge, Oliver

1947   *Santa Eulalia: The Religion of a Cuchumatán Indian Town*. Chicago: University of Chicago Press.

LaFarge, Oliver, and Douglas Byers

1931   *The Yearbearer's People*. Middle American Research Series, Publication No. 3. New Orleans: Department of Middle American Research, Tulane University.

Land, Hugh C.

1970   *Birds of Guatemala*. Wynnewood, Pa.: Livingston.

Landa, Diego de

1978   *Relación de las Cosas de Yucatán*. 9th ed. Mexico City: Editorial Porrúa.

Laughlin, Robert M.

1975   *The Great Tzotzil Dictionary of San Lorenzo Zinacantan*. Smithsonian Contributions to Anthropology, No. 19. Washington D.C.: Smithsonian Institution Press.

1977   *Of Cabbages and Kings: Tales from Zinacantan*. Smithsonian Contributions to Anthropology, No. 23. Washington, D.C.: Smithsonian Institution Press.

1988   *The Great Tzotzil Dictionary of Santo Domingo Zinacantan: 1: Tzotzil-English*. Smithsonian Contributions to Anthropology, No. 31. Washington, D.C.: Smithsonian Institution Press.

Law, Danny, John Robertson, and Stephen Houston

2006   Split Ergativity in the History of the Ch'olan Branch of the Mayan Language Family. *International Journal of American Linguistics* 72, no. 4 (October 2006): 415–50.

Lenkersdorf, Carlos

1979   *B'omak' Umal Tojol Ab'al–Kastiya 1 diccionario tojolabal-español*. Mexico City: Editorial Nuestro Tiempo.

León, Juan de

1945   *Mundo quiche miscelánea*. Guatemala City: Taller Tipográfico "San Antonio."

1954   *Diccionario quiché–español*. Guatemala City: Editorial Landivar.

León Portilla, Miguel

1973   *Time and Reality in the Thought of the Maya*. Charles L. Boilès and Fernando Horcasitas, trans. Boston: Beacon Press.

Liman, Florence F., and Marshall Durbin

1975   Some New Glyphs on an Unusual Maya Stela. *American Antiquity* 40(3): 314–20.

Lincoln, Jackson Steward

1942 The Maya Calendar of the Ixil of Guatemala. In *Contributions to American Anthropology and History*, no. 30, vol. 7, pp. 97–128. Publication 528. Washington, D.C.: Carnegie Institution of Washington.

n.d. Field Notes on the Ixil Indians of the Guatemalan Highlands, 1939–1941. MS notebooks. 6 vols. Tozzer Library, Harvard University, Cambridge, Mass.

Looper, Matthew G.

2003 *Lightning Warrior*. Austin: University of Texas Press.

Lounsbury, Floyd

1980 Some Problems in the Interpretation of the Mythological Portion of the Hieroglyphic Text of the Temple of the Cross, Palenque. In *Third Palenque Round Table, Part 2*, ed. Merle Greene Robertson, pp. 99–115. Austin: University of Texas Press.

1982 Astronomical Knowledge and Its Uses at Bonampak, Mexico. In *Archaeoastronomy in the New World*, ed. A. F. Aveni, pp. 143–68. New York: Cambridge University Press.

1989 The Ancient Writing of Middle America. In *The Origins of Writing*, ed. Wayne M. Senner, pp. 203–37. Lincoln: University of Nebraska Press.

MacLeod, Barbara, and Dennis E. Puleston

1979 Pathways to Darkness: The Search for the Road to Xibalbá. In *Tercera Mesa Redonda de Palenque, 1978*, ed. Merle Greene Robertson and Donnan C. Jeffers, pp. 71–77. Pre-Columbian Art Research. Monterey, Calif.: Herald Printers.

Macri, Martha J., and Matthew G. Looper

2003 *The New Catalog of Maya Hieroglyphs, Volume One: The Classic Period Inscriptions*. Norman: University of Oklahoma Press.

Macri, Martha J., and Gabrielle Vail

2009 *The New Catalog of Maya Hieroglyphs, Volume Two: The Codical Texts*. Norman: University of Oklahoma Press.

Maldonado Andrés, Juan, Juan Ordóñez Domingo, and Juan Ortiz Domingo

1983 *Diccionario de San Ildefonso Ixtahuacan Huehuetenango, mam-español*. Hanover: Verlag für Ethnologie.

Maler, Teobert

1901 *Researches in the Central portion of the Usumatsintla Valley, Part II (Piedras Negras). Memoirs, Volume 2, No. 1*. Cambridge, Mass.: Peabody Museum of Archaeology and Ethnology, Harvard University.

Malmström, Vincent H.

1997 *Cycles of the Sun, Mysteries of the Moon*. Austin: University of Texas Press.

Martin, Simon

1997 The Painted King List: A Commentary on Codex-Style Dynastic Vases. In *The Maya Vase Book, Volume 5*, ed. Barbara Kerr and Justin Kerr, pp. 846–67. New York: Kerr Associates.

Martin, Simon, and Nikolai Grube

2000 *Chronicle of the Maya Kings and Queens*. London: Thames and Hudson.

Mathews, Peter, and David M. Pendergast

1979   The Altun Ha Jade Plaque: Deciphering the Inscription. *Contributions of the University of California Archaeological Research Facility, Studies in Ancient Mesoamerica*, 4:197–214. Berkeley: Archaeological Research Facility, Department of Anthropology, University of California.

Mathews, Peter, and Linda Schele

1974   Lords of Palenque: The Glyphic Evidence. In *Primera Mesa Redonda de Palenque, Part I*, ed. Merle Greene Robertson, pp. 63–76. Pebble Beach, Calif.: Pre-Columbian Art Research, Robert Louis Stevenson School.

Maudslay, Alfred Perceval

1974   *Biologia Centrali-Americana*. 2 vols. F. Ducans Godman and Osbert Salvin, eds. Reprint of the original of 1889–1902 published by the same company. New York: Milpatron.

Mayer, Karl Herbert

1980   *Maya Monuments: Sculpture of Unknown Provenance in the United States*. Sandra L. Brizee, trans. Ramona, Calif.: Acoma Books.

1987   *Maya Monuments: Sculpture of Unknown Provenance, Supplement 1*. Berlin: Verlag von Flemming.

1989   *Maya Monuments: Sculpture of Unknown Provenance, Supplement 2*. Berlin: Verlag von Flemming.

Mayers, Marvin K.

1956   *Vocabulario pocomchi-español español-pocomchi*. Guatemala City: Instituto Lingüístico de Verano.

1958   *Pocomchi Texts*. Norman, Okla.: Summer Institute of Linguistics.

1960   The Linguistic Unity of Pocomam-Pocomchi. *International Journal of American Linguistics* 26(4): 290–300.

McArthur, Carolina, and Ricardo McArthur, eds.

1995   *Diccionario pocomam y español*. Guatemala City: Instituto Lingüístico de Verano de Centroamérica.

McClaran Stefflre, Marlys

1972   Lexical and Syntactic Structures in Yucatec Maya. Ph.D. dissertation. Harvard University, Cambridge, Mass.

McGee, R. Jon

1990   *Life, Ritual and Religion among the Lacandon Maya*. Belmont, Calif.: Wadsworth Publications.

McQuown, Norman A.

1976a   Vocabulario chol de la Cueva. MS (1949). Microfilm Collection of Manuscripts on American Cultural Anthropology, No. 156 (XXIX). Chicago: University of Chicago Library.

1976b   *Vocabulario kanjobal*. Microfilm Collection of Manuscripts on American Cultural Anthropology, No. 195. Chicago: University of Chicago Library.

Mediz Bolio, Antonio
1985   *El libro de Chilam Balam de Chumayel*. Mexico City: Secretaría de Educación Pública.

Mengin, Ernst, ed.
1972   *Bocabulario de mayathan*. Graz: Akademische Druck- und Verlagsanstalt.

Mexico Air Navigation Map: Merida: N-F16-South
1938   Fort Humphreys, D.C.: Engineer Reproduction Plant, U.S. Army.

Michelon, Oscar, ed.
1976   *Diccionario de San Francisco*. Graz: Akademische Druck- und Verlagsanstalt.

Milbrath, Susan
1999   *Star Gods of the Maya*. Austin: University of Texas Press.

Miles, Suzanne W.
1957   The Sixteenth-Century Pokom-Maya: A Documentary Analysis of Social Structures and Archaeological Setting. *Transactions of the American Philosophical Society* (Philadelphia) 47, part 4:731–81.

Miller, Mary Ellen
1986   *The Art of Mesoamerica*. New York: Thames and Hudson.

Miller, Mary, and Karl Taube
1993   *The Gods and Symbols of Ancient Mexico and the Maya*. London: Thames and Hudson.

Miram, Helga-Maria
1988a   *Maya Texte II: Transkriptionen/Transcriptions/Transcripciones der/of the/de los Chilam Balames*, vol. 1: *Ixil Chumayel Nah*. Hamburg: Toro Verlag.
1988b   *Maya Texte II: Transkriptionen/Transcriptions/Transcripciones der/of the/de los Chilam Balames*, vol. 3: *Codice Perez*. Hamburg: Toro Verlag.

Molina, Fray Alonso de
1977   *Vocabulario en lengua castellana y mexicana, y mexicana y castellana*. 2nd ed. Mexico City: Editorial Porrúa.

Mondloch, James P.
1978   *Basic Quiche Grammar*. Publication No. 2. Albany: State University of New York.
1980   K'e̊š: Quiché Naming. *Journal of Mayan Linguistics* 1(2): 9–25.

Montgomery, John
2001   *Tikal: An Illustrated History of the Ancient Maya Capital*. New York: Hippocrene Books.

Morán, Francisco
1935   *Arte y diccionario en lengua choltí*. Maya Society, Publication 9. Baltimore: Maya Society.

Morán, Pedro
1720   Bocabulario de nombres que comiençan en romance, en lengua pocomam de Amatitlán. MS. Fonde Americaine No. 50: Pokoman 642, R. 7491. Bibliothèque nationale, Paris.

1725   Bocabulario de sólo los nombres de la lengua pokoman. Photographic copy of MS. Latin American Library, Tulane University, New Orleans.

Morley, Sylvanus Griswold
1920   *The Inscriptions at Copan.* Publication No. 219. Washington, D.C.: Carnegie Institution of Washington.
1938   *The Inscriptions of Peten: Volume IV.* Publication No. 437. Washington, D.C.: Carnegie Institution of Washington.

Nachtigall, Horst
1978   *Die ixil: Maya-indianer in Guatemala.* Marburger Studien zur Völkerkunde, Vol. 3. Berlin: Dietrich Reimer Verlag.

Narciso, Victor A.
1897   Pokomchi Month Names. In Sapper 1906:412.
1906   Pokomchi Month Names. In Termer 1930:394–95.
1914   Pokomchi Month Names. In Gates 1932b:75–76.

National Almanac Office, U.S. Naval Observatory
1997   *Astronomical Almanac for the Year 1997.* Washington, D.C.: U.S. Government Printing Office.

Navarrete Cáceres, Carlos A.
1984   *Guía para el estudio de las ruinas de Chinkultic, Chiapas, México.* Mexico City: Universidad Nacional Autónoma de México.

Newman, James R., ed.
1967   *The Harper Encyclopedia of Science.* New York: Harper and Row.

Nuttall, Zelia
1928   *La observación del paso del sol por el zenit por los antiguos habitantes de la América tropical.* Mexico City: Talleres Gráficos de la Nación.

Owen, Michael Gordon, III
1968   The Semantic Structure of Yucatec Verb Roots. Ph.D. dissertation. Yale University, New Haven. Ann Arbor, Mich.: University Microfilms.

Pacheco Cruz, Santiago
1939   *Lexico de la fauna yucateca.* 2nd ed. Mérida, Yucatán, Mexico: Imprenta Oriente.
1958   *Diccionario de la fauna yucateca.* 1st ed. Mérida, Yucatán, Mexico: n.p.
1960   *Usos, costumbres, religión, y supersticiones de los mayas.* 2nd ed. Mérida, Yucatán, Mexico: n.p.

Pagden, A. R., trans. and ed.
1975   *The Maya: Diego de Landa's Account of the Affairs of Yucatan.* Chicago: J. Philip O'Hara.

Page, J. L.
1933   The Climate in the Yucatan Peninsula. In *The Peninsula of Yucatan: Medical, Biological, Meteorological and Sociological Studies*, ed. George Cheever Shattuck, pp. 409–22. Publication 431. Washington, D.C.: Carnegie Institution of Washington.

Pearse, A. S.

1977   La fauna (1945). In *Enciclopedia yucatanense*, 1:109–271. 12 vols. 2nd ed. Mexico City: Edición Oficial del Gobierno de Yucatán.

Pereira, Dionicio

1723   *Arte de la lengua tzotzlem or tzinacanteca con explicación del año solar y un tratado de las quentas de los indios en lengua tzotzlem, lo todo escrito en 1688 . . . sacadas a luz por el P. Fr. Juan de Rodaz . . . y ahora trasladadas nuevamente por el padre fray Dionycio Pereyra . . . 1723*. In Ruz 1989:87–168.

Pérez González, Benjamín, and Santiago de la Cruz

1998   *Diccionario chontal: Chontal-español, español-chontal*. Villa Hermosa, Mexico: Instituto Nacional de Antropología e Historia, Fondo Estatal para la Cultura y las Artes de Tabasco.

Pérez Martínez, Vitalino, Federico García, Felipe Martínez, and Jeremías López

1996   *Diccionario del idioma ch'orti'*. Antigua, Guatemala: Proyecto Lingüístico Francisco Marroquín.

Pérez Toro, Augusto

1942   *La milpa*. Mérida, Yucatán, Mexico: Publicaciones del Gobierno de Yucatán.

Peterson, Roger Tory, and Edward L. Chalif

1973   *A Field Guide to Mexican Birds*. Boston: Houghton Mifflin.

Pineda, Emeterio

1845   *Descripción geográfica del departamento de Chiapas y Soconusco*. Tuxtla Gutiérrez, Mexico: Consejo Estatal para la Cultura de las Artes de Chiapas.

Pineda, Vicente

1887   *Gramática de la lengua tzeltal* (bound with 1888 *Historia*).

1888   *Historia de las sublevaciones indígenas habidas en el estado de Chiapas: Gramática de la lengua tseltal y diccionario de la misma*. Bound with *Diccionario de la lengua tzeltal*. San Cristobal de las Casas, Chiapas, Mexico: Tipografía del Gobierno en Palacio.

Pinkerton, Sandra, ed.

1976   *Studies in K'ekchi*. Texas Linguistic Forum 3. Austin: Department of Linguistics, University of Texas.

Pío Pérez, Juan

1843   Ancient Chronology of Yucatan; or, A True Exposition of the Method Used by the Indians for Computing Time [translated by Stephens]. In John Lloyd Stephens, *Incidents of Travel in Yucatan* [1843], 1:278–303. 2 vols. New York: Dover Publications, 1963.

1864   Cronología antigua de Yucatán y exámen del método con que los indios contaban el tiempo, sacado de varios documentos antiguos. In Brasseur de Bourbourg 1864:366–429.

1866–77   *Diccionario de la lengua maya*. Mérida, Yucatán, Mexico: Imprenta Literaria de Juan F. Molina Solis.

1898   *Coordinación alfabética de las voces del idioma maya*. Mérida, Yucatán, Mexico: Imprenta La Ermita.

Pohl, Mary, and John Pohl
1983   Ancient Maya Cave Rituals. *Archaeology* 36:28–51.

Pollock, H. E. D.
1980   *The Puuc*. Memoirs of the Peabody Museum, Vol. 19. Cambridge, Mass.: Peabody Museum of Archaeology and Ethnology, Harvard University.

Po'ot Yah, Eleuterio, and Victoria R. Bricker
1981   *Yucatec Maya Verbs (Hocaba Dialect)*. New Orleans: Tulane University Center for Latin American Studies.

Quirke, Stephen, and Jeffrey Spencer, eds.
1992   *The British Museum Book of Ancient Egypt*. London: Thames and Hudson.

Rand McNally: The New International Atlas
1986   New York: Rand McNally.

Rätsch, Christian, and K'ayum Ma'ax
1984   *Ein Kosmos im Regenwald: Mythen und Visionen der Lakandonen-Indianer*. Diedrichs Gelbe Series 48. Cologne: Eugen Diedrichs Verlag.

Recinos, Adrián
1947   *Popol Vuh: Las antiguas historias del quiché*. Mexico City: Fondo de Cultura Económica.
1954   *Monografía del departamento de Huehuetenango*. 2nd ed. Guatemala City: Editorial del Ministerio de Educación Pública.

Recinos, Adrián, and Delia Goetz, trans.
1953   *The Annals of the Cakchiquels*. Norman: University of Oklahoma Press.

Recinos, Adrián, Delia Goetz, and Sylvanus Morley
1950   *Popol Vuh: The Sacred Book of the Ancient Quiché Maya*. Norman: University of Oklahoma Press.

Redfield, Robert, and Alfonso Villa Rojas
1934   *Chan Kom: A Maya Village*. Publication No. 448. Washington, D.C.: Carnegie Institution of Washington.
1939   *Notes on the Ethnography of Tzeltal Communities of Chiapas*. Publication No. 509. Contributions to American Anthropology and History, No. 28. Vol. 5, Nos. 24–29. Washington, D.C.: Carnegie Institution of Washington.
1962   *Chan Kom: A Maya Village*. Chicago: University of Chicago Press/Phoenix Books.

Reina, Ruben E.
1966   *The Law of the Saints*. New York: Bobbs-Merrill.

Richards, Julia Becker
1985   Vowel Variability in a Linguistic Transition Zone. *International Journal of American Linguistics* 51(4): 549–53.

Ridpath, Ian, ed.
1979   *The Illustrated Encyclopedia of Astronomy and Space*. New York: Thomas Y. Crowell.

Ringle, William
1985   Notes on Two Tablets of Unknown Provenance. In *Fifth Palenque Round Table, 1983*, ed. Virginia M. Fields, pp. 151–58. San Francisco: Pre-Columbian Art Research Institute.
1988   *Of Mice and Monkeys: The Value and Meaning of T1016, the God C Hieroglyph.* Research Reports on Ancient Maya Writing, No. 18. Washington, D.C.: Center for Maya Research.

Rivera Dorado, Miguel
1994   Notas de arqueología de Oxkintok. In *Hidden among the Hills: Maya Archaeology of the Northwest Yucatan Peninsula*, ed. Hanns J. Prem, pp. 44–58. Acta Mesoamericana, 7. Möckmühl, Germany: Verlag von Flemming.

Rivera y Rivera, Roberto
1977   *Los instrumentos musicales de los mayas.* Mexico City: Instituto Nacional de Antropología e Historia.

Robertson, John S.
1977   A Proposed Revision in Mayan Subgrouping. *International Journal of American Linguistics* 43(2): 105–20.
1992   *The History of Tense/Aspect/Mood/Voice in the Mayan Verbal Complex.* Austin: University of Texas Press.
1998   A Ch'olti'an Explanation for Ch'orti'an Grammar: A Postlude to the Language of the Classic Maya. *Mayab* (Sociedad Española de Estudios Mayas) 11:5–11.

Robertson, John S., Danny Law, and Robbie A. Haertel
2010   *Colonial Ch'olti': The Seventeenth-Century Morán Manuscript.* Norman: University of Oklahoma Press.

Robertson, Merle Greene
1991   *Sculpture of Palenque, Vol. IV: The Cross Group, the North Group, the Olvidado, and Other Pieces.* Princeton: Princeton University Press.

Robertson, Merle Greene, ed.
1974   *Primera Mesa Redonda de Palenque, Part I.* Pebble Beach, Calif.: Robert Louis Stevenson School, Pre-Columbian Art Research.

Robicsek, Francis, and Donald Hales
1981   *The Maya Book of the Dead: The Ceramic Codex.* Charlottesville: University of Virginia Art Museum.
1982   *Maya Ceramic Vases from the Classic Period (The November Collection of Maya Ceramics).* Charlotte, N.C.: Maya Publishing.

Robles Urribe, Carlos
1966   *La dialectología tzeltal y el diccionario compacto.* Mexico City: Instituto Nacional de Antropología e Historia.

Rodríguez Guaján, Demetrio, Jesús Leopoldo Tzian Guantá, and José Obispo Rodríguez Guaján
1990   *Ch'uticholtzij: Maya-kaqchikel: Vocabulario kaqchikel-español, vocabulario español-kaqchikel.* Guatemala City: Coordinadora Kaqchikel de Desarrollo Integral.

Rodríguez Sánchez, Juan, Efraín Poma Sambrano, Teresa Pérez Toma, and Francisco
Castro Osorio
1995   *Aq'b'al Elu'l Yol Vatzsaj diccionario ixil.* Programa de Rescate Cultural Maya-Ixil. Guate-
        mala City: Cholsamaj.

Roys, Ralph L.
1931   *The Ethnobotany of the Maya.* Middle American Research Series, Publication No. 2. New
        Orleans: Tulane University.
1940   *Personal Names of the Maya of Yucatan.* Publication 523, Contribution 31. Washington,
        D.C.: Carnegie Institute of Washington.
1949   *The Prophecies for the Maya Tuns or Years in the Books of Chilam Balam of Tizimin and
        Mani.* Contributions to American Anthropology and History, No. 51. Washington, D.C.:
        Carnegie Institution of Washington.
1967   *The Book of Chilam Balam of Chumayel.* 2nd ed. Norman: University of Oklahoma Press.

Ruyán Canú, Déborah, and Rafael Coyote Tum
1991   *Diccionario de cakchiquel central y español.* Guatemala City: Instituto Lingüístico de
        Verano.

Ruz, Mario Humberto
1982   *Los legítimos hombres: Aproximación antropológica al grupo tojolabal.* Vol. 2. Mexico City:
        Universidad Nacional Autónoma de México.

Ruz, Mario Humberto, ed.
1989   *Las lenguas del Chiapas colonial: Manuscritos en la Biblioteca Nacional de París.* Vol. 1.
        Mexico City: Universidad Nacional Autónoma de México.

Sáenz de Santa María, Carmelo
1940   *Diccionario cakchiquel-español.* Guatemala City: Impresa de Tipografía Nacional.

Sam Juárez, Miguel, Ernesto Chen Cao, Crisanto Xal Tec, Domingo Cuc Chen,
and Pedro Tiul Pop
2001   *Diccionario q'eqchi' molob'aal aatin q'eqchi-español.* Antigua, Guatemala: Proyecto
        Lingüístico Francisco Marroquín.

Santo Domingo, Thomas de
1693   Vocabulario en la lengua cakchiΣel. Fonde Américaine No. 44 (R7503). Paris: Biblio-
        thèque nationale.

Sapper, Karl
1897   *Das nördliche Mittelamerika nebst einem Ausflug nach dem Hochland von Anahuac:
        Reisen und Studien aus den Jahren 1888–1895.* Braunschweig, Germany: Druck und
        Verlag von Vieweg und Sohn.
1906   Sitten und Gebräuche der Pokonchi-Indianer, nach Sr. Vicente A. Narciso. *International
        Congress of Americanists, Stuttgart* (1904) 2:403–17.
1907   Choles y chorties. In *Proceedings of the International Congress of Americanists, XVth Ses-
        sion,* 1:423–65. 3 vols. Quebec: Dussault and Proulx.

Saquic Calel, Felipe Rosalio
1989   *Idioma quiché: Inglés/quiché/español.* Guatemala City: Editorial Piedra Santa.

Sarles, Harvey B.

1961 Monosyllabic Dictionary of the Tzeltal Language. MS in possession of the Estate of Linda Schele.

Schele, Linda

1985 *Notebook for the Maya Hieroglyphic Workshop at Texas, March 9–10, 1985*. Austin: Institute of Latin American Studies, University of Texas at Austin.

1987a *A Possible Death Date for Smoke-Imix-God K*. Copan Note 26. Austin, Tex.: Copán Acropolis Project and Instituto Hondureño de Antropología e Historia.

1987b *Some Ideas of the Protagonist and Dating of Stele E*. Copan Note 25. Austin, Tex.: Copán Acropolis Project and Instituto Hondureño de Antropología e Historia.

1990 *The Early Classic Dynastic History of Copan, Interim Report*. Copan Note 70. Austin, Tex.: Copan Acropolis Project and Instituto Hondureño de Antropología e Historia.

Schele, Linda, and David Freidel

1990 *A Forest of Kings*. New York: William Morrow.

Schele, Linda, and Nikolai Grube

1994 *Notebook for the XVIIIth Maya Hieroglyphic Workshop at Texas, March 12–13, 1994*. Austin: University of Texas Press.

1995 *Notebook for the XIXth Maya Hieroglyphic Workshop at Texas, March 9–18, 1995*. Austin: University of Texas Press.

Schele, Linda, and Matthew Looper

1996 *Notebook for the XXth Maya Hieroglyphic Forum at Texas, March 9–10, 1996*. Austin: University of Texas Press.

Schele, Linda, and Peter Mathews

1979 *The Bodega of Palenque, Chiapas, Mexico*. Washington, D.C.: Dumbarton Oaks.

1997 *Notebook for the XXIst Maya Hieroglyphic Forum at Texas*. Austin: University of Texas Press.

1998 *The Code of Kings*. New York: Scribner.

Schele, Linda, and Mary Ellen Miller

1986 *The Blood of Kings*. Fort Worth: Kimbell Art Museum.

Schoenhals, Alvin, and Louise C. Schoenhals

1965 *Vocabulario mixe de Totontepec*. Serie de Vocabularios y Diccionarios Indígenas Mariano Silva y Aceves, No. 14. Mexico City: Instituto Lingüístico de Verano.

Schoenhals, Louise C.

1988 *A Spanish-English Glossary of Mexican Flora and Fauna*. Mexico City: Summer Institute of Linguistics.

Scholes, Frances B., and Eleanor B. Adams

1960 *Relaciones histórico-descriptivas de la Verapaz, el Manché y Lacandón, en Guatemala*. Guatemala City: Editorial Universitaria.

Schuller, Rodolfo

n.d. Vocabularios comparativos: Vocabulario Castellano-Tz'eltal II. MS. Howard-Tilton Memorial Library, Tulane University, New Orleans.

Schultze Jena, Leonhard
1972   *Popol Vuh: Das Heilige Buch der Quiché-Indianer von Guatemala.* Berlin: Verlag W. Kohlhammer.

Schulz, Ramón P. C.
1942   Apuntes sobre algunas fechas del templo de la cruz de Palenque. *Actos del 28. Congreso International de Americanistas (Mexico City 1939)*: 1:352–55.
1953   Nuevos datos sobre el calendario tzental y tzotzil de Chiapas. *Yan: Ciencias Antropológicas* 2:114–16.

Schumann, Otto
1971   *Descripción estructural del maya itzá del Petén, Guatemala C.A.* Centro de Estudios Maya, 4. Mexico City: Universidad Nacional Autónoma de Mexico.
1973   *La lengua chol, de Tila (Chiapas).* Centro de Estudios Mayas, 8. Mexico City: Universidad Nacional Autónoma de México, Coordinación de Humanidades.

Sedat, Guillermo S.
1955   *Nuevo diccionario de las lenguas kʼekch y español.* Chamelco, Guatemala: Instituto Indigenista Nacional.

Seler, Eduard
1899   Mittelamerikanische Musikinstrumete. *Globus* 76(7): 109–12.
1902   Der Festkalender der Tzeltal und der Maya von Yucatan. In *Gesammelte Abhandlungen zur Amerikanischen Sprach- und Alterthumskunde.* 5 vols. Berlin.: A. Asher.

Sharer, Robert J.
2006   *The Ancient Maya.* 6th ed. Stanford: Stanford University Press.

Silva Leal, Felipe
1875   Diccionario de los dialectos que se hablan en los pueblos del centro y de los altos de la República de Guatemala o sean los llamados kachiquel [*sic*] y quiché. Microfilm copy of MS in Lehmann Library of Ibero-Amerikanisches Institut, Berlin.

Slocum, Marianna
1948   Tzeltal (Mayan) Noun and Verb Morphology. *International Journal of American Linguistics* 14(2): 77–86.
1953   *Vocabulario tzeltal-español.* Mexico City: Instituto Lingüístico de Verano/Dirección General de Asuntos Indígenas de la SEP.

Slocum, Marianna, and Florencia L. Gerdel
1965   *Vocabulario tzeltal de Bachjón.* Serie de Vocabularios y Diccionarios Indígenas Mariano Silva y Aceves, No. 13. Mexico City: Instituto Lingüístico de Verano (second printing 1971).

Smailus, Ortwin
1975   *Textos mayas de Belice y Quintana Roo: Fuentes para una dialectología del maya yucateco.* Berlin: Gebr. Mann Verlag.
1989a   *Gramática del maya yucateco colonial.* Wayasbah Publikation 9. Hamburg: Wayasbah Verlag.

1989b *Vocabulario en lengua castellana y guatemalteca que se llama cakchiquel chi*: *Análisis gramatical y lexicológico del cakchiquel colonial según un antiguo diccionario anónimo* (Bibliothèque nationale de Paris, Fond Américaine No. 7). 3 vols. Hamburg: Wayasbah Verlag.

Smith, A. Ledyard
1932a *Two Recent Ceramic Finds at Uaxactun*. Contributions to American Archaeology, Vol. 2, No. 5. Publication No. 436: 1–25. Washington, D.C.: Carnegie Institution of Washington.
1932b *Two Recent Ceramic Finds at Uaxactun*. Publication No. 456. Contributions to American Archaeology 3(19): 189–231. Washington, D.C.: Carnegie Institution of Washington.

Smithe, Frank B.
1966 *The Birds of Tikal*. Garden City, N.Y.: Natural History Press.

Solís Alcalá, Ermilo
1949 *Diccionario español-maya*. Mérida, Yucatán, Mexico: Editorial Yikal Maya Than.

Sosa, John R.
1985 The Maya Sky, the Maya World: A Symbolic Analysis of Yucatec Maya Cosmology. Ph.D. dissertation. State University of New York, Albany.

Sper, S.
1970 Results of Dialectological Research for an Atlas of Lake Atitlan Cakchiquel. *Papers from the Sixth Regional Meeting: Chicago Linguistic Society* 6:36–56.

Stadelman, Raymond
1940 Maize Cultivation in Northwestern Guatemala. Publication 523. Contributions to American Anthropology and History, Vol. 6, No. 33:83–263. Washington, D.C.: Carnegie Institution of Washington.

Standley, Paul C., and Julian A. Steyermark
1958 *The Flora of Guatemala*. Fieldiana: Botany, Vol. 24, Part I. Chicago: Chicago Natural History Museum.

Starr, Frederick
1902 *Notes upon the Ethnography of Southern Mexico, Part II*. Proceedings of the Davenport Academy of Sciences, Vol. 9. Davenport, Iowa: Putnam Memorial Publication Fund.

Stauder, Jack
1966 Algunos aspectos de la agricultura zinacanteca en tierra caliente. In *Los zinacantecos: Un pueblo tzotzil de los altos de Chiapas*, ed. Evon Z. Vogt, pp. 145–62. Colección de Antropología Social, 7. Mexico City: Instituto Nacional Indigenista.

Stewart, Stephen
1980 *Gramática kekchí*. Guatemala City: Editorial Académica Centro Americana.

Stoll, Otto
1887 *Die Sprache der Ixil-Indianer: Ein Beitrag zur Ethnologie und Linguistik der Maya-Völker*. Leipzig: F. A. Brockhaus Verlag.

1888 *Die Maya-Sprachen der Pokom-Gruppe, Erster Theil: Die Sprache der Pokomchi-Indianer.* Vienna: Alfred Hölder.

1889 Die Ethnologie der Indianerstämme von Guatemala. *Internationales Archiv für Ethnologie.* Supplement to vol. 1, pp. 1–112. Leyden: Verlag von P. W. M. Trap.

1896 *Die Maya-Sprachen der Pokom-Gruppe, Zweiter Theil: Die Sprache der K'ekchi-Indianer— Nebst einem Anhang: Die Uspanteca.* Leipzig: K. F. Köhler.

1938 Vocabulario comparado de los idiomas mayances. In *Etnografía de la República de Guatemala,* trans. Antonio Goubaud Carrera 46–84. Guatemala City: Cosmos.

Stone, Andrea J.
1995 *Images from the Underworld: Naj Tunich and the Tradition of Maya Cave Painting.* Austin: University of Texas Press.

Stone, Andrea, and Marc Zender
2011 *Reading Maya Art.* London: Thames and Hudson.

Stuart, David
1985 The Yaxha Emblem Glyph as Yax-ha. *Research Reports on Ancient Maya Writing, No. 1.* Washington, D.C.: Center for Maya Research.

1987 *Ten Phonetic Syllables.* Research Reports on Ancient Maya Writing, 14. Washington, D.C.: Center for Maya Research.

1990 "Directional Count Glyphs" in Maya Inscriptions. *Ancient Mesoamerica* 1:213–24.

1998 Fire Enters His House: Architecture and Ritual in Classic Maya Texts. In *Function and Meaning in Classic Maya Architecture,* ed. Stephen D. Houston, pp. 373–425. Washington, D.C.: Dumbarton Oaks Research Library and Collection.

2004 The Entering of the Day: An Unusual Date from Northern Campeche. *Mesoweb:* www.mesoweb.com/stuart/notes/EnteringDay.pdf.

2005 *The Inscriptions from Temple XIX at Palenque: A Commentary.* San Francisco: Pre-Columbian Art Research Institute.

2012 On Effigies of Ancestors and Gods (dated 20 January 2012, in Maya Decipherment): https://decipherment.wordpress.com/2012/01/20/on-effigies-of-ancestors-and-gods/.

Stuart, David, and Linda Schele
1986 *Interim Report on the Hieroglyphic Stairs of Structure 26. Copan Note 17.* Copán: Copán Mosaics Project.

Stuart, George
1988 *Glyph Drawings from Landa's Relación: A Caveat to the Investigator.* Research Reports on Ancient Maya Writing, No. 19. Washington, D.C.: Center for Maya Research.

Sullivan, Thelma
1988 *Compendium of Nahuatl Grammar.* Salt Lake City: University of Utah Press.

Tate, Carolyn E.
1985 Summer Solstice Events in the Reign of Bird Jaguar III of Yaxchilan. *Estudios de Cultura Maya* 16:85–112.

1992 *Yaxchilan: The Design of a Maya Ceremonial City.* Austin: University of Texas Press.

Taube, Karl A.

1988a   A Prehispanic Maya Katun Wheel. *Journal of Anthropological Research* (Albuquerque) 44(2): 183–203.

1988b   A Study of Classic Maya Scaffold Sacrifice. In *Maya Iconography*, ed. Elizabeth P. Benson and Gillett G. Griffin, pp. 331–51. Princeton: Princeton University Press.

Taube, Karl A., and Bonnie L. Bade

1991   *An Appearance of Xiuhtecuhtli in the Dresden Venus Pages*. Research Reports on Ancient Maya Writing, No. 35. Washington, D.C.: Center for Maya Research.

Tedlock, Barbara

1982   *Time and the Highland Maya*. Albuquerque: University of New Mexico Press.

1985   Hawks, Meteorology and Astronomy in Quiché-Maya Agriculture. *Archaeoastronomy* 8(1–4): 80–88.

1992a   Mayan Calendars, Cosmology, and Astronomical Commensuration. In *New Theories on the Ancient Maya*, ed. Elin C. Danien and Robert Sharer, pp. 217–27. University Museum Monograph 77. University Museum Symposium Series, Vol. 3. Philadelphia: University Museum, University of Pennsylvania.

1992b   The Road of Light: Theory and Practice of Mayan Skywatching. In *The Sky in Mayan Literature*, ed. Anthony Aveni, pp. 18–42. New York: Oxford University Press.

Tedlock, Dennis

1985   *Popol Vuh: The Mayan Book of Creation*. New York: Simon and Schuster.

1993   *Breath on the Mirror: Walking the World of Mayan Myths*. New York: HarperCollins.

Teletor, Celso Narciso

1959   *Diccionario castellano-quiché y voces castellano-pocomam*. Guatemala City: Tipografía Nacional.

Termer, Franz

1930   Zur Ethnologie und Ethnographie des Nördlichen Mittelamerika. *Ibero-Amerikanisches Archiv* 4(3): 303–492.

Thompson, John Eric Sidney

1925   The Meaning of the Maya Months. *Man* (London) 25:121–23.

1932   A Maya Calendar from the Alta Vera Paz, Guatemala. *American Anthropologist* 34:449–54.

1950   *Maya Hieroglyphic Writing: An Introduction*. Publication 589. Washington, D.C.: Carnegie Institution of Washington.

1962   *A Catalog of Maya Hieroglyphs*. Norman: University of Oklahoma Press.

1970   *Maya History and Religion*. Norman: University of Oklahoma Press.

1971   *Maya Hieroglyphic Writing: An Introduction*. 3rd ed. Norman: University of Oklahoma Press.

1972   *A Commentary on the Dresden Codex*. Philadelphia: American Philosophical Society.

1973   The Painted Capstone at Sacnikte, Yucatan, and Two Others at Uxmal. *Indiana* 1:59–64.

Townsend, Paul G., ed.

1986   *Guatemalan Maya Texts*. Guatemala City: Summer Institute of Linguistics.

Townsend, Paul G., Te'c Cham, and Po'x Ich'
1980   *Ritual Rhetoric from Cotzal (Ixil).* Guatemala City: Instituto Lingüístico de Verano.

Tozzer, Alfred M., trans. and ed.
1941   *Landa's Relación de las cosas de Yucatán.* Papers of the Peabody Museum of American Anthropology and Ethnology, Harvard University, 18. Cambridge, Mass.: Peabody Museum. Reprinted New York: Kraus Reprint, 1978.
1957   *Chichen Itza and Its Cenote of Sacrifice: A Comparative Study of Contemporaneous Maya and Toltec.* Memoirs of the Peabody Museum of Archaeology and Ethnology, Harvard University, Vol. 12. Cambridge, Mass.: Peabody Museum.

Tozzer, Alfred M., and Glover M. Allen
1910   *Animal Figures in the Maya Codices.* Papers, Vol. 4, No. 3. Cambridge, Mass.: Peabody Museum of Archaeology and Ethnology, Harvard University. Reprint New York: Kraus Reprint, 1967.

Trik, Helen W., and Michael E. Kampen
1983   *The Graffiti of Tikal.* University Museum Monograph 57. Philadelphia: University Museum.

Ulrich, Matthew, and Rosemary Dixon de Ulrich
1976   *Diccionario bilingüe: Maya mopán y español, español y maya mopán.* Guatemala City: Instituto Lingüístico de Verano.

Vail, Gabrielle, and Anthony Aveni, eds.
2004   *The Madrid Codex: New Approaches to Understanding an Ancient Maya Manuscript.* Mesoamerican Worlds Series. Boulder: University Press of Colorado.

Varea, Francisco
1600   Calepino en lengua kaqchikel. 1699 copy by Francisco Zerón. MS. Philadelphia: American Philosophical Society.

Vico, Domingo de
1555   Vocabulario de la lengua cakchiquel, con advertencia de los vocablos de las lenguas quiché y tzutohil: se trasladó de la obra compuesta por el Ill.mo Padre el Venerable Fr. Domingo de Vico. MS. Fonde Américain No. 47, R. 7507. Bibliothèque nationale, Paris.

Villacorta C., J. Antonio, and Carlos A. Villacorta
1977   *Códices Maya.* 2nd ed. Guatemala City: Tipografía Nacional Guatemala.

Villa Rojas, Alfonso
1946   *Notas sobre la etnografía de los indios tzeltales de Oxchuc.* Microfilm Collection of Manuscripts on American Cultural Anthropology, Series 1, No. 7. Chicago: University of Chicago Library.
1986   *Estudios etnológicos: Los mayas.* Instituto de Investigaciones Etnológicas, Etnología: Serie Antropológica: 32. Mexico City: Universidad Nacional Autónoma de México.

Vogt, Evon
1966   *Los Zinacantecos: Un pueblo tzotzil en los altos de Chiapas.* Mexico City: Instituto Nacional Indigenista.

1969   *Zinacantan: A Maya Community in the Highlands of Chiapas*. Cambridge, Mass.: Harvard University Press.

1976   *Tortillas for the Gods: A Symbolic Analysis of Zinacantan Rituals*. Cambridge, Mass.: Harvard University Press.

1992   Cardinal Directions in Mayan and Southwestern Indian Cosmology. In *Antropología mesoamericana: Homenaje a Alfonso Villa Rojas*, ed. V. Esponda Jimeno et al., pp. 105–28. Tuxtla Gutiérrez, Mexico: Instituto Chiapaneco de Cultura, Gobierno del Estado de Chiapas.

1997   Zinacanteco Astronomy. *Mexicon* 19(6): 110–17.

Vogt, Evon, and Victoria R. Bricker

1996   The Zinacanteco Fiesta of San Sebastian: An Essay in Ethnographic Interpretation. *Res* 29/30:203–22.

Whelan, Frederick G., III

1967   The Passing of the Years. MS. Harvard Chiapas Project, Cambridge, Mass.

Whittaker, Arabelle, and Viola Warkentin

1965   *Chol Texts on the Supernatural*. Publication No. 13. Norman, Okla.: Summer Institute of Linguistics.

Wichmann, Søren, ed.

2004   *The Linguistics of Maya Writing*. Salt Lake City: University of Utah Press.

2006   Mayan Historical Linguistics and Epigraphy: A New Synthesis. *Annual Review of Anthropology* 35:279–94.

Wick, Stanley A., and Remigio Cochojil-González

1966   *Spoken Quiché Maya*. 3 vols. Microfilm Collection of Manuscripts on American Cultural Anthropology, No. 70, Series 11. Chicago: University of Chicago Library.

Williams, Kenneth L.

n.d.   Chuj-San Sebstián Coatán Dialect. MS. University of Chicago, Chicago.

Wirsing, Pablo

n.d.   Kekchi-German Dictionary. MS. Howard-Tilton Library, Tulane University, New Orleans.

Wisdom, Charles

1940   *The Chorti Indians of Guatemala*. Chicago: University of Chicago Press.

1949   *Materials on the Chortí Language*. Microfilm Collection of Manuscripts on American Cultural Anthropology, Series 5, No. 28. Chicago: Joseph Regenstein Library, University of Chicago.

Ximénez, Francisco

1985   *Primera parte del tesoro de las lenguas ΣaΣchiquel, K'iché y 4,utuhil, en que las dic[h]as lenguas se traducen en la nuestra española*. Carmelo Sáenz de Santa María, ed. Publicación Especial, No. 30. Guatemala City: Academia de Geografía e Historia de Guatemala.

Yucatán: Atlas cultural de México, Cartográfico II
1987   Mexico City: Secretaría de Educación Pública, Instituto Nacional de Antropología e Historia, Grupo Editorial Planeta.

Zender, Marc
2006   Teasing the Turtle from Its Shell: AHK and MAHK in Maya Writing. *PARI Journal* 6(3): 1–14.

Zimmerman, Günther
1954   La lista de los meses quichés según Domingo de Basseta. *Yan: Ciencias Antropológicas* 3:60–61.

Zinn, Raymond, and Gail Zinn
n.d.   Dictionary Pocomam-Oriental: San Luis Jalapa, Guatemala Dialect. MS. Copy in possession of the author.

Index

Page numbers in italic type refer to tables.